The Biology of Polar Regions

THE BIOLOGY OF HABITATS SERIES

This attractive series of concise, affordable texts provides an integrated overview of the design, physiology, and ecology of the biota in a given habitat, set in the context of the physical environment. Each book describes practical aspects of working within the habitat, detailing the sorts of studies which are possible. Management and conservation issues are also included. The series is intended for naturalists, students studying biological or environmental science, those beginning independent research, and professional biologists embarking on research in a new habitat.

The Biology of Polar Regions

David N. Thomas
Bangor University, U.K.

G.E. (Tony) Fogg
Deceased

Peter Convey
British Antarctic Survey, U.K.

Christian H. Fritsen
Desert Research Institute, Navada, USA

Josep-Maria Gili
Institut de Ciències del Mar, Barcelona, Spain

Rolf Gradinger
University of Fairbanks, Alaska, USA

Johanna Laybourn-Parry
University of Tasmania, Australia

Keith Reid
CCAMLR, Tasmania, Australia

David W.H. Walton
Emeritus, British Antarctic Survey, U.K.

OXFORD
UNIVERSITY PRESS

OXFORD
UNIVERSITY PRESS

Great Clarendon Street, Oxford OX2 6DP

Oxford University Press is a department of the University of Oxford.
It furthers the University's objective of excellence in research, scholarship,
and education by publishing worldwide in

Oxford New York

Auckland Cape Town Dar es Salaam Hong Kong Karachi
Kuala Lumpur Madrid Melbourne Mexico City Nairobi
New Delhi Shanghai Taipei Toronto

With offices in

Argentina Austria Brazil Chile Czech Republic France Greece
Guatemala Hungary Italy Japan Poland Portugal Singapore
South Korea Switzerland Thailand Turkey Ukraine Vietnam

Oxford is a registered trade mark of Oxford University Press
in the UK and in certain other countries

Published in the United States
by Oxford University Press Inc., New York

© Oxford University Press 2008

The moral rights of the authors have been asserted
Database right Oxford University Press (maker)

First published 2008

British Library Cataloguing in Publication Data

Data available

Library of Congress Cataloging in Publication Data

The biology of polar regions / David N. Thomas... [et al.].
    p. cm.
    Rev. ed. of: The biology of polar habitats / G.E. Fogg. 1998.
    Includes bibliographical references and index.
    ISBN 978-0-19-929813-6 (alk. paper)—ISBN 978-0-19-929811-2 (alk. paper)
    1. Ecology—Polar regions. 2. Natural history—Polar regions. I. Thomas,
David N. (David Neville), 1962– II. Fogg, G. E. (Gordon Elliott), 1919– Biology
of polar habitats.
    QH541.5.P6F64 2007
    578.0911—dc22   2007047118

Typeset by Newgen Imaging Systems (P) Ltd., Chennai, India
Printed in Great Britain
on acid-free paper by
Biddles Ltd, King's Lynn, Norfolk

ISBN 978-0-19-929811-2 (Hbk.)   978-0-19-929813-6 (Pbk.)

10 9 8 7 6 5 4 3 2 1

To the memory of Tony Fogg (1919–2005), who wrote the first version of this book on his own and which has taken eight of us to update.

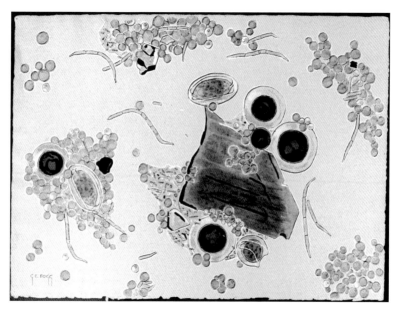

Watercolour painting by Tony Fogg of red snow algae, green algae, and cyanobacteria on Signy Island in the Antarctic during his visit in 1966 (see colour plate).

Preface

This book is a revision of G.E. (Tony) Fogg's *Biology of Polar Habitats* published in 1998. The fact that a major revision is needed in such a short period of time, and that an additional eight authors are needed to update the material, is a reflection of just how fast progress is being made in our exploration of polar regions and in our understanding of how life survives in these extremes. There are still surprisingly many true outposts on our planet that are not influenced by humans: unclimbed mountains, unexplored cave systems, extensive regions of desert and isolated islands. We can only presume that there is a wealth of undiscovered biology in these remote places and as-yet-unimagined and unique adaptations within the biology found there. It is true that few new mammals and birds are being discovered, but among the other groups of plants and animals, *organisms new to science* are recorded routinely as we push forward our access to these remote places.

Most staggering are the discoveries being made in the micobial world, where it has only been in the past decade or so that we have developed the tools capable of giving us an insight into the myriad of organisms just a few micrometres (µm) in size (1 µm is one thousandth of a millimetre). Within this microbial world, especially the bacteria and archaea, there are organisms capable of growth in extremes of temperature (lower limit –20°C, upper limit 113°C), acidic and alkaline pHs, high pressure, low water content, high ionizing radiation, and high salt concentrations. Collectively these organisms are referred to as *extremophiles*, and they often require the extreme condition for growth and reproduction to take place. The last few decades have seen a forging ahead of extremophile research as technological advances enable us to sample the organisms without compromise and, once isolated, to investigate their biology under simulated realistic environmental conditions.

But after centuries of exploration it is still the frozen wastelands of the polar regions that most readily captivate the attention of a wide public audience. For many the introduction is beguiling tales of adventure, but especially in the last 50 years a diverse group of scientists and non-specialists alike have been intrigued by the discoveries of a fascinating biology thriving in what were perceived for so long to be wastelands devoid of any life.

Despite all the advances many of the sentiments expressed by Tony Fogg in his Preface to the first edition of the book are worth repeating here.

A great attraction in studying polar habitats is that, although it may involve sophisticated biochemical or electronic techniques, it still takes one into wild, physically challenging, and hauntingly beautiful places. Fortunately for those who want to indulge themselves in this way there are sound scientific reasons to support their applications for the necessary funding. Apart from the interest in finding out how living organisms manage to exist under the apparently adverse conditions, polar habitats provide particularly favourable opportunities for investigating basic ecological relationships. Compared with the communities of temperate and tropical regions, those in polar habitats are of recent origin and, as a result, relatively simple with few species. This makes it somewhat easier to identify the critical factors operating in the environment to determine the presence and abundance of individual species, interrelations between species, cycling of nutrients, and energy flow. A further feature adding to the interest is that although the Arctic and Antarctic are both cold, with nights and days which last for months, they have inverse patterns of distribution of land and sea. This makes for differences in climate and ocean circulation which have had profound effects on the invasion of plants and animals from temperate zones so that the respective ecologies of the two regions are different. Not unrelated to this, there are radical differences in political regimes which have resulted in the support and organization of science in the Arctic and Antarctic being remarkably dissimilar.

Knowledge of polar habitats is valuable from the practical as well as the academic point of view. The polar regions, taken as including the sub-Arctic and sub-Antarctic as well as the high Arctic and Antarctic, occupy a great part of the earth's surface and with increasing pressure from human populations it is advisable to know what resources and potentialities these relatively unused lands and seas might provide. Proper management of polar fisheries, mineral exploitation, oil drilling, and human settlement all depend on an understanding of the ways in which habitats work. Tourism, also, burgeoning in both regions, needs an input of this understanding if visitors are to get the best out of the experience with the least damage to the environment. Beyond this, polar ecology has world-wide significance. It is increasingly evident that polar ecosystems intermesh into global processes and that they play key roles in the regulation of the environment which is vital for the well-being of all mankind.

We are grateful to a host of colleagues and collaborators for sharing material with us for publication here, including many acknowledged by Tony in *The Biology of Polar Habitats*. We are thankful to the funding agencies, colleagues, logistic support, and opportunities that we have all enjoyed while conducting work in the Arctic and Antarctic. It is these rich experiences that have resulted in our being able to contribute to a book such as this. In all of our various visits to the frozen climes we have been gifted unique glimpses into a world dominated by extremes. Our chosen field sites are still largely unspoilt by mankind, and despite some recent media reports they will surely remain so, simply because they are so vast and inaccessible. For anybody, scientist or tourist, who has the opportunity to visit any part of the polar regions, even the margins, there is a sense of privilege of experiencing the harsh, desolate, awesome beauty that has driven the spirit of exploration and adventure ever since the frozen oceans and lands were first described.

The authors of this book have all participated in scientific expeditions, most of them multi-national ventures that take many years of planning. A highlight of our work is the close friendships and companionship that such research fosters. None of our work is possible without the strength that comes from working within a team, be it 50 people on a remote ice camp or a research vessel filled with 50 scientists and 50 crew. As individual researchers we are driven to expand our own knowledge and experience, and we certainly know much about our individual research specializations. However, working within a team forged by a common desire for discovery and to see scientific understanding furthered by our modest exploits is one of richest rewards a career as a scientist can provide.

Increasingly our work in polar regions has become more routine with permanent bases, ice-breaking research ships, and the latest satellite communications. Although still working at extreme temperatures, modern clothing and technology make even remote field campaigns quite safe and relatively comfortable. But the lure of polar research is just as great as ever before. In no small part this is driven by the tremendous changes that we are recording in the polar regions attributed to global climate change. At the time of writing the Intergovernmental Panel on Climate Change (www.ipcc.ch) is about to launch its fouth assessment report *Climate Change 2007*. There can be no doubt that in regions such as the Arctic and at least in parts of the Antarctic such as the Peninsula region in the past 50 years or so dramatic changes in the frozen landscapes have been induced by rising temperatures. It is recognized that the effects of a changing global climate will be most prominent in the polar regions, and for this reason there has to be a concentration of scientific endeavour if we are to understand the biology of the regions and interpret how the changing climate may influence the biology of these frozen realms.

This revision is produced during another major highlight in polar research, The International Polar Year 2007 to 2008 (IPY; www.ipy.org), which follows on from similar campaigns in 1882, 1932 and 1957. During the period March 2007 to March 2009 there will be an internationally co-ordinated effort to further our understanding of the Arctic and Antarctic and their respective roles in the Earth system (see special editions of the journals *Nature* published on 8 March 2007 and *Science* published on 16 March 2007). Scientists from 60 nations are joining the IPY in over 200 major projects, and it is very likely that the fourth IPY will generate a considerable wealth of knowledge that will surely necessitate this edition being revised substantially within the next decade.

This book was first conceived by Tony Fogg, a keen and influential communicator of the history of Antarctic exploration and scientific endeavour in the polar regions. Chapters 1 and 2, although updated, are largely unchanged from his first edition, since they describe the basic physical and biological constraints restricting survival polar regions. The subsequent chapters are updated substantially to reflect advances made over the last decade. However, it is a great testament to Tony that the concluding chapter is also largely unchanged: he was a perceptive scientist who was able to see the much bigger picture. As in the first edition, it is impossible to give references for everything covered in the text, but the selected citations together with the further reading list will enable the reader to locate all of the sources used.

While updating the information we have tried to maintain the same spirit of enthusiasm for the subject that was a hallmark of Tony's writing. This is not an easy task for a disparate group of authors, but we trust we have done enough to arouse curiosity and stimulate further research for a new generation of visitors to the frozen waters and lands of the Arctic and Antarctic.

David N. Thomas,
July 2007

Scientist working on Arctic pack ice in the Fram Strait in February at temperatures below −30°C (photograph David N. Thomas).

Contents

1 Introduction to the polar regions

1.1 Introduction

Freezing temperatures, ice, snow, continuous daylight, and long periods with no light at all. The polar regions are notorious for being frozen deserts at the ends of the Earth where nothing can survive. As this book will outline, nothing could be further from the truth. Both on the land and in the oceans of these frozen realms there is a wealth of biology that is adapted to the strong seasonality of light and temperature extremes. This extends from viruses and bacteria through to the charismatic mammals and birds that capture a wide-ranging popular interest. However, in order to understand the acclimations and adaptations of Polar biology, be it a bacterium growing on the surface of a glacier, or a phytoplankton cell encased in a frozen sea ice floe, it is essential to have an understanding of the physical forces structuring these regions.

Although the examples given in this book will point to many similarities between life in the two polar regions it is important to establish from the outset that despite being very cold the Arctic and Antarctic are very different. Much of the Arctic region is a land-locked ocean, covered by pack ice that can persist for several years. The Arctic has large areas of tundra and permafrost and several very large river systems. It contains Greenland, covered by the massive Greenland ice sheet which is on average 2 km thick. In contrast, the Antarctic is made up of a land mass almost entirely covered by the huge East and West Antarctic ice sheets (2–4 km thick) that are separated by the Transantarctic Mountains. The Antarctic continent is completely surrounded by the Southern Ocean, in which 16 million km^2 freezes over every year, effectively doubling the area of the frozen Antarctic. Whereas the Arctic has land connections with other climate zones, the Antarctic is effectively cut off from the rest of the world due to the barrier of the Southern Ocean (the shortest distance is the 810-km Drake Passage between South America and the Antarctic

Peninsula). Together the Greenland and Antarctic ice sheets account for more than 90% of the Earth's fresh water and if they were to melt sea level would increase by about 68 m (61 m from the Antarctic ice sheets and 7 m from the Greenland ice sheet).

One definition of the *polar regions* is that they are the areas contained within the *Arctic* and *Antarctic Circles*. These are parallels of latitude at 66°33′ north and south, respectively, corresponding to the angle between the axis of rotation of the Earth and the plane of its orbit around the Sun. They are the furthest latitudes from the North and South Poles where there is at least one day when the Sun does not fall below the horizon in the summer or does not rise above it in winter.

The areas encompassed by the Polar Circles total some 84 million km², 16.5% of the surface of the Earth. However, these circles mark no sharp transitions, either in climate or in flora and fauna, and they have little ecological significance. In places inside the Arctic Circle there are forests and thriving towns. Within the Antarctic Circle there is nothing but sea, ice, and sparse exposures of rock at the present day (Fig. 1.1). However, 60 million years ago, in the early Cenozoic, forests flourished despite Antarctic winter darkness. The polar regions are defined in various other ways by climatologists, terrestrial ecologists, marine biologists, geographers, and lawyers, but these will be considered later.

Another term that is frequently used in conjunction with polar regions is the *cryosphere* (derived from the Greek *kryos* for cold). The term collectively describes regions of the planet where water is in its solid form: it includes sea ice, lake ice, river ice, snow, glaciers, ice caps, and ice sheets, as well as frozen ground and permafrost. There is no doubt that the greatest proportion of the cryosphere is found in the polar regions, but not exclusively since high-altitude habitats clearly also store frozen water.

1.2 The energy balances of the polar regions

1.2.1 Solar irradiance

About half the solar energy entering the atmosphere consists of visible radiation; most of the rest is infrared, and a small fraction is ultraviolet. Mean values for total direct radiation from the Sun penetrating to the Earth's surface at various latitudes in the northern hemisphere are shown in Fig. 1.2. Because they have the Sun for all or most of the 24 h, polar situations actually receive more radiation around midsummer than do those on the equator. Nevertheless, the low angular height of the Sun, even in midsummer, and its disappearance in winter result in the total radiation per unit of surface area delivered during the year at the North Pole being

Fig. 1.1   Contrasting climates north and south in *summer* situations inside the Polar Circles. (a) Longyearbyen, Svalbard (78°N) in July (photograph by David N. Thomas) and (b) Scott Base (78°S) in January (photograph by Jean-Louis Tison). Note that despite the comparable time of the year that both photographs were taken, the Antarctic landscape is dominated by sea ice cover off the coast whereas the northern seas at these latitudes are ice-free (see colour plate).

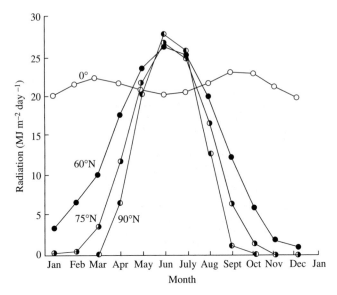

Fig. 1.2   Estimates of total direct radiation, with corrections for variations in atmospheric turbidity on the 15th day of each month at sea level at various latitudes north. From Hutchinson (1957).

less, by about 43%, than that at the equator. Direct solar radiation is augmented by scattered light from the sky to a variable extent, usually about 20% of the total. When the Sun is obscured by cloud much of its radiation is reflected back into space so that the values in Fig. 1.2, although not including scattered radiation from the sky, are likely to be overestimates.

These generalizations apply to both polar regions but there are differences: Since the Earth is closest to the Sun in the austral summer but most distant in the boreal summer, 7% more energy enters the Antarctic than the Arctic. Furthermore, the Antarctic atmosphere has less radiation-absorbing dust and pollutants, and, the continent having a higher elevation, the atmospheric mass to be penetrated by incoming radiation is less. Together, these factors result in the Antarctic getting 16% more energy. Nevertheless, the Antarctic is the colder region. The reasons for this will become apparent in the following sections.

1.2.2 Reflection and absorption of solar radiation

Incident radiation falling on a body may be reflected, transmitted, or absorbed. Absorbed radiation is changed into thermal energy in the absorbing material. The ratio of reflected to incident radiation is known as the *albedo* and has the value of 1.0 for complete reflection and of 0.0 for complete absorption. Snow and ice have high albedos, water a low albedo, and rocks

Fig. 1.3   Snow-covered ice shelf and sea ice with high albedo compared with open water with low albedo (photograph by David N. Thomas).

Table 1.1 Albedos of various natural surfaces

Surface	Albedo
Snow-covered sea ice	0.95
Fresh snow	0.8–0.85
Melting snow	0.3–0.65
Quartz sand	0.35
Granite	0.15
Bare earth	0.02–0.18
Coniferous forest	0.10–0.14
Water	0.02
The Earth as a whole	0.43

are intermediate (Perovich *et al.* 2002; Fig. 1.3 and Table 1.1). The mean albedos of the polar regions vary seasonally but are always higher than that of the Earth as a whole. That of the Arctic is lower than that of the Antarctic, 0.65 compared with 0.90, as a result of loss of reflective snow cover and relatively greater ice melt in summer. In the Arctic ice covers about 2 million km^2 of land and sea ice extends over 7 million km^2 at its minimum and 14–16 million km^2 at its maximum in late February or March. In the Antarctic, land ice extends over 12.6 million km^2 and sea ice over 4 million km^2 at its minimum, increasing to about 20 million km^2 at

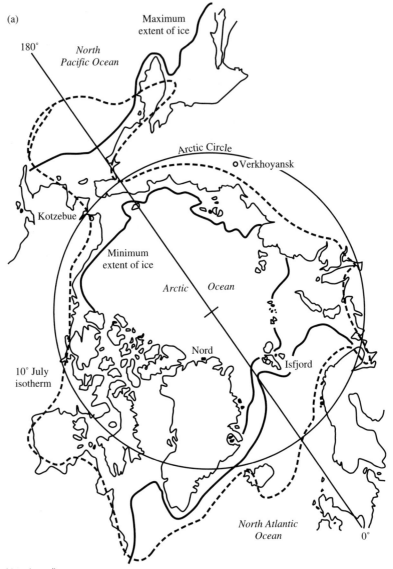

(a)

Fig. 1.4   (Continued)

its maximum in September, when it goes well north of the Antarctic Circle all round the continent and effectively more than doubling its size (Fig. 1.4). The total area of high albedo in the summer is sufficient at both poles to reflect much of the incident radiation back into space and thus reduce heating of land and sea. This resolves the paradox that the area which receives the maximum monthly input of solar energy of any on Earth, the ice sheet of East Antarctica, is also the coldest on Earth.

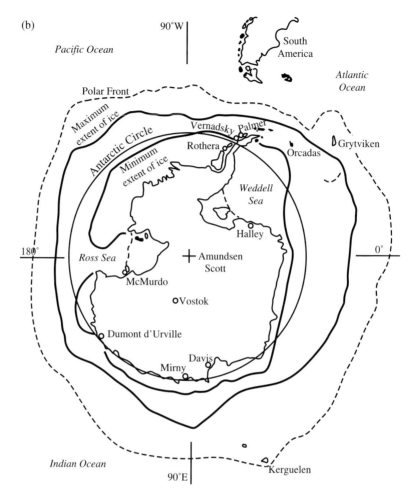

Fig. 1.4   Extents of sea ice and positions of Polar circles in the (a) Arctic and (b) Antarctic. The weather stations indicated are the same that appear in Fig. 1.5.

Spots of low albedo within polar regions can absorb large amounts of heat. The Russian station, Mirny (66°33′S 93°01′E), is on snow-covered ground whereas the nearby Oazis station (66°30′S 101°E) has bare rock around it. At Mirny, most of the incoming radiation is reflected and little heat is accumulated in the ground even at midsummer; because of its high albedo, snow cover tends to persist once established. At Oazis, the rock surface heats up in summer and soil temperatures rise to 10°C or so above that of the ambient air. A water surface, which also has a low albedo except at low angles of incidence, behaves similarly. Since water has a high specific heat, lakes and seas act as particularly effective heat stores. Not only does water transmit radiation into its depths but heat can be carried downwards by its turbulence. Were it not for extensive and persistent high-albedo snow cover, Polar climates could be temperate, as, indeed, they have been in the past.

1.2.3 Long-wave radiation from terrestrial sources and its absorption in the atmosphere

Thermal energy acquired by absorption of solar radiation is lost by emission of radiation of a longer wavelength, infrared radiation. The amount of energy re-radiated is a function of the infrared emission characteristics of the surface and the fourth power of its absolute temperature. Most natural surfaces have emission characteristics in the same range: snow, ice, rock, and water all having similar high values at the same temperature. The Earth's surface, to a good approximation, can be regarded as having perfect infrared emissivity at a temperature of 285 K in the waveband 4.5–50 μm, with a peak at about 10 μm.

Whereas the atmosphere is highly transparent to solar radiation it absorbs terrestrial radiation because of the presence of clouds, water vapour, and certain gases, all of which show high absorption within the waveband just specified. These gases, which include carbon dioxide and methane, have achieved notoriety as so-called greenhouse gases because of their increasing concentrations in the atmosphere, leading to increasing interception of infrared radiation, resulting in global climate warming. Without the blanketing effect of the atmosphere the Earth's surface temperature would be 30–40°C lower than it is and would vary between greater extremes of heat and cold. Just now the blanket is becoming oppressively thick (see Chapter 10).

Liquid water in the form of clouds is nearly opaque to terrestrial radiation even though its concentration may be only $1\,g\,m^{-3}$, equivalent to a thickness of $0.001\,mm\,m^{-1}$. Clouds of ice or snow are similarly highly absorbing. Water vapour present in the clear atmosphere also has high absorption for most wavelengths in the terrestrial emission spectrum but has a window in the region of 10 μm, the region of maximum terrestrial emission. The frost which often accompanies a cloud-less night is a familiar example of the heat loss that this allows.

1.2.4 Long-wave radiation emission in the atmosphere

The absorbing agents in the atmosphere heat up and in their turn emit long-wave radiation according to their temperature and emission characteristics (Law and Stohl 2007). Some of this, about 75%, will return to the Earth's surface and there be reabsorbed. The net loss of energy from this surface will thus be the radiation which it emits itself less that of the back radiation received from above. When a layer of warm cloud overlies cold ground the balance becomes positive. The rest of the long-wave radiation emitted by atmospheric components escapes into space. There is an overall heat loss via long-wave radiation from the Earth and its atmosphere because their temperatures are higher than that in space.

Net heat loss is greater in the Antarctic than in the Arctic. The reason for this lies partly in the greater prevalence of clouds in the Arctic, particularly at its periphery, as compared with the Antarctic, the high continental plateau of which is generally cloud-free. Furthermore, the atmosphere over the plateau is more transparent to long-wave radiation because of its thinness and dryness. An added complication in both polar regions is that of temperature inversions, that is to say increases, rather than decreases as normally found, of temperature with height above ground. These arise as a result of snow surfaces beneath clear skies reflecting nearly all incoming radiation, so that air near the surface becomes chilled and dense. Above it, between 200 and 1000 m, is less-dense air containing more moisture which intercepts some outgoing long-wave radiation and remains warmer. Inversions are prevalent over the Antarctic plateau for most of the year and account for the fact that temperatures on the plateau, after a rapid fall in autumn, scarcely decrease thereafter so that the winter is *coreless* (Fig. 1.5). Variations in refraction associated with inversions produce the optical phenomena, such as mirages, which are characteristic of polar regions (Pielou 1994).

1.2.5 Transport and global balance of thermal energy

Around the poles there is net loss of energy by radiation over the year whereas in equatorial latitudes there is a net gain. Losses balance gains and as a whole the Earth and its atmosphere neither warm up nor cool down. The loss of heat from the polar regions is made good by a flow of excess heat carried in currents of air or water from lower latitudes. The poles are sinks for thermal energy and the equatorial regions the source. Air transports not only sensible heat but also latent heat in the water vapour it contains. Some is released when the vapour condenses to form clouds ($539 \, \text{cal} \, \text{g}^{-1}$) and some more when the liquid water freezes ($79.8 \, \text{cal} \, \text{g}^{-1}$).

If Ptolemy's geocentric theory were correct and the Earth remained motionless while the Sun went around it, warm air carrying water vapour with it would rise in equatorial regions and flow towards each pole along meridional paths. Cooling on the way, this air would eventually sink and return towards the equator, again along meridional tracks, at a lower level. In the oceans, sea water, concentrated by evaporation in low latitudes, would become saltier, more dense, and sink, likewise flowing polewards and carrying heat along meridional pathways, assuming no obstruction by land masses. In high latitudes it would rise as it encountered water which, although less salty, would be denser because it was colder. However, because the Earth rotates, flows of air and water are deflected, by the Coriolis

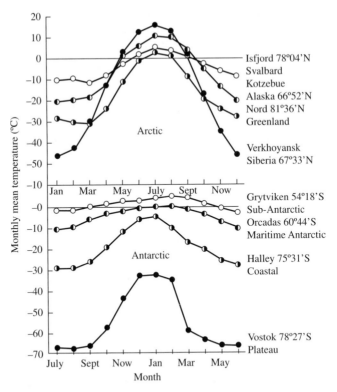

Fig. 1.5   Mean monthly temperature at different locations (see Fig. 1.4) in the Arctic and Antarctic. N. B. at Vostok the temperature from April to September remains more or less level — the winter is *coreless*, unlike that in the Arctic, which has a sharp minimum. Data from Stonehouse (1989).

force, to the right of the direction of movement in the northern, and to the left in the southern hemisphere. The Coriolis force is zero at the equator and maximal at the poles.

In the atmosphere, thermal energy is mostly transported in the lower layer, the *troposphere*, which is about 10 km thick and separated by a temperature minimum, the *tropopause*, from the *stratosphere*. The basically meridional two-way traffic of warm air polewards and cool air equatorwards is obscured, not only by the Earth's rotatiffon, but also by the different thermal effects of continents and oceans. The result is a complex pattern of zonal and cellular circulation. Salient features are that, between latitudes 30 and 60°, both north and south, there are zones of predominantly low pressure and westerly winds, and polewards of 60° there are zones of high pressure with north-easterly and south-easterly winds, respectively, north and south.

Water currents are also subject to the Coriolis force but are obstructed and deflected by land masses and irregularities in the seabed. A further complication is that atmosphere and ocean interact. The drag of winds on the sea surface induces currents and, also, by setting up slopes in the surface, winds produce other currents in response to the pressure gradients that arise. Wind-induced currents may be temporary, varying with local weather, but the major oceanic circulations correspond roughly to the pattern of the prevailing winds. Such currents are largely superficial, and, for present purposes, deep-water currents, the direction of which need show no relation to those at the surface, are more important. The temperature at great depths in the oceans is everywhere near to freezing point. This cold bottom water comes from two main sources, one in the Greenland Sea, the other in the Weddell Sea in the Antarctic, where surface water becomes cold and dense enough to sink to the bottom and flow equatorwards. Other deep-water currents carry warm salty water from equatorial regions polewards in replacement. These currents are of enormous volume but move slowly, perhaps around 1 or 2 km per month, and so the water in them stays below the surface for many hundreds of years.

1.2.6 Heat influx and balance in the polar regions

Against this general background we can look more specifically at the paths by which thermal energy reaches the polar regions. First, a radical geographic difference between Arctic and Antarctic is of key importance. Whereas the Arctic centres on a sea of some 14 million km² enclosed by islands and the northern stretches of continents, the other is a continent of 13.3 million km²—larger than Europe but smaller than South America—surrounded by a belt of ocean which separates it by 800 km from an outlier of the nearest major land mass (Fig. 1.6). This difference has profound consequences for their respective climates, biology, and importance in the regulation of the global environment (Walton 1987).

1.2.7 The Arctic

The major input of heat is provided by northward-moving warm air which interchanges with cold polar air in cyclones associated with low pressure along the atmospheric Polar Front in the region of 60°N. Variations in surface topography introduce complications and a regular succession of cyclones is frequently obstructed by well-developed stationary regions of high pressure: so-called anticyclonic blocking. From there, the warm air travels high in the troposphere to subside around the pole. It then returns as surface winds away from the pole, the Coriolis force giving these an easterly direction. This is an anticyclonic situation. Over the sea ice of the Arctic Ocean the motion of these winds is imparted by pressure differences. On the Greenland ice cap, winds become related to topography, dense cold air flowing downslope. Such *katabatic* winds are intermittent. The air accelerates as it

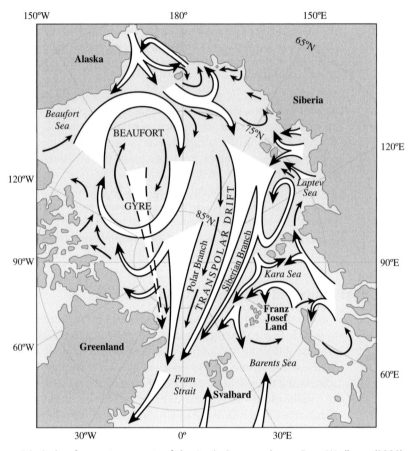

Fig. 1.6   Principal surface water currents of the Arctic Ocean and seas. From Wadhams (2000).

descends, becoming compressed by the higher pressure at the lower levels and developing heat equivalent to the work done.

Heat is also contributed by the great Siberian rivers (Fig. 1.7), which introduce fresh water at a rate of about 3500 km³ year⁻¹, mainly in summer when its temperature may get up to between 10 and 15°C. An additional 1500–2000 km³ year⁻¹ enters as a freshwater fraction of the Bering Strait inflow. The main ocean current flowing into the Arctic Ocean (Carmack in Smith 1990) is the West Spitsbergen Current, a northward-flowing extension of the Norwegian Atlantic Current, passing through the Fram Strait (approximately 80°N 0°; Fig. 1.6). This follows a deep trench leading to the Arctic Ocean. The access via the Barents Sea is partially obstructed by shallows. The water in this current is warm (above 3°C) and relatively saline (salinity* greater than 34.9). The amount of water transported is uncertain: estimates

---

* Salinity is defined as the grams of salt dissolved per kilogram of water. There are no units.

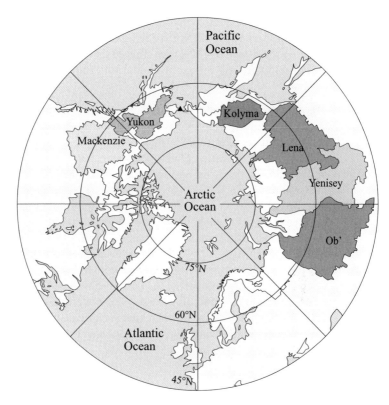

Fig. 1.7   Map showing major river catchments discharging into the Arctic Ocean (image courtesy of R.M. Holmes).

vary between 2 and 8 Sv (1 Sv or Severdrup = $10^6 \, \mathrm{m}^3 \mathrm{s}^{-1}$), or 60 000 and 250 000 $\mathrm{km}^3 \, \mathrm{year}^{-1}$, but it is possible that as much as half of this circulates in the vicinity of the Fram Strait without entering the Arctic Ocean.

Alongside the West Spitsbergen Current to the west is the East Greenland Current. This is the main current out of the Arctic Ocean and carries cold (below 0°C), relatively fresh (less than 34.4 salinity), water southwards. The flow of this current is between 3 and 30 Sv (91 000–910 000 $\mathrm{km}^3 \, \mathrm{year}^{-1}$) and it carries with it some 4 million MT of drift ice to lower latitudes. The Bering Strait is narrow (85 km) and shallow (50 m), allowing a small (about 0.8 Sv or 25 000 $\mathrm{km}^3 \, \mathrm{yr}^{-1}$) northerly flow. Within the Arctic Ocean the Transpolar Drift, a surface current, flows from the Siberian to the Greenland side, where it feeds into the East Greenland Current. It was trusted by Nansen to carry his ship, the *Fram*, beset in the ice, across the Arctic Ocean into the vicinity of the North Pole. A less heroic demonstration of the Transpolar Drift is being provided by a consignment of 29 000 plastic floating bath toys including ducks, frogs, and turtles, lost from a container ship in the North Pacific at 44°N 178°E in October 1992. It was proposed that the floating toys would be transported by the pack ice across the Arctic Ocean into the North Atlantic. To the delight of oceanographers tracking these toys in 2003, they showed up on North Atlantic coastlines and findings are still being reported in July 2007.

The pack ice, up to 12 m thick, plays an important part in conserving heat. The sea water itself, with a mean depth of about 1200 m and a volume of about 17 million km³, provides an immense heat reservoir. Its ice cover reduces heat transfer to the atmosphere by one or two orders of magnitude compared with that from open water. Furthermore, largely because of the inflow of river and Bering Strait waters, a layer of low salinity water floats on top of the denser water in the Arctic basin, producing a marked *halocline* at between 30 and 60 m which limits the convection that would otherwise mix the whole water column and promote heat loss. The deep water consequently remains between −0.5 and −0.9°C, appreciably above its freezing point of −2.0°C. These various factors contribute to the generally higher temperatures of the Arctic compared with the Antarctic. Plans to divert the southward part of the flow of some Siberian rivers to alleviate water shortages in the south perhaps need not create too much alarm. On present evidence it seems unlikely that such diversions would have major effects on circulation or sea ice distribution in the Arctic Ocean.

1.2.8 The Antarctic

Although the heat exchanges of Antarctica are still incompletely understood, it presents a simpler situation for analysis, having an approximately circular ice-covered land mass forming a dome, without too many topographical irregularities, nearly centred on the geographical pole and surrounded by a continuous belt of deep ocean (Walton 1987). This allows the collection of meaningful data and the construction of realistic mathematical models of circulatory processes.

The chief agent of imput of thermal energy is again atmospheric circulation. The zone of westerly winds produces a succession of cyclones which satellite images show as a regular procession of cloud spirals around the continent at latitudes 60–65°S. The Antarctic thus contrasts with the Arctic in that anticyclonic blocking is infrequent. The cyclones often swing south into the Ross Sea area but rarely depart from their circumpolar track to carry warm air into the centre of the continent. As in the Arctic, cyclones provide a mechanism for exchanging cold polar air for warm moist air from lower latitudes. The water vapour is again a source of thermal energy. This air travels south at an intermediate height in the troposphere and sinks in the high-pressure region over the summit of the polar plateau. The ice dome favours the initiation of katabatic winds, which under the influence of the Coriolis force follow a north-westerly track. This layered system of air movements is seen in dramatic form when smoke from the volcano Erebus (3785 m, 77°40′S 167°20′E) is carried polewards whereas at sea level a blizzard may be blowing from the south. If these low-level winds are sufficiently strong, surface irregularities such as wind-produced ridges in the snow (*sastrugi*) cause turbulence which disturbs inversions and mixes in warmer air and moisture from above. The speed of katabatic winds increases with slope and so turbulent heat exchange is about four times greater around the edge of the continent than it is in the interior.

There are no permanent rivers to contribute to the heat budget of the Antarctic. In the sea (Carmack in Smith 1990) there is meridional transport which is, however, deflected by the wind-driven Antarctic Circumpolar Current. This current, being deep-reaching and confined only at the Drake Passage and south of the Australasian land mass, where it has to pass through a deep channel connecting the Indian and Pacific Oceans, is an enormous flow of about 130 Sv (4 million km^3 year^{-1}). This is no mean barrier to meridional transport and is regarded as one of the major factors contributing to the exceptionally frigid state of the Antarctic. Nevertheless, considerable southward transport of heat, mainly from the Indian Ocean, takes

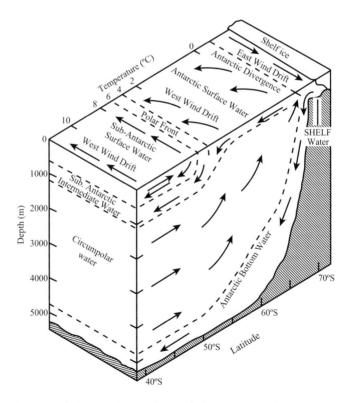

Fig. 1.8   Three-dimensional diagram showing the circulation patterns and water masses in the Southern Ocean.

place in the Circumpolar Deep Water (Fig. 1.8), which has temperatures between 0 and 1.8°C and maximum salinity of around 34.76. This wells to the surface at the Antarctic Divergence, about 70°S, and spreads north and south. The north-flowing fraction mixes with fresh water released by ice melt, giving temperatures of around −0.4°C and salinities of about 34.20. It forms a layer, some 200 m in depth, separated by a sharp density gradient (*pycnocline*) from the Circumpolar Deep Water. Meeting warmer Subantarctic Surface Water, it plunges beneath this at the Antarctic Polar Front, a feature also known by the not-quite-synonymous name of Antarctic Convergence (Fig. 1.8). The Polar Front remains in a surprisingly constant position, extending all the way round the continent within the Circumpolar Current (see Fig. 1.4b). It is a boundary of great biogeographical importance and is easily detected by abrupt changes in temperature in both surface sea water and air. The upwelling water from the Circumpolar Deep Water which continues on south becomes colder and mixes with water, from the continental ice shelves, charged with brine formed by freezing of sea water. Being both cold (0.4–1.3°C) and saline (salinity of 34.66–34.72) it sinks and then spreads northwards across the sea bottom (Jacobs 2004). The Weddell Sea and to a lesser extent the Ross Sea are the major centres for the production of this Antarctic Bottom Water. Traces of it have been found as far beyond the equator as 17°N in the Atlantic Ocean. The scale of events in the Weddell Sea is shown by the estimate of between 76 and 97 Sv (2.4–3.0 million km^3 year^{-1}) for water transport in the gyre occupying its basin. This vastly exceeds that in Arctic waters.

The area of the Southern Ocean covered by sea ice increases five- or six-fold each winter (Fig. 1.4b) but with great year-to-year variation in timing, extent, and distribution. This interannual variability is linked to atmospheric processes and the flow rates and directions of ocean currents (Murphy *et al.* 1995). Unlike Arctic ice, which is constrained by land, Antarctic ice is free to spread over deep ocean, almost anywhere that wind and tide take it. The heat exchanges of the Southern Ocean are modified correspondingly. Ice formation is most active in coastal regions subject to cold katabatic winds: further offshore turbulence retards ice formation. Both pack ice and icebergs are carried in a generally northern direction and, since freezing involves release of latent heat, whereas the eventual melting requires supply of heat, this implies a net poleward transport of heat against an export of fresh water. The release of heat by freezing at the beginning of winter and its uptake on melting in summer work to buffer temperatures. The high albedo and insulating properties of the ice also minimize heat exchanges.

The Arctic and Antarctic are the two great heat sinks which between them determine the patterns of both atmospheric and oceanic circulations and are thus key areas in regulating the global environment. Of the two, the Antarctic is the dominant.

1.3 Climate

1.3.1 The climatic boundaries of the polar regions

Another simple definition of the polar regions is that they are those areas in the vicinity of the poles, where the mean temperature of the warmest month is less than 10°C. The 10°C summer isotherm (Fig. 1.4a) usually coincides with the limits of tree growth (Aleksandrova 1980). Isolines of radiation balance give a better match to tree line, although they sometimes deviate by as much as 160 km. The position of the tree line depends on both latitude and altitude. The transition from tall forest to dwarf, shrubby, vegetation, which marks the tree line, is sometimes strikingly sharp, largely because single exposed trees, not being able to ameliorate their environment, tend to be eliminated. A closed canopy affords some protection from wind stress but since forest vegetation is penetrated by large-scale eddies temperatures of the above-ground tissues are closely coupled to those in the air. In contrast, shrubby vegetation is aerodynamically smoother and dissipates heat less readily, experiencing tissue temperatures and microclimates that, on average, are warmer than the air. Consequently, dwarf shrubs can succeed in polar climates in which trees fail to grow and reproduce.

In the Arctic the 10°C summer isotherm undulates around the Polar Circle, going well south of it in the regions of the Bering Strait and the north-west Atlantic where the Kamchatka and Labrador Currents, respectively, bring cold surface water down from the north. The isotherm goes north along the coast of Norway because of the warm North Atlantic Drift (Fig. 1.6). Around the Antarctic the 10°C summer isotherm runs well north of the Polar Circle, at about 50°S, and almost entirely over ocean, only touching land at the tip of South America. There the coast is mostly forested. The sub-Antarctic islands south of the isotherm have vegetation which resembles Arctic tundra and are treeless. The Falkland Islands (approximately 53°S 58°W) lie just on the cold side of the isotherm and have no native trees but their grasslands, dwarf shrub heaths, and *fell fields* (with discontinuous cover of cushion plants) are scarcely sub-Antarctic in character and will not be dealt with here.

1.3.2 The Arctic climates

Within the confines of the 10°C summer isotherm conditions are generally cold, dry, and windy (Sugden 1982, Stonehouse 1989) but there are variations which are not easy to classify. The Arctic can be divided into the central maritime basin and the areas peripheral to it, in which can be distinguished the ice caps, polar maritime climates (located principally around the Atlantic and Pacific coastlines), and the polar continental climates as in north Alaska, Canada, and Siberia. There are no fixed meteorological stations in the central maritime basin but observations have been made from a succession of stations on drifting ice islands. This is a climatically stable area with a strong central anticyclone, clear skies, and light centrifugal winds during the winter. Because of the reservoir of heat in the ocean, temperatures do not fall to extremely low levels, averaging −30°C offshore and −26 to −28°C in the coastal regions during the depth of winter. When the Sun returns the anticyclone weakens and there are incursions of depressions, bringing moist air, fog, cloud, snow, rain, and strengthening winds. Temperatures, except in a small central area, rise above freezing point so that the periphery of the pack ice melts and large areas of open water appear along the Alaskan and Eurasian coasts during June and July.

There are ice caps on the more northerly islands, for example, about 58% of Svalbard (approximately 79°N 15°E) is permanently ice-covered. That on Greenland, which covers most of the island and rises to 3000 m, is by far the most massive. It is fed by snow borne by year-round south-western airstreams, which deposit some 100 cm rain equivalents annually in the south but only about 20 cm in the north, parts of which are consequently almost ice-free. Temperatures on the plateau of the ice cap fall to −40 to −45°C in winter, rising to −12°C in summer.

Arctic continental climates are dominated by anticyclonic conditions in winter with low temperatures, light winds, and little precipitation. Mean monthly temperatures rise to freezing point around May and can get well above it in the short summer (Fig. 1.5). Weakening of the anticyclone allows incursions of depressions, bringing warm, moist, oceanic air with precipitation that favours the development of tundra on low ground and ice caps and glaciers higher up. The most extreme continental conditions are found around Verkhoyansk (67°33′N 133°25′E), the 'pole of cold', in eastern Siberia, where an intense winter anticyclone spreads cold, dry, air in all directions. Being well away from the sea it has variations in temperature from −67.8°C in winter to 36°C in summer (Fig. 1.5). Precipitation is mainly in the form of summer rain and amounts to only 15 cm per annum.

Maritime climates are ameliorated by the sea, especially where there are warm currents. In the Canadian Arctic, the worst climate is encountered in the Hudson Strait area (approximately 63°N 70°W), which is dominated by open water and frequent cyclonic activity, giving the highest average temperature, but the heaviest snowfall, highest average wind speeds, and greatest number of summer fogs in this sector. This is the region in which many of the early seekers after the Northwest Passage came to grief. The south-west of Greenland is warmed by an offshoot of the North Atlantic Drift (see Fig. 1.5) and is comparatively free of sea ice. Its mild climate allows sheep farming on luxuriant tundra within a short distance of the ice cap. Parts of Iceland likewise have a mild climate and forests of birch and spruce in the south. Its northern shores have pack ice drifted in by the East Greenland Current. The same current keeps the east coast of Greenland cold, even in summer.

The North Atlantic Drift passes Iceland to give Svalbard (Fig. 1.5) and the north-western tip of Europe remarkably temperate climates with the tree line going far north. Only the northern part of Svalbard remains ice-bound in summer. The effect of the same current persists along the Eurasian coast of the Arctic Ocean as far east as Novaya Zemlaya (approximately 75°N 60°E), keeping the Barents Sea open in summer. Depressions bring abundant summer rain as well as winter snow. The great Siberian rivers have some ameliorating influence but further east still the coastal climate becomes harsher with short, cold, summers and frigid winters. Precipitation decreases and Kotel'niy (75°59'N 138°00'E) on Ostrova Novosibirskiy, which has only 13 cm per year, can be described as desert.

1.3.3 The Antarctic climates

The array of Antarctic climates is simpler (Sugden 1982, Stonehouse 1989). The central feature here is the enormous, high, continental plateau, usually dominated by a high-pressure system. With its 'coreless' winter goes a 'pointed' summer, lasting only a few weeks (Fig. 1.5). When planning their attempts on the South Pole, neither Amundsen nor Scott had any idea of this state of affairs. Amundsen was lucky to arrive just before the peak of summer and Scott desperately unlucky to arrive just after it. Temperatures depend on altitude as well as latitude and the Russian base Vostok at 78°28'S and 3400 m above sea level holds the world record for low temperature, −89.5°C (Fig. 1.5). Wind speeds are generally low and precipitation extremely low. Direct measurement of snowfall is imprecise at best and the prevalence of drift on the continent makes it difficult or impossible. Between 3 and 7 cm rain equivalents seems to be likely—less than in most tropical deserts—but it is the extremely low moisture content of the air which makes the Antarctic plateau so highly desiccating.

As the slope of the ice cap steepens towards the coast, a different type of climate predominates, with strong and persistent katabatic winds averaging around 11 m s^{-1} (39 km h^{-1}) and occasionally reaching 300 km h^{-1} (Fig. 1.9). The coast itself has milder temperatures, dropping to around −20°C in winter and rising to near zero in summer (Fig. 1.5). Over the sea the cold air from katabatic winds rises and is dissipated in turbulence, leaving conditions at the surface more tranquil. Cyclonic activity sometimes penetrates landwards, bringing strong winds and precipitation. The weather is very dependent on topography, which affects the incidence of katabatic winds and the amount of sea ice insulating the coast from the relatively warm

Fig. 1.9   Leaning on the wind while collecting ice for the kitchen, Adelie Land, Antarctica. From the original lantern slide taken by Frank Hurley taken during the Australasian Antarctic Expedition between 1911 and 1914. Courtesy of the Scott Polar Research Institute.

sea. The US station, McMurdo, at 78°S on the Ross Sea has persistent sea ice but sunny summers with little snow so that the rocks are exposed for 3 or 4 months each year. The effect of topography is particularly marked on the Antarctic Peninsula. On its west coast the Ukrainian station, Vernadsky (previously the British station, Faraday, at 65°15′S 64°15′W) has a mild maritime climate. At the same latitude on its east coast frigid conditions are maintained year round by cold water brought from higher latitudes by the Weddell gyre and an ice shelf extends out from the shore. Temperatures are some 4–6°C colder than on the west coast.

Topography also produces oases, or dry valleys, which are a special feature of Antarctica. These are ice- and snow-free areas are found at various points around the continent. The dry valleys of Victoria Land, accessible from McMurdo Station, have been investigated intensively, as have those of the Bunger Hills in the vicinity of Mirny. Dry valleys (Fig. 1.10; see also Chapters 3 and 5) exist where loss of snow and ice by *ablation* (i.e. removal by sublimation or run-off of melt water) exceeds addition by precipitation and movement of ice into the area. The configuration of the land surface must be such as to divert the flow of ice elsewhere and also to provide a precipitation shadow. The effect on the radiation balance of the resulting lowering of albedo has been mentioned and temperature fluctuations from around −38°C to as much as +15.6°C have been recorded. Winds are generally light but strong katabatic winds blow occasionally and wind-eroded rocks, *ventifacts*, are a striking feature of the dry valleys. The bare area tends to extend along the direction of the prevailing wind since debris is carried downwind and, being deposited on snow, decreases its albedo,

Fig. 1.10   Aerial view of Lake Fryxell in the McMurdo Dry Valleys, Antarctica (photograph by Dale Anderson) (see colour plate).

promoting melting and exposure of bedrock. Extreme desiccation is a major factor for living organisms; the mummified remains of the occasional seals and penguins which stray into these cold deserts may remain for centuries. The annual precipitation is around 4.5 cm rain equivalents.

The maritime Antarctic, taken as the zone from 70°S northwards to 55°S, including the Antarctic Peninsula and its associated islands together with adjacent archipelagos, falls in the domain of cyclones. Vernadsky Station has mean temperatures of around −10°C in the winter rising to near zero in summer. There is more cloud than on the coasts of the main continent and winds are stronger. Sea ice usually disperses in early spring and reforms in autumn. Signy Island (60°43′S 45°36′W), in the South Orkney, has much the same sea ice conditions and temperature range but has cloud for 80% of the summer and 60–80% of the winter. Annual precipitation, mostly snow but sometimes rain, amounts to 40 cm rain equivalents.

All these islands are well south of the Polar Front. Those in the vicinity of the front are termed *sub-Antarctic*. South Georgia (approximately 54°30′S 37°00′W) is some 350 km south of the front. It has sub-zero temperatures every month of the year (Fig. 1.5) and permanent ice fields, although temperatures of 15°C are not uncommon. It receives an almost continuous series of atmospheric depressions and its climate may be summed up as generally cold, wet, and cloudy, with strong winds, subject to abrupt change but without great seasonal variation. Katabatic winds, caused by cold air spilling down valleys, give rise to sudden squalls and whirlwinds, known as williwaws, which are frequent in some of the harbours. The south-west coast, exposed to the prevailing westerly winds, has a more rigorous climate than the sheltered north-east. These winds, heavily laden with water vapour, are forced to rise over the steep 3000-m spine of the island and in doing so expand, cool, and deposit their moisture as rain or snow. The then relatively dry air descending on the leeward side of the mountains is compressed and so warms up, producing rises in temperature of as much as 10°C. These *Föhn* winds are an outstanding feature of South Georgia. Icebergs are common about its coasts but pack ice reaches it rarely. The French archipelago of Kerguelen (approximately 49°S 70°E) is about as far south as Paris is north, but its situation just on the Polar Front ensures a sub-Antarctic climate similar to that of South Georgia. Temperatures are again rather uniform, only falling a little below freezing in winter and rising little above 10°C in summer but at 1050 m it has an ice cap giving rise to numerous glaciers. Like South Georgia it has strong westerly winds, fogs, rain, and snow. Icebergs are occasionally seen but its coastal waters are always free of ice.

1.4 Thermohaline circulation

Despite their very different characterstics and the vast expanses covered by the oceans of the world, they are all interconnected by a large-scale movement of water that is referred to as the meridional overturning circulation (MOC), or *global thermohaline circulation* or *global ocean conveyor belt* (Broecker, 1997, Broecker *et al.* 1999, Clark *et al.* 2002; Fig 1.11). The basis of

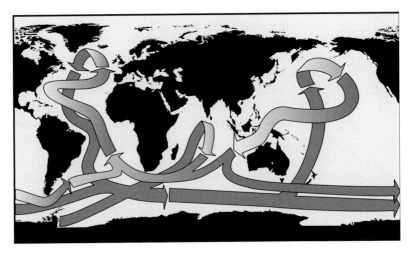

Fig. 1.11   Illustration of the global thermohaline circulation or meridional overturning circulation (MOC) (sometimes referred to as the *global ocean conveyor belt*).

themohaline circulation is that a kilogram of water that sinks from the surface into a deeper part of the ocean displaces a kilogram of water from the deeper waters. As sea water freezes in the Arctic and Antarctic and ice sheets consolidate, cold, highly saline brines are expelled from the growing ice sheet (see Chapter 7) increasing the density of the water and making it sink.

In the conveyor-belt circulation, warm surface and intermediate waters (0–1000 m) are transported towards the northern North Atlantic, where they are cooled and sink to form North Atlantic Deep Water that then flows southwards. In southern latitudes rapid freezing of sea water during ice formation also produces cold high-density water that sinks down the continental slope of Antarctica to form Antarctic Bottom Water. These deep-water masses move into the South Indian and Pacific Oceans where they raise towards the surface. The return leg of the conveyor belt begins with surface waters from the north-eastern Pacific Ocean flowing into the Indian Ocean and then into the Atlantic Ocean.

It is not just the temperature and salinity of the deep-water formation in the polar regions that is crucial to the ocean circulation. These water masses are rich in oxygen, and so are fundamental for transporting oxygen to the ocean depths where respiration by deep-sea organisms consumes oxygen. The transport of dissolved organic matter and inorganic nutrients is also governed fundamentally by this transport, increasing the nutrients being remineralized during the transfer of the deep-water masses. Therefore water rising at the end of the conveyor belt in the north-eastern Pacific has higher nutrient loading, and lower oxygen concentrations than North Atlantic waters at the beginning of the conveyor belt (Sarmiento *et al.* 2004).

1.5 El Niño Southern Oscillation

The El Niño Southern Oscillation (ENSO) is the largest climate oscillation on Earth to influence ocean currents and surface temperatures. El Niño is the term used to refer to unusually warm surface temperatures in the equatorial region of the Pacific. In contrast La, Niña is the state when there are abnormally cold ocean surface temperatures in the region. During non-El Niño and non-La Niña conditions sea surface temperatures are about 5–8°C warmer in the western than in the eastern tropical Pacific, and the trade winds blow to the west across the region. The sea level is also higher in the western tropical Pacific because of the prevailing winds.

During an El Niño period the sea surface temperatures increase significantly in the eastern tropical Pacific, and the trade winds either slacken or reverse direction, also moving less water from east to west, greatly affecting the physical characteristics of the waters in the regions. However, the effects of El Niño are far more widespread than the Pacific Ocean, and weather patterns and ocean circulation patterns throughout the world are influenced by these events.

The Southern Oscillation part of ENSO refers to the east–west atmospheric circulation pattern characterized by rising air above Indonesia and the western Pacific and sinking air above the eastern Pacific. The strength of this circulation pattern is defined by the Southern Oscillation Index (SOI) which is a measure of the monthly differences in surface air pressure between Tahiti and Darwin. During an El Niño period the surface air pressure is higher in the western tropical Pacific than in the eastern tropical Pacific the SOI has a negative value, and the reverse is true for La Niña periods (Fig. 1.12).

ENSO events generally happen every 4–7 years and can last between 1 and 2 years; however, it seems as though in the 1980s and 1990s El Niño events were more frequent and lasted longer than recorded previously. There was a very protracted El Niño from 1990 to 1995 and exceptionally strong ENSO events in 1992/1993 and 1997/1998.

There is substantial evidence that the SOI is correlated closely to climate anomalies in certain sectors of the Southern Ocean, and that year-to-year variation in sea ice cover in these regions is linked to recent ENSO events. When the SOI is in a positive phase there are generally lower sea-level pressure, cooler surface air, and sea surface temperatures in the Bellingshausen, Amundsen, and Ross Seas, with the potential for greater ice growth. In

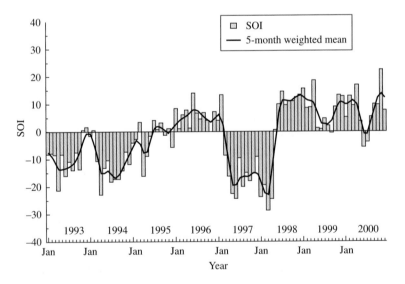

Fig. 1.12   Southern Oscillation Index (SOI) from 1993 to 2000. Redrawn from data presented by the Australian Government's Bureau of Meteorology. Copyright Commonwealth of Australia reproduced by permission.

contrast, during El Niño events (negative SOI) the reverse is true and declines in the ice extent of these regions have been noted. In particular the ENSO years of 1983, 1988, 1992, and 1998 show very good correlations with lower sea ice extents in the Ross Sea in particular. ENSO links to sea ice distribution are not confined to these sectors but are also reported for cyclical sea ice dynamics in other regions of the Southern Ocean.

1.6 Arctic and North Atlantic Oscillations

Over the past few decades it has become clear that many oceanographic trends in the northern hemisphere are closely linked to the North Atlantic Oscillation (NAO). This is one of the most dominant modes of climate variability following El Niño, although there is very little connection between the two. The NAO links the atmospheric pressure distribution between the region of Greenland–Iceland and the subtropical central North Atlantic in the Azores. The NAO index is defined as the difference between the Icelandic low and the Azores high in winter (December to March).

A positive NAO index is characterized by a strong Icelandic low and Azores high pressure with a corresponding strong north–south pressure gradient. When this is the case the pressure differences result in stronger and more frequent storms crossing the Atlantic Ocean towards a more northerly routing. This results in warmer and wetter winters in Europe and the eastern USA in conjunction with cold, dry winters in northern Canada and Greenland (Marshall *et al.* 2001).

During negative phases the pressure gradient is weak with an Icelandic high and Azores low and a south–north pressure gradient. This results in fewer and weaker winter storms crossing in a more-or-less west–east trajectory. They bring moist air into the Mediterranean and cold air to northern Europe and the east coast of the USA, bringing about cold, snowy weather. Greenland has milder winters during these phases.

The *Arctic Oscillation* (AO; sometimes referred to as the *Northern Hemisphere Annual Mode*, or NAM) is thought to be highly linked to the NAO, and in fact some researchers say that the NAO is rather a component of the larger-scale AO, and so often the two terms are interchanged, especially since the variations described by the two are highly correlated. From the 1950s until 1979 a negative phase dominated, after which a more positive phase has predominated. There are years when these general trends are reversed, such as in 1995–1996 when there was a very abrupt reversal of the index (Fig. 1.13).

There has been general warming of surfaces in the Arctic over the past 100 years. However, the increases in the past 20 years has been increasing

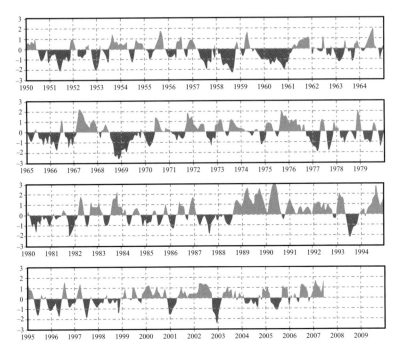

Fig. 1.13   Three-month running mean of the Arctic Oscillation Index from 1950 to present day (May 2007) Image redrawn from data presented by National Weather Service, Climate Prediction Service, National Oceanic and Atmospheric Administration (NOAA).

at a rate eight times higher than the longer 100-year trend which indicates that there has been a rapid acceleration in the warming process linked to global warming processes (Chapter 10). It has also been noted that these rapid warming trends are associated with increasing positive phase in the AO/NAO. Such AO/NAO trends in wind, storm, and warming events will have great influence on the Arctic climate, as is highlighted by interannual variations in sea ice distribution in the Arctic basin (Serreze *et al.* 2007); for example, when the NAO index shifted from positive to negative during the winter of 1995 to 1996 the sea ice export through the Fram Strait is estimated to have been reduced by half.

1.7 Magnetic and electrical phenomena

Investigation of the Earth's magnetic field in the vicinity of the poles was the principal attraction of the Arctic and Antarctic for early scientific expeditions. At the beginning of this century it began to be realized that electrical phenomena in the atmosphere have some relationship with the magnetic field and, in recent years, polar studies have made major contributions to the concept of geospace. Cusps in the magnetosphere, one over each pole, allow the charged particles of the solar wind to penetrate deeply into the polar atmospheres. One manifestation of this is in the aurora (Walton 1987, Pielou 1994). In the present state of our knowledge it seems that this uniquely Polar situation is not of great significance for the life of these regions.

There is a possibility that the fixation of molecular nitrogen produced by the greater electrical activity may support marginal increases in biological productivity. It is also well established experimentally that some organisms react to magnetic fields, and that variations in magnetic fields can affect the time course of plant growth. Birds and mammals may orientate themselves with reference to the Earth's magnetic field so that migrations in the vicinity of the poles may have interesting features. There is a suggestion that magnetic disturbances around the poles may interact with brain activities and interfere with sleep in humans. It may therefore be premature to dismiss the special geophysical characteristics of the polar regions as of no interest to the biologist.

2 Stress, adaptation, and survival in polar regions

2.1 Introduction

This chapter deals with the effects of polar conditions on living systems in general, the ways in which microorganisms, plants, and animals are able to adapt to the stresses imposed, and how they come to be in these habitats.

Stress is a difficult concept to define in precise physiological terms. Radical change in photoperiod is one stress but cold—both in itself and through its secondary effects—is the factor which we think of first as inflicting greatest stress in polar regions. The impact is rather different on small organisms, termed *ectotherms*, whose internal temperature conforms to that of their environment, and on larger plants, which to a limited extent can achieve higher temperatures than their surroundings, and warm-blooded animals. Among ectotherms the term *psychrophile* denotes an organism which is able to grow at or below 0°C, with an optimum growth temperature at less than 15°C and an upper limit at 20°C. It is thought that some microorganisms, especially bacteria and archaea, can metabolize at –20°C and are even predicted as surviving temperatures around –40°C.

Psychrotolerant organisms (sometimes referred to in the literature as *psychrotrophs*) can also grow at around zero but show optimum growth at temperatures greater than 15°C and have upper limits at about 40°C. The distinction between a psychrophiles and a *mesophile* (growth optima between 25 and 40°C) is a bit unclear since microorganisms classified as mesophiles may, given suitable conditions, grow below their supposed lower limit of 5°C (Fig. 2.1).

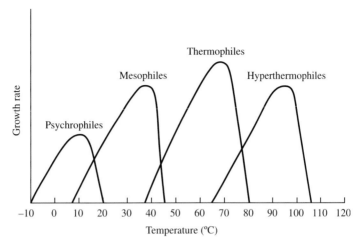

Fig. 2.1   Idealized growth curves comparing psychrophilic with mesophilic thermophilic and hypothermophilic organisms.

It is has been suggested that this terminology is somewhat misleading and out of tune from commonly used ecological terminology, which commonly utilizes the prefixes *eury-* and *steno-* to denote broad and narrow respectively. Instead the terms *eurypsychrophile* and *stenopsychrophile* should be used (Cavicchioli 2006). A stenopsychrophile (true psychrophile) is an organism with a restricted growth temperature range that cannot tolerate higher temperatures for growth. In contrast, eurypsychrophiles (formally psychrotolerants or psychrotrophs) grow best in permanently cold temperatures, but can tolerate a wide range of temperatures extending into the mesophilic range of temperatures.

There are also some questions about using growth rates and temperature optima as the best measures of defining how well an organism is adapted to cold environments. Cavicchioli (2006) cites the examples of two Antarctic archaea to illustrate this point: *Methanoccoides burtonii* grows three times faster than *Methanogenium frigidum* at 4°C. However, the former has an upper growth temperature limit of 28°C compared to 14°C for *M. frigidum*. The thermal regulation of the former has been shown to be regulated through complex gene-expression events that only the use of molecular techniques will elucidate (Goodchild *et al.* 2004).

The adaptations to cold shown by higher plants and animals, as well as by microorganisms, may be *genotypic*; that is, inherited. *Phenotypic* adaptation, resulting from interaction between the genetic constitution of the individual and its environment, is termed *acclimatization* if it is a response in the field to a whole complex of cold-related factors. *Acclimation* refers to laboratory-induced adaptation to a single variable.

2.2 Life at low temperatures

2.2.1 Effects of low temperatures on cell physiology

As temperature falls the proportion of molecules that are in a sufficiently activated state to undergo reaction diminishes and the rate of chemical transformation is reduced. The Arrhenius equation, which expresses this, provides a fundamental approach to the study of the effect of temperature on an isolated chemical reaction and can be applied to enzyme-catalysed processes. However, it is inappropriately applied to complicated living systems and for these it is more convenient to use an empirical temperature coefficient, Q_{10} [where $Q_{10} = (k_1/k_2)^{10/(t_1-t_2)}$], derived from the rates of a process, k_1 and k_2, measured at temperatures of t_1 and t_2, respectively, in degrees Centigrade. This can be applied to any sort of activity. The effects of chilling on the processes going on in an organism are complicated by biochemical interactions but it is usually found that Q_{10} values are between 2 and 3, so that rates at 0°C are only a half, or less, of what they are at 10°C. So long as liquid water is available, which it may be down to −40°C because of supercooling (or, more correctly, *undercooling*), biochemical processes should continue at lower, but still perceptible, rates at temperatures below zero.

Nevertheless, cold may kill and few tropical or temperate plants or animals can survive polar temperatures. Broadly speaking, this lethal effect arises because different processes have different temperature coefficients so that the balance between them becomes distorted as temperature falls. The efficiency with which enzyme reactions are coupled together may be reduced and adenosine triphosphate (ATP) production may become insufficient to meet the energy requirements of the cell. Enzymes themselves undergo changes and become inactivated. Denaturation of proteins at high temperatures is familiar to anyone who has boiled an egg but it may equally happen at low temperatures, although this has been much less studied. The particular folding pattern of its peptide chain, on which the catalytic activity of the enzyme protein depends, is determined by hydrophobic interactions, which decrease in strength, and hydrogen bonding, which increases in strength, on chilling. The relation between them becomes disturbed, and the molecule unfolds and becomes inactive (Jaenicke, in Laws and Franks 1990).

One way of countering the effects of low temperature is to compensate for the reduced activity of a key enzyme by producing more of it. This happens in some Antarctic fish, especially with enzymes involved in aerobic respiration (Eastman 1993). High rates of photosynthesis in Arctic flowering plants (and some Southern Ocean phytoplankton microalgae) are made possible by higher-than-usual concentrations of the rate-limiting photosynthetic enzyme, ribulose bisphosphate carboxylase (Rubisco).

The leaves of Arctic plants have, as a consequence, higher nitrogen contents than those of temperate plants. Rubisco is an enzyme of very high molecular mass and in plants and algae it can amount to 50% of the protein in the cell. It is the most abundant protein on the planet (estimated at 40 million t). The carboxylase reaction catalysed by Rubisco is as follows:

$$\text{Ribulose 1,5-bisphosphate} + CO_2 \rightarrow 2(\text{3-phosphoglycerate})$$

Then follows a complex sequence of reactions, from which the sugars glucose ($C_6H_{12}O_6$) and sucrose ($C_{12}H_{22}O_{11}$) are common initial products.

Rubisco is unusual as an enzyme in that it has a second and quite different function as an *oxygenase*. In this reaction, rather than adding CO_2 to ribulose 1,5-phosphate, oxygen is added:

$$\text{Ribulose 1,5-bisphosphate} + O_2 \rightarrow \text{3-phosphoglycerate} + \text{phosphoglycolate}$$

This reaction, known as *photorespiration*, results eventually in the loss as carbon and the formation of CO_2. It is an important loss reaction in plants, especially tropical plants. The balance between the two alternative reactions is controlled by the ratio of O_2 and CO_2 concentrations at the enzyme:

- the carboxylase (CO_2-fixing) reaction is *high at high* CO_2/O_2 ratios;
- the oxygenase (CO_2-releasing) reaction is *high at low* CO_2/O_2 ratios.

Genotypic modifications in the molecular configurations of enzymes in polar organisms, giving them greater efficiency at low temperatures, have also been found (Russell, in Laws and Franks 1990). An enzyme in a psychrophilic organism may have the same qualitative properties as its isozyme in a mesophile but differ in quantitative characteristics. Thus, Rubisco isolated from some species of Antarctic diatoms has the same activation energy as that from a species living in a temperate habitat, but, surprisingly, the rate of catalysis by the psychrophile enzyme reaches a maximum at 50°C compared with 40°C with that from the mesophile. However, the affinity of the psychrophilic enzyme for its substrate is greatest at 4.5°C as compared with 20°C for the mesophilic form and it is this which is evidently of biological advantage in cold waters. The expectation that enzymes of ectotherms adapted to cold conditions should have lower activation energies than those from warmer climates has been verified for some higher-plant enzymes, fish and bacteria enzymes, and for protein-cleaving enzymes of planktonic and benthic crustaceans in Antarctic waters (Committee on the Frontiers of Polar Biology 2003).

Another damaging effect of chill is derangement of the cell membrane, the integrity of which is essential for the control of exchange of substances between the cell and its surroundings. This membrane is a liquid lipid layer modified by the presence of protein. The principal lipid molecules are arranged at right angles to the plane of the membrane to form a bimolecular layer with the non-polar hydrophobic groups on the inside. Some 15–25% of the total lipid does not form bilayers but instead the molecules are grouped into aggregates. The protein consists of various enzymes which porter substances across the membrane and also play a part in stabilizing it. Few general principles relating to membrane stability have been agreed but it seems that a usual occurrence when temperature is lowered is a transition from a liquid crystal structure to that of a gel, with damage showing itself by the leakage of solutes from the cell (Williams, in Laws and Franks 1990).

The lipid composition of whole cells of ectotherms changes, particularly with a shift from saturated to unsaturated fatty acids, when they are grown at near-zero temperatures rather than in warmer conditions. This phenotypic change parallels genetic differences. Psychrophiles, as a group, have higher proportions of unsaturated fatty acids (52%) than do mesophiles (37%) and thermophiles (10%). Psychrophiles also have more short-chain fatty acids. Both these changes could lower the temperature at which the transition from the liquid crystal to the gel phase of the membrane occurs and perhaps might contribute to its chill resistance. However, the weak interactions between the different membrane components are sensitive to a complex of factors, including pH, ionic concentrations, and hydration, as well as temperature, and it is not possible at present to be sure that lipid composition is of major importance. One of the effects of cryoprotectants—substances such as glycerol and proline which protect against freezing—is to increase membrane stability and hence chill resistance. They seem to do this by acting as water-replacement agents that maintain a balance between membrane components similar to that which exists under normal physiological conditions.

It had long been supposed that polar *poikilotherms*—cold-blooded animals—adapt to cold by having elevated basic (*routine*) metabolic rates as compared with those of similar animals from warmer environments. This would diminish the amount of energy available for growth and reproduction and so conveniently account for the slow growth rates, delayed maturation, and prolonged gametogenesis characteristic of polar organisms. However, the data on which this theory was based seem to have been obtained by faulty experimental procedures. There is now much physiological and biochemical evidence to refute it (Clarke 1983). With Antarctic marine invertebrates, for example, under properly controlled experimental conditions there is no detectable elevation of the rate of routine metabolism above that to be expected at low temperatures. The slow rates of growth and reproduction are puzzling but presumably are due to an overall reduction in energy utilization in response to conditions in the environment (Johnston, in Laws and Franks 1990).

2.2.2 Effects of freezing and freeze resistance

The effects of freezing are distinct from those of low temperature per se. Chill does not necessarily involve the separation of ice nor the changes in water-soluble components which inevitably accompany freezing. Ice is formed in the extracellular phase before freezing occurs inside the cell. This releases latent heat and helps to buffer the cytoplasm against a fall in temperature and freezing. Freezing requires the presence of nuclei for formation of ice crystals. These can form spontaneously in the course of the random movements of molecules in the liquid, but, except in the region of −40°C, usually decay before crystal growth is initiated. In living systems nucleation is catalysed by particulate matter presenting particular molecular configurations. It is unfortunate for invertebrates that nucleators are present in food materials so that freezing conditions are best faced with an empty gut.

When ice forms in the immediate vicinity of unicellular organisms or in the intercellular fluids of multicellular organisms, solutes are excluded from the crystals and so their concentration in the remaining liquid is increased. The osmotic stress which results is the immediate and most injurious consequence of freezing. Water is drawn from the cell, the magnitude of the efflux depending on the excess solute concentrations in the extracellular phase and the permeability of the cell membrane. Under field conditions cooling rates are usually slow and high membrane permeability to solutes ensures that osmotic equilibrium is maintained as the extracellular fluid becomes gradually more concentrated. The reduction in cell volume as water is withdrawn may lead to injury by impairing the resilience of the membrane so that the cell bursts on rewarming. Under more extreme conditions cell dehydration and perhaps internal freezing may cause mechanical deformation and leakage of cytoplasmic solutes. Another possibility is that as substrates become more concentrated the rates of enzyme reactions are speeded up and a deleterious imbalance of metabolic pattern set up.

These injurious effects may be minimized in various ways. Cold-acclimatized microorganisms, plants, and animals produce substances which counter the effects of freezing. Osmotic imbalance can be reduced by the production within the cell of 'compatible' solutes, a process known as osmoregulation. These substances, which include free amino acids and low-molecular-mass polyhydroxy compounds such as sugars, must, of course, be soluble and metabolically inactive. Additionally, they appear to have a role in stabilizing proteins. Production of compatible solutes may take a few minutes in microorganisms but freeze resistance takes several days to develop in insects and weeks in large plants. Changes in membrane elasticity may occur during cold acclimation, making cells more able to withstand contraction and expansion stresses. Plant protoplasts, for example, can develop tolerance to stretching as much as three times greater than that which they possess normally. In certain yeasts and snow algae this correlates with the appearance, as seen by electron microscopy, of numerous folds in the membrane, which are not present in unadapted cells and which seem as though they should allow expansion or contraction to be taken up without stress. Effective water management is a basic requirement in all strategies for surviving the cold (Franks *et al.*, in Laws and Franks 1990).

2.2.3 Avoidance of chill and keeping warm

Complete avoidance or inhibition of nucleation at sub-zero temperatures prevents freezing. This is a thermodynamically unstable state; hickory trees are able to withstand cooling down to about −45°C but then freezing happens suddenly and they are killed. Marine fish in polar regions have blood freezing points of between −0.9 and −1.0°C but can live in deep water at a temperature of −1.8°C because the absence of ice nuclei allows them to remain in the supercooled state. If ice is introduced into their vicinity they instantly freeze as nuclei diffuse from ice to fish. Antifreezes occur universally in Antarctic fish and less commonly in Arctic fish, some of which only produce them during the winter. They are not specific to particular taxonomic groups. Antarctic fish have had more time, about 25 million years, to evolve the necessary biochemical mechanisms, as compared with only 2–3 million years for those in the Arctic (Johnston, in Laws and Franks 1990, Eastman 1993). Antifreeze substances are also present in some plants and some insects. They include glycerol, polyhydric alcohols, and glycoproteins. The action of the low-molecular-mass substances, such as glycerol, is *colligative*, that is to say proportional to the molecular concentration of the solute. In terms of the fresh weight of the organism a colligative antifreeze may reach a concentration of 3–6%, or even as high as 14%. The high-molecular-mass substances have low osmotic activity and their action is non-colligative. They act by becoming adsorbed on ice crystals and inhibiting their growth. The long straight fronts in the growing crystal are interrupted by the adsorbed molecules and divided into smaller fronts. These become curved, causing an increase in the free surface energy of the crystal and hence a lower temperature is required for freezing (Eastman 1993).

Form and function plays a part in ameliorating temperature for plants. A cushion or turf not only provides an insulating layer protecting the lower parts of the plants (and, incidentally, a better environment for animals) but also acts as a trap for radiant energy. Absorbed radiant energy which is not used for photosynthesis or evaporation is re-emitted at longer wavelengths but, because of the structure of the plants, much is re-absorbed by them. In the sunshine, the temperature inside a cushion or turf may rise to 10°C or more above that of the ambient air. Cloches have a similar effect (Figs 2.2, 2.3, and 2.4). Phototropism may keep a plant organ orientated so as to receive the maximum amount of radiant energy. The glacier buttercup (*Ranunculus glacialis*) has flowers which swivel on their stems to face the Sun as it crosses the sky. The flower itself acts as a parabolic reflector, directing the rays to its centre and, presumably, providing a more favourable temperature for the development of ovaries and stamens as well as attracting pollinating insects.

Insects are the largest metazoan animals to tolerate low body temperatures and to survive freezing. The larger marine invertebrates are not exposed to temperatures below −1.9°C in their normal habitats. Arctic insects are mostly darker, smaller, and hairier than their temperate counterparts, these being features maximizing radiative warming. Their activity patterns take advantage of favourable microclimates and unfavourable periods are passed in dormant condition. Bumblebees, large insects with dense insulating 'fur', can raise their body temperature to as much as 35°C by shivering of their flight muscles. This enables them to remain active in cool weather and to hasten development of eggs and larvae in the nest.

Mammals and birds avoid the problems faced by ectotherms by using heat released by metabolic processes, in conjunction with insulation and various physiological and behavioural mechanisms which conserve heat, to maintain body temperatures between 36 and 41°C. Heat has often to be

Fig. 2.2 Experimental manipulation of Antarctic systems using cloches to ameliorate temperature at Leonie Island near Rothera on the Antarctic Peninsula. Courtesy of British Antarctic Survey.

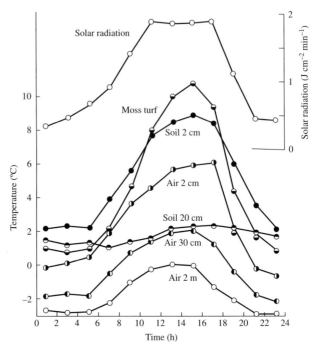

Fig. 2.3   Mean temperatures at 2-h intervals over 5 days in, above, and below a moss turf (*Bryum argenteum*) on Ross island, Antarctic. Redrawn after Longton (1988).

Fig. 2.4   Antarctic moss cushions. Courtesy of British Antarctic Survey.

conserved against steep temperature gradients and requires a correspond-ingly ample supply of food or internal, energy-yielding, reserves. An extreme example is the male emperor penguin (*Aptenodytes forsteri*; Fig. 2.5), which survives 105–115 days of winter, without feeding, at temperatures which may fall to −48°C.

Aquatic mammals are in a better situation because the sea does not fall below freezing point and, since it supports a greater bulk of body, it allows low surface area/volume ratios to reduce heat loss. Hibernation, by allow-ing body temperature to approach that of the surroundings, enables large savings in energy-yielding reserves but is not typically resorted to by polar animals. Perhaps this is simply because it is only possible in frost-free nooks and these are scarce. Breeding female polar bears (*Ursus maritimus*) enter a limited winter sleep in which body temperature falls for a few days at a time, but this is not true hibernation. Many birds and mammals avoid the hazard of a polar winter and find a more congenial habitat for breed-ing by migrating to warmer places. This, too, may involve considerable expenditure of energy.

There is some tendency for polar birds and mammals to be larger than their temperate or tropical counterparts but factors other than heat con-servation by reducing the surface area/volume ratio are involved (e.g. the

Fig. 2.5 The emperor penguin (*Aptenodytes forsteri*) is a supremely well-adapted ectotherm (photograph by David N. Thomas) (see colour plate).

need for completing the reproductive cycle as quickly as possible makes small size desirable), and there are many exceptions to this rule. Finally, it should not be overlooked that conservation of body temperature carries the danger of occasional overheating. Animals may sleek their fur or feathers, pant, employ circulatory bypasses to their extremities, or have bare or lightly insulated areas which they can expose, to cool themselves when necessary.

Thawing may not be the end of an organism's suffering, as this too has its hazards. Cell membranes may be in an unstable condition and not able to tolerate sudden change in osmotic stress. Freeze–thaw cycles are powerful in disintegrating rock and organisms cannot altogether escape mechanical disruption. In Signy Island soils there is a spring peak in microbial activity in response to organic substances released by thawing from damaged cells.

2.2.4 Wind-chill

Some effects of wind—mechanical damage, snow drift, abrasion by suspended ice crystals or rock particles, increased evaporation—perhaps do not require much discussion here, but the concept of wind-chill does. The physiological effects of cold depend on the rate of loss of heat from the organism. Loss by convection/conduction, as distinct from radiative loss, is determined by the temperature gradient between the organism and the ambient air. This gradient is steepened if the air is moving. For a given air temperature the loss will be greater as wind speed increases. An index of this effect is based on experiments on heat loss from dry, uninsulated surfaces. This *wind-chill factor* is expressed either as equivalent temperature reduction (Fig. 2.6) or as additional heat loss per unit time. It is used extensively but has defects from the physiological point of view. If the surface in question is damp then more heat will be lost than the index shows, because of evaporation of water, but the temperature of the surface cannot fall below the wet-bulb temperature of the air; below this point heat is added by condensation of water. During storms, the effective cooling power of the wind is greatly increased by the ice and snow it carries. The index also takes no account of radiation or insulation.

Animals with wind-resistant plumage or fur, and humans in wind-proof clothing, may be scarcely affected, even in strong winds. There is little correlation between freezing of uninsulated finger skin and wind-chill index because varying degrees of supercooling, cold-induced vasodilation, and skin moisture complicate the situation. Nevertheless, the wind-chill index, used with discretion, remains a useful guide to polar conditions, especially as they affect bird mortality and the comfort of humans. Among emperor penguins wind-chill in the period between leaving the brood pouch and learning to huddle seems to be the major cause of chick mortality, accounting for 80% of the loss.

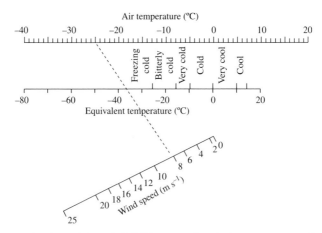

Fig. 2.6   A nomogram for determining wind-chill in terms of equivalent temperature. Equivalent temperature is read off by drawing a line between actual air temperature and the wind speed, as illustrated by the dashed line in the figure which shows an air temperature of −25°C and wind speed of 8 m s^{-1}, giving an equivalent temperature of −36°C. Redrawn from Rees (1993).

An index of survival time outdoors in extreme cold, based on body–atmosphere energy budget modelling procedures, is more informative about effects on humans than the wind-chill factor. The index is the calculated time for a fall in core temperature, in an inactive healthy subject in full polar clothing, from 37 to 27°C. The most severe Antarctic conditions become life-threatening after 20 min. At most stations on the continent, a similar point would be reached in 2 h of outdoor exposure in winter. Conditions at all coastal stations in summer are mild enough to allow core temperature to be maintained (de Freitas and Symon 1987).

2.3 Desiccation

For the maintenance of life, exchange of substances, including water, across the cell membrane is essential. If the water potential is less outside (e.g. if the concentration of osmotically active substances is greater than it is inside), then water is inevitably lost from a cell. The gradient in water potential between a cell grown in fresh water then transferred to sea water is equivalent to an osmotic stress of around 23 bar (1 bar = 10^5 Pa ≈ 1 standard atmosphere). Organisms exposed to the air may be subjected to much greater stresses. When the air is 90% saturated with water vapour at 20°C the difference in water potential between it and a moist surface at the same temperature is 140 bar. When it is half saturated, as it may be in an English summer, it is as much as 924 bar. Relative humidities in the dry valleys of Antarctica are usually low, 30–40% in winter, rising to 80% in summer, or even to 100% in the very rare showers of light rain. Relative humidity falls to less than 10% when katabatic winds, already containing little moisture, warm up as they blow down-valley. This would result in a water-potential deficit of about 800 bar at 0°C. Freshwater habitats present no problem but water remaining unfrozen in polar regions is often highly saline, with water-potential deficits up to 200 bar.

2.3.1 Effects of desiccation on cell physiology

Lack of water is generally more limiting for terrestrial life than is low temperature, although resistance to desiccation has been little investigated at the cellular level. As might be expected from the crucial involvement of water relations in resistance to freezing, there is a general correlation of the abilities to withstand drought and low temperatures. This is seen in the tolerance to both these stresses, which is generally characteristic of the cyanobacteria and the chlorococcalean green algae, in contrast to the sensitivity of other groups of photosynthetic microorganisms. There is a genetic element in resistance to desiccation; Antarctic cyanobacteria and unicellular green algae appear to have a greater tolerance to freeze-drying than their counterparts in temperate latitudes. The conditions under which drying and rehydration take place have a great effect on whether cells will survive. The stage of growth, the rate of drying, and whether it takes place in light or dark, all seem important for one or other species but it is difficult to discern any general pattern.

Metabolism is retarded in desiccated cells; cyanobacteria and lichens show no perceptible photosynthesis when dry, although they remain viable and may resume photosynthetic activity quickly on rewetting. The respiration of *Phormidium*, unlike that of another cyanobacterium, *Nostoc*, continues after photosynthesis is inhibited, which must deplete carbon reserves and possibly it is this which slows down recovery on rewetting of mats consisting predominantly of *Phormidium* (Hawes *et al.* 1992). A lipid pellicle is sometimes produced around microbial cells on drying and mucilaginous sheaths or thick cell walls, acting as reservoirs, will also reduce the rate of water loss from the cells themselves, giving time for physiological adaptation. Total lipid increases in water-stressed microbial cells but this does not seem to be related to desiccation resistance, although individual lipid components may be concerned in providing resistance.

2.3.2 Effects of desiccation on whole organisms

Dehydration, whether brought about by low relative humidity, freezing, or high salinity, is something the microflora of polar regions must withstand for long periods. The most severe test, no doubt, is during the colonization stage when cells are transported in the air, but viable microorganisms do arrive by this route.

Plants are particularly affected by water deficiency since exposure of extensive areas of water-imbibed cell surface is required for carbon dioxide intake from the air for photosynthesis. With frozen roots, and shoots exposed to Sun and wind, a plant quickly becomes water-stressed. Flowering plants in the sub-Arctic have morphological characteristics, such as reduced or fleshy leaves, usually associated with growth in water-deficient conditions (i.e. they are *xeromorphic*). Examples are some *Saxifragas* and crowberries (*Empetrum* spp.). A capacity to adapt metabolism to water stress is essential for the existence of most terrestrial plants. Provided that withdrawal of water is gradual, osmotic adjustment may be achieved via concentration of inorganic ions and organic solutes. Mosses and lichens, which are the predominating components of polar land floras are, as classes, characteristically desiccation-resistant and the polar forms remarkably so. Thallus morphology is important in reducing water loss and crustose or foliose habits in lichens and aggregation of mosses into mats or cushions (Fig. 2.4) decrease evaporation. Net photosynthesis occurs in polar lichens and mosses over a water content range of 100–400% of dry weight (Longton 1988). Some mosses and lichens remain viable after years at water contents of 5–10%. Tolerance to desiccation is related to habitat; many mosses growing in wet places are killed by drying for only a few hours. There is also seasonal variation.

Terrestrial invertebrates become inactivated if desiccated and their distribution is largely determined by water availability. The high degree of cold tolerance which polar forms have suggests a parallel resistance to desiccation. Indeed, the effects of desiccation and low temperature are closely intertwined. The springtail *Cryptopygus antarcticus* can survive a reduction in total body water from 60 to 40% of fresh weight, representing a fivefold increase in solute concentration assuming no osmoregulation. In the dehydrated state it has a remarkable ability to take in water but this is done at some sacrifice of cold tolerance. The Antarctic mite *Alaskozetes antarcticus* (Fig. 2.7) is very resistant to desiccation; on drying to a water content of 60% of fresh weight there is a significant accumulation of glycerol, which is an antifreeze. A degree of mobility and a capacity for encystment or dormancy also help in evading drought conditions. Birds and mammals are able to search out liquid water and presumably do not have a desiccation-resistant type of cell physiology.

Desiccation in an arid environment may be avoided by occupying an enclosure within which water can be conserved. Continued life in such a niche is only possible if there is an input of energy and cycling of materials; that is, there must be an association of photosynthetic activity, producing organic matter and oxygen and assimilating carbon dioxide, with

Fig. 2.7   The Antarctic mite, *Alaskozetes antarcticus*, which at only about 0.2 mm in length is still one of the largest terrestrial invertebrates in Antarctica. Courtesy of W. Block and the British Antarctic Survey.

heterotrophy, consuming organic matter and oxygen and producing carbon dioxide. This was achieved in the sealed containers containing living ferns which graced the dimly lit Victorian drawing room. Water, transpired by the foliage, condensed on the glass and ran back into the soil which sustained enough heterotrophic activity to balance the low net photosynthesis. Analogous microcosms, almost self-contained communities of algae and lichens with yeasts and bacteria, are found in the micropores of rocks in the dry valleys of Antarctica.

2.4 The effects of radiation

2.4.1 Light

Light is, strictly speaking, radiation visible to the human eye, but the word can be used without too much inaccuracy to denote radiation visible to other animals and also photosynthetically active radiation. The quality and range in intensity and the reactions of plants and animals to these are much the same in polar regions as elsewhere. However, the relations of photosynthetic activity to light intensity are of particular importance in many polar habitats and must be outlined. The rate of arrival of radiant energy per unit area of surface is properly termed *irradiance*, although *intensity* is generally used, in error, to mean the same thing. Irradiance is measured in joules ($J\,m^{-2}\,s^{-1}$), Watts ($W\,m^{-2}$)—that is, in terms of energy—or most usually, photon flux density (moles photons or einsteins $m^{-2}\,s^{-1}$)—that

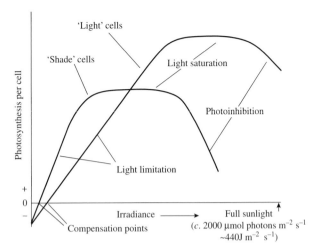

Fig. 2.8   Relationship between the rate of photosynthesis with irradiance in light- and shade-adapted photoautotrophs (plant, algae, and photosynthetic bacteria).

is, in number of quanta. The energy of a quantum is dependent on its wavelength but for full sunlight a conversion factor of $1\,J\,m^{-2}\,s^{-1} = 4.6\,\mu mol$ photons $m^{-2}\,s^{-1}$ is a reasonable approximation. Photon flux density will generally be used in this book and it will be helpful in visualizing light conditions to remember that the photosynthetically active component of full sunlight amounts to roughly $2000\,\mu mol$ photons $m^{-2}\,s^{-1}$.

The relation of rate of photosynthesis, usually measured as oxygen output or carbon dioxide uptake, to irradiance is shown in Fig. 2.8. In darkness, the uptake of oxygen and release of carbon dioxide in respiration are the dominant exchanges. With low levels of irradiance these exchanges diminish and, at a particular level—the *compensation point*—become zero as photosynthesis and respiration cancel each other out. Beyond this point the rates of oxygen output and carbon dioxide uptake increase linearly with irradiance. Light is at first the limiting factor and variation in concentration of carbon dioxide has no effect on these rates. As irradiance is increased further, carbon dioxide supply becomes limiting and the rate of photosynthesis approaches a plateau in which it is independent of the level of irradiance. The level of this plateau can be raised by increasing carbon dioxide or lowered by diminishing it. The rate of light-limited photosynthesis is normally independent of temperature, photochemical reactions being temperature-insensitive, but when it is light-saturated it becomes temperature sensitive because it is now limited by ordinary chemical reactions. At high irradiances, approaching full sunlight, the rate of photosynthesis may fall.

This *photoinhibition* happens because the absorption of light energy exceeds the capacity of the chemical reactions of photosynthesis to utilize it and excess energy becomes diverted to destructive oxidative processes. The irradiances at which the compensation point, light saturation, and photoinhibition occur depend on the physiological state of the plant and particularly on the light levels to which it has previously been exposed. In *shade* plants and algae, which usually contain more photosynthetic piments, these points occur at lower irradiances and the rate of photosynthesis per cell (but not per unit amount of chlorophyll) is higher at low irradiances than in *sun* plants and algae (Fig. 2.8). Unicellular algae and photosynthetic bacteria may adapt from one state to the other within hours, but for higher plants it can take longer. In polar plants and algae the effects of low irradiance and low temperature are interrelated, both lowering the compensation point. The lower rate of respiration enables net photosynthesis and growth to take place at lower light levels than would otherwise be possible and by economizing on use of reserves enables the plant to survive for longer in darkness.

2.4.2 Ultraviolet radiation

Ultraviolet (UV) radiation is a meteorological element of considerable biological importance which has received much attention by researchers in polar regions (Karentz 1991). Anyone visiting the Arctic or Antarctic in spring and summer will be aware of the strong light and burning UV radiation (Fig. 2.9). The solar radiation which impinges on the stratosphere has about 1% of its energy contained in the wavelengths 400 down to 200 nm (i.e. the ultraviolet). This can be damaging to life since DNA

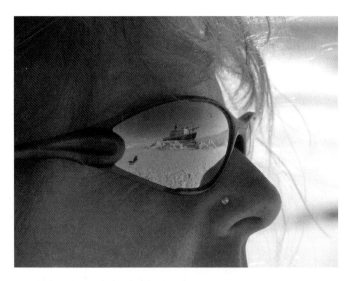

Fig. 2.9   Because of strong ultraviolet radiation in polar regions, adequate protection for skin and eyes is essential (photograph by David N. Thomas).

absorbs maximally at 260 nm and wavelengths between 280 and 320 nm are capable of disrupting its structure and producing mutations. Other cell components, such as proteins and photosynthetic pigments, are also damaged by the shorter wavelengths of UV.

The atmosphere is efficient in filtering out wavelengths below 286 nm, largely because of the presence of ozone, which absorbs between 320 and 225 nm (UV-B down to 280 nm, UV-C below this), while allowing the less-harmful longer wavelengths to penetrate. If the ozone in the atmosphere were concentrated into a discrete layer, this would have a thickness of 2.4–2.6 mm at the equator and 3.1–4.3 mm at 70°N, both at standard temperature and pressure. This seems, and actually is, a flimsy shield against a serious danger for living organisms. The concentration of ozone in the stratosphere, where it is most abundant, depends on a balance between its production, by the action of UV of wavelengths below 190 nm on oxygen, and its destruction, also by photochemical processes involving UV. It may be noted that these processes themselves utilize the energy of UV and result in warming of the stratosphere. Different balances between production and destruction explain the variation in ozone concentration with latitude.

In the 1970s it was realized that because chlorine has a catalytic effect on the destructive process, the release into the atmosphere of chlorofluorocarbons (Freons or CFCs), much used in industry as aerosol propellants and refrigerants, might affect ozone levels following transport of these substances into the stratosphere and their photolysis to release active chlorine. The use of CFCs was restricted and more extensive and accurate monitoring of atmospheric ozone was put in train. This showed that between 1979 and 1986 there was a fall in the global mean total ozone of about 5%, an appreciable but not catastrophic decline. However, scientists reported a dramatic decrease in total ozone over Halley Station, lasting over some 6 weeks in the austral spring of 1984. Values were down by about a third from those in 1957–1977. At other times of the year the amounts were normal. This local 'hole' in the ozone layer had been missed by a sophisticated satellite-based worldwide monitoring system because this was programmed to discard low values which might be due to instrumental error. The existence of the hole was confirmed on re-examination of the satellite data (Jones and Shanklin 1995).

The ozone depletions are now tracked estensively by scientists worldwide and bulletins from the World Meteorological Organization (WMO) enable long-term records to be coordinated: the ozone depletions continue and the ozone hole recorded in 2000 was the largest on record. The 2006 event, although not quite as large in surface area as the one 2000, actually suffered the most deficit, and so there was less ozone over the Antarctic than ever measured previously. A similar, but smaller and less-consistent, hole is present in the Arctic. The annually averaged loss of ozone over the Arctic has been about 7% since 1979, although the average springtime loss is between 10 and 15% in the same period (Arctic Climate Impact Assessment 2005).

Such rapid, massive, and localized depletion of ozone was completely unexpected. The explanation lies in the pattern of circulation in the stratosphere in the polar regions during winter (Pyle *et al.*, in Drewry *et al.* 1992, McIntyre, in Wadhams *et al.* 1995). In both regions, strong westerly winds set up low-pressure vortices in the stratosphere. The Antarctic vortex, centred over the pole, is the more stable whereas that in the Arctic is variable and its centre usually does not lie over the pole. The strong wind gradients around the vortices act as barriers to horizontal mixing and cut off masses of cold air from replenishment with ozone from outside. Since during the winter there is no UV falling on the polar regions to stimulate local production of ozone, processes eventually resulting in its destruction predominate. The declines in total ozone over Halley and the southern hemisphere were correlated with increases in CFC concentrations (now due to be phased out completely by 2030 by the Montreal Protocol that was implemented in 1989 and last revised in 1999) but low temperatures play an important part by providing ice crystals, on the surfaces of which the ozone-destroying reactions take place more readily. These ice crystals are visible as the nacreous clouds characteristic of the polar stratosphere. The increases in intensity at ground level of UV-B, to be expected from the falls in ozone concentration in the stratosphere, have been observed. When dynamical breakdown of the vortices occurs in the spring, the barrier to transport disappears, ozone levels are restored and the UV intensity falls to normal levels.

Naturally there is much debate as to whether following the Montreal Protocol ozone depletions will recover. *The Scientific Assessment of Ozone Depletion* published by WMO and the United Nations Environment Programme (World Meteorological Organization 2007) says that the ozone layer over latitudes 30–60° north and south should recover by 2049, 5 years later than anticipated by their previous assessment in 2002. Because of the unique atmospheric conditions at southern latitudes ozone over the Antarctic is predicted to recover by 2065, 15 years later than earlier predictions.

These findings have accentuated public anxiety about the global environment and have led to some international action to reduce industrial pollution of the atmosphere. The biological impacts of UV radiation are discussed in the next section but a few further points may be made at this stage. One is that cloud reduces the intensity and impact of damaging UV radiation but it has no effect on the spectral composition of that radiation. Snow provides good protection to organisms covered by it by reflecting radiation and rapidly attenuating that which does penetrate. Clear ice and moderate depths of water do not afford protection to the same extent.

2.4.3 Biological responses to UV radiation

The potential damage to biological systems from shorter-wavelength UV is illustrated by the estimate that radiation of wavelength 295 nm has a thousand times the potency in sunburning human skin as does the wavelength 320 nm. The impact on polar organisms depends on their level of organization and particular habitat and there are various general mechanisms for countering the challenge (Karentz 1991).

Damage to DNA may have mutagenic or lethal effects but repair is possible. The capacity to repair is determined genetically and microorganisms and humans deficient in this respect are killed by low doses of UV that scarcely affect repair-efficient strains. A DNA-repair-deficient strain of the bacterium *Escherichia coli* has been used as a biological dosimeter to measure, for example, the penetration of lethal radiation into the sea. Apart from genetic damage, absorption of UV can disrupt RNA, proteins (including enzymes and hormones), pigments, and other metabolically important molecules. Whether or not a cell is able to make good such damage and survive depends, among other things, on its physiological condition.

Protection is provided by substances that absorb UV without damage being caused. Such substances are the melanin produced by humans, mycosporine-like amino acids (Shick and Dunlap 2002) contained in some Antarctic invertebrates and algae, and flavonoids in mosses and flowering plants. There is great variation, even between related species, in ability to produce these substances. Again, the ability is genetically determined. In animals, behavioural characteristics are important. Most animals avoid full sunlight as far as possible: only 'mad dogs and Englishmen go out in the midday sun'. Marine birds and mammals that breed ashore may not be able to avoid it but are largely protected by feather or fur. Eyes are at risk but Antarctic penguins and skuas have higher thresholds for UV damage of the cornea than domesticated fowls. Tears of the southern elephant seal (*Mirounga leonina*) show strong absorption of wavelengths shorter than 300 nm. Many insect species have vision extending to 300nm. Increase in radiation of less than 320 nm caused by reduction in the ozone shield would be perceived by them with possible effects on behaviour patterns.

It may be that increased exposure to UV-B will have little effect on the total biological productivity of polar regions but because protection and repair are genetically determined there may well be selective effects on species in the course of time. Such changes should first be evident among microorganisms with short generation times.

As an example, there has been considerable efforts by biological oceanographers to understand the effects of UV radiation on the productivity of photoautotrophs and bacteria. Clearly UV radiation damage and subsequent repair is the result of a complex suite of cellular mechanisms including protection, repair, cell size, and growth rates (Vincent and Neale, in de Mora *et al.* 2000). However, the ratio of UV-B to longer wavelengths is very important (Vincent and Roy 1993), and for this reason shade-adapted phytoplankton are more susceptible to UV damage (Prézelin *et al.* 1998, Buma *et al.* 2006, van de Poll *et al.* 2006).

Normally, most of the photoinhibition near the water surface is caused by excess absorption of visible light. Additionally, there may be inhibition by UV radiation. In transparent oceanic waters about 1% of surface UV-B irradiance reaches 60 m depth, whereas in inshore polar waters significant amounts, say 2.5%, may reach 15 m depth and biological effects may be detected at 20 or 30 m. In well-mixed waters this may not be important since individual cells are near the surface for short times only and have ample time for repair in the shade of deeper water, so effects of UV may be overestimated. It is a different matter in stratified water, as in the marginal ice zone, where actively growing phytoplankton is held in a 10–20 m surface layer in its season of maximum growth, just when the ozone hole is open. Antarctic studies have shown, that shallow mixing greatly increases the photoinhibition of photosynthesis by UV-B (Villafañe *et al.* 2003).

In the marginal ice zone of the Bellingshausen Sea, at around 64°S 72°W, in the austral spring of 1990 (when the edge of the ozone hole touched the tip of South America), observations were made both inside and outside the area under the hole (Smith *et al.* 1992). The sampling areas were selected on the basis of daily transmissions of ozone concentrations from the NASA Nimbus 7 satellite. The ratio of UV-B to total irradiance at the sea surface increased under the hole and there was an estimated minimum of 6–12% reduction in primary productivity, although this should be viewed in relation to a ±25% year-to-year variation which usually occurs. Numbers of *Phaeocystis* in the surface waters were positively correlated to ozone concentrations, the decline in numbers being seemingly due to cell lysis caused by increased exposure to UV-B. Diatoms were not affected in this way.

Interspecific differences in the responses to this complex of factors will dictate any changes in phytoplankton. However, the most likely scenario due to this UV-radiation-induced environmental change is that there will be a shift in species composition or successional patterns and ecosystem function (Karentz 1991, Vincent and Roy 1993, Mostajir, in de Mora *et al.* 2000). One of the few generalizations that it seems possible to make with reference to susceptibility to UV damage is that smaller cells are more vulnerable (Karentz 1991, Buma *et al.* 2001, 2006). This is because of the increased surface area to volume ratios, and the low effectiveness of UV-screening pigments in small cells.

The prymnesiophyte *Phaeocystis antarctica* plays a large role in the seasonal dynamics of many parts of the worlds oceans, in particular in coastal waters and waters of the marginal ice zone in polar oceans (Chapter 7). The colonial stage of *Phaeocystis* contains high concentrations of UV-B-absorbing compounds which are highly efficient in protecting *Phaeocystis* colonies from UV damage. In contrast, the flagellate stages of the *Phaeocystis* life cycle suffer severe mortality from UV-A (Davidson and Marchant 1994). Riegger and Robinson (1997) concluded that many phytoplankton species have the potential to respond to the increases in UV radiation by increasing cellular levels of mycosporine-like amino acids, which have been shown to be the case in other Antarctic diatoms during both short- and long-term photoacclimation to UV radiation (Hernando *et al.* 2002). It is not just microalgae that have been studied in response to UV, but macroalgae (seaweeds) are also prone to damage in shallow coastal waters. UV effects can influence macroalgal life histories and community structure (Wiencke *et al.* 2007, Hanelt *et al.* 2007).

Bacterioplankton can also suffer DNA damage from UV, although there can be very quick recovery of this (Jeffrey and Mitchell 2001). In temperate waters UV-A and UV-B radiation lead to only minor alterations in bacterioplankton species composition, since only approximately 10% of the species there are sensitive to UV radiation (Winter *et al.* 2001). In Antarctic waters Helbing *et al.* (1995) and Buma *et al.* (2001) showed that significant UV radiation reduced the viability of natural bacterial assemblages. Davidson and van der Heijden (2000) also found significant inhibition of bacterial growth with increasing UV irradiance, although UV-induced damage was repaired rapidly (see also Hernandez *et al.* 2006).

2.5 Biological rhythms in the polar environment

2.5.1 Photoperiodism

Rhythms of activity, development, behaviour, and reproduction are usually geared to periodicities in the environment, particularly to those in light and dark, heat and cold, and tides. Some degree of synchrony between these biological and physical cycles must occur if organisms are to succeed. Timing, as, for example, in producing progeny when the physical conditions make food available, is a crucial factor in the life-cycle strategies of polar animals. Many species have timing mechanisms to spread development over several years, rather than risk concentrating it all in one short favourable period.

An overriding periodicity, since it limits the time during which the primary production by photosynthesis upon which everything else depends takes place, is the cycle of light and dark. The mechanism by which organisms keep track of the progress of the seasons by response to duration of day and night was first studied in plants. These fall broadly into three classes according to their requirements for flower initiation, namely short-day, long-day, and day-neutral plants. If they are to flower, short-day plants must have an uninterrupted dark period of at least a certain length, usually about 8 h in the 24. Most spring- and autumn-flowering plants of temperate regions are of this kind. Long-day plants, which include all summer-flowering plants of temperate regions, do not need a dark period but require 12 h or more of light in the daily cycle. Little work has been done on the photoperiodicity of polar plants but it seems to be of the long-day type. Vegetative growth and dormancy also show photoperiodic effects and there are indications that short days induce cold-hardening. Parallel responses to those in flowering plants are found in ferns, mosses, and seaweeds. The inability of some mosses to maintain photosynthesis under continuous light may account for their absence in high latitudes (Kallio and Valanne 1975).

Photoperiodic effects are found in some Arctic insects, diapause (suspension of development) being brought about by short days. They are also shown in the migrations, reproduction, and changes in coat or plumage of mammals and birds. The change to white in winter in the Arctic hare (*Lepus arcticus*) and in ptarmigans (*Lagopus mutus* and *Lagopus lagopus*) is photoperiodic and can be induced out of season by exposure to short days. In reindeer (*Rangifer tarandus*; also known as caribou), levels of growth hormone and testosterone show seasonal variations which seem ultimately related to external factors such as photoperiod. Northern-hemisphere reindeer introduced into the Antarctic adapted well to a reversal of the photoperiodic cycle.

Photoperiodism is often complicated by temperature effects. *Colobanthus subulatus*, a pearlwort native in South Georgia, is a long-day plant but needs cold pretreatment for flower initiation. It seems as though the very early inflorescences which it produces in September must have been initiated in the previous season with the cold pretreatment happening in the winter before that. In *homoiothermic*—warm-blooded—animals the complicating effects of ambient temperature are less than they are for plants and illumination, which acts through eye stimulation releasing hormones under pituitary control, has more clear-cut influence.

2.5.2 Circadian rhythms

Circadian rhythms in metabolism (i.e. variations in activity with a periodicity of around 24 h) seem to be involved in the biochemical mechanisms which produce photoperiodic response, as a means of counting the passage of time. These rhythms, which are found at all levels of organization, are endogenous (i.e. continuing to a large extent independently of environmental changes), although the clock may be stalled and reset by sudden alterations in conditions such as temperature.

In plants and many algae, photosynthesis commonly shows a circadian rhythm. Transferred from a light/dark regime to continuous light, plants may show for some days afterwards a minimum in photosynthesis at the time corresponding to that of the original dark period. However, it is difficult to establish that similar endogenous rhythms are important under field conditions. Measurement of carbon dioxide fixation by the marsh plant *Caltha palustris* on the Murmansk coast, around 69°N 35°E, during the continuous daylight of the Arctic summer showed periodicity with a maximum near noon and a minimum near midnight. However, the variations were correlated with radiation intensity and there was no evidence of an endogenous rhythm (Pisek 1960). Diurnal variations in phytoplankton photosynthesis in polar regions seem likewise related to irradiance rather than to a biological clock.

Among animals, Arctic bees show diurnal rhythms. Birds might be expected to show activity throughout the 24 h of continuous daylight but, in fact, usually retain some degree of diurnal rhythm (Sage 1986). The resting period of light active passerines—song birds—shifts further into the evening with lengthening light period put there is usually a resting period around midnight. An overriding effect of a circadian rhythm is evident in the behaviour of the snow buntings (*Plectrophenax nivalis*) in the narrow valley of Longyearbyen (78°12′N 15°40′E) in Svalbard. Because of its orientation between high mountains, this valley only receives direct sunlight around midnight and is otherwise in shadow. Nevertheless, the resident birds sleep at midnight as do those in open tundra. Different species can, however, show different timings of activity and inactivity. Non-passerines, such as ducks, auks, gulls, and terns, are capable of continuous activity although they show circadian rhythms in other respects.

Circadian rhythms occur in humans and the polar regions are obvious places for study of the effects of unusual schedules or time-zone shifts, without confinement of subjects to controlled environment rooms. Interest in this began in the Arctic as far back as 1910. It is a general experience that the photoperiods of high latitudes have an upsetting effect, resulting for some in serious impairment of sleep cycles and sleep structure, peaking at the times of continuous daylight or the middle of the polar night. Anomalous responses to medication may be induced. In an elaborate study at the South Pole sleep and activity patterns, haematological and cardiopulmonary data, and the acute effects of oxygen shortage (the pole is at an altitude of 2912 m) were examined. A surprising finding was that stage 4 (slow-wave) sleep was almost completely lost by the end of the austral winter and failed to return 6 months after the subject was back in the USA. Nothing similar was found to happen at the British station at Halley, further north. The discrepancy may be related to differences in group behaviour—such as having fixed mealtimes, a British rather than an American habit—which may act as synchronizers to adapt circadian rhythms to a 24-h day (Wortmann 1995).

2.5.3 Season anticipators and responders

The sublittoral macroalgae (seaweeds) can survive dark periods for as long as a year and grow at low irradiance levels, light saturation occurring between 4 and 20 μmol photons m^{-2} s^{-1} (Wiencke et al. 2007). They become photoinhibited under high irradiance levels in summer but have a photoprotective mechanism which dissipates excess energy as heat. Growth occurs only at temperatures below 10°C (5°C for *Desmarestia* spp.) and upper survival levels are usually less than 17°C (Wiencke 1996).

Several species, including the large brown seaweed, *Laminaria solidungula*, from the Arctic and the red *Palmaria decipiens* from the Antarctic, actually begin to grow during periods of darkness at the end of the winter, when the overlying water is still covered by sea ice. These species have not found a way of growing without light; instead, they begin to grow by using the starches and other metabolites that they built up in the previous year's growth period (Weykam *et al.* 1997). The new tissues produced are ready to begin photosynthesis as soon as light becomes available when the ice breaks up. This kick-start maximizes the growth period during the short summer months when light conditions are more favourable. The development of new blades in the dark is probably controlled by inherent rhythms within the biochemistry of the seaweeds that govern seasonal growth patterns. This hypothesis is supported by experiments to measure the seaweed growth under constant daylength or in the dark that indicate that there are free-running growth rhythms that are independent of the light conditions. These are *season anticipators*, in contrast to the opportunist *season responders* characteristic of the littoral zone (Kain 1989).

The red seaweed, *Myriogramme mangini*, common in the sublittoral of Signy Island, is a good example of a season responder, not growing in winter. Detailed studies *in situ* on whole fronds showed that photosynthesis was saturated between 10.5 and 18 μmol photons m^{-2} s^{-1} in early spring and autumn respectively, with compensation occurring from 2.5 to 2.8 μmol photons m^{-2} s^{-1}. From the photosynthesis measurements and records of solar radiation and light attenuation in the water over the year, annual net production can be estimated as 1.5 and 0.007 g carbon g dry weight^{-1} year^{-1} at 5 and 20 m depth, respectively. From this a maximum depth of occurrence of this species can be predicted as 22.9 m, at 2.7% of surface radiation. That it is, in fact, only occasionally found below 14 m may be attributed to competition with other algae (Brouwer 1996).

3 Periglacial and terrestrial habitats in polar regions

3.1 Introduction

This chapter introduces the terrestrial ecosystems of the Arctic and Antarctic. First to be considered are some of their physical and chemical characteristics and then subsequent subsections detail the ecosystems and their component biota based by practical means of *regionalization* within these two areas. It is self-evident that both encompass far too wide a geographical extent and different levels of biological complexity to be usefully treated as a single whole. Throughout, similarities and differences between the two regions are drawn out.

Climates operate over wide areas but what is crucial for an organism is the combination of physical factors in its immediate vicinity (Walton, in Laws 1984). These can differ enormously from those in the general environment. Warm sheltered spots and exposed prominences can be within short distances of each other. Slope is of obvious importance for the receipt of direct radiation but is of little significance in overcast conditions. Habitats in soil, among rocks or vegetation, provide irradiances, temperatures, humidities, and wind speeds quite different from those outside. Within short spaces of time conditions in a microhabitat may shift with shafts of sunlight or as thawing or freezing occur in its vicinity. Larger animals and plants are to varying extents able to select or modify their microenvironments but small organisms survive, or perish, according to the conditions in their immediate surroundings. How we define microclimate depends on the size of the organisms under consideration.

Measurements made by standard meteorological procedures give little, if any, information about these conditions. An example is provided by the two indigenous flowering plants of the Antarctic (see below) which are found in the maritime climatic zone but in very limited areas, within an otherwise ice-bound landscape, on sheltered north-facing slopes where they can get sufficient warmth and liquid water to be able to complete their reproductive cycles in a short summer. Correlation of their distribution with macroclimatic data is poor. Only measurements made in the air immediately surrounding the plants, among their leaves, and in the soil underneath can give useful information about the conditions they need.

3.2 Substrata

3.2.1 Exposed rock surfaces

A rock surface offers some hospitality to life, and those organisms able to take advantage of it are termed *epilithic*. Rock surfaces are present on the most northerly islands in the Arctic Ocean (approximately 84°N) and most southerly nunataks within the Antarctic continent (approximately 87°S), with both being colonized by biota. Conditions experienced on these surfaces depend very much on their aspect: on a clear day near the summer solstice the maximum radiation flux on a north-facing surface in the continental Antarctic Dry Valleys is high, at about $1050\,\mathrm{J\,m^{-2}\,s^{-1}}$, which is a large proportion of the maximum levels typical of much lower latitudes. Indeed, if radiation receipt is integrated over a 24-h period in midsummer, the total is greater than that of a tropical location. As snow and ice have a high *albedo* (see Chapter 1), radiation reflected from neighbouring glaciers and snowfields may further increase flux for some terrestrial habitats. Conversely, at these same latitudes during winter, the sun remains well below the horizon, and radiation receipt is virtually zero. Under these conditions, exposed surfaces will experience temperature regimes following that of the air, with averages in the range of −40 to −60°C, and extreme minima in the Antarctic below −89°C. In contrast, in summer the temperature of a dark surface perpendicular to incoming radiation may rise to as much as 42°C, with lesser inceases at other orientations, and diurnal ranges of up to 25°C (Nienow and Friedmann, in Friedmann 1993). Snow and ice cover can ameliorate winter conditions. A thin ice sheet provides a warmer atmosphere below it, as in a greenhouse. Snow both insulates against extreme temperatures and provides a filter for incoming and potentially damaging ultraviolet (UV) radiation (Cockell *et al.* 2002).

The supply of water, arriving as precipitation, dew, frost, or run-off, is also affected by slope and aspect. Water can penetrate the surface layers of some rock types, providing a resource for endolithic biological communities (discussed below). In coastal and maritime situations, melt water may trickle over rock surfaces for much of the summer. In inland or high-altitude locations, water may only be available to biota for a few days each year, or not at all in some years. Epilithic crustose or foliose growth will retain water. Mineral nutrients may arrive in airborne dust, precipitation, run-off, or splash, or derive from the rock itself. Near bird colonies the nutrient supply may be so copious as to be toxic.

An overriding consideration for biota colonizing rock surfaces is that they should be able to attach to it. If a rock weathers grain by grain or if it *exfoliates* (i.e. weathers by flaking of the surface), it may not provide a sufficiently permanent substrate. Different species may have preferences for different rocks, based on aspects of their chemistry. On Signy Island in the South Orkney Islands, quartz-mica-schist, an acidic rock, is colonized by the moss genus *Andreaea* and the lichens *Omphalodiscus* and *Usnea*, whereas adjacent marble, which is basic, supports mosses such as *Syntrichia* spp. and the lichens *Caloplaca* and *Leptogium* (Table 3.1).

Orientation relative to the prevailing wind is important, growth being favoured on the leeward side. Abrasion by wind-blown rock debris or ice crystals is an important and strong force of erosion (Fig. 3.1), and may completely polish off any vegetation. Snow accumulation on the leeward side of boulders and in crevices provides a longer-term source of moisture, whereas to the windward side evaporation is increased so that the time available for active metabolism is reduced. Wind speed decreases logarithmically as a surface is approached, the distance above the surface at which it becomes virtually zero being determined by an aerodynamic roughness factor, equivalent to about 0.001 cm over smooth ice, 0.1 cm for vegetation up to 1 cm high, and 0.5 cm for smooth fresh snow. Cracks in rocks (*chasmolithic* habitats) and their undersurfaces (*sublithic*) may provide shelter for small invertebrates and even contain entire communities. Sublithic habitats where some light transmission is possible (such as quartz pebbles in the Victoria Land Dry Valleys) permit the inclusion of autotrophic elements (algae and mosses) and, indeed, may be the only places where these are found in these otherwise extremely hostile locations.

A critical element required for epilithic lichen survival is their ability to create and maintain a secure attachment to the rock surface layers. In creating this, lichens penetrate into the rock, a process facilitated by their secretion of, often species-specific, so-called lichen compounds (which are frequently acids). In exfoliating rocks, lichen growth may both take place beneath a plate and help its eventual detachment, thereby providing a biological contribution to weathering, initial soil formation, and nutrient release.

3.2.2 The endolithic habitat

The rock matrix itself may be inhabited by biota, known as an *endolithic* habitat. These range from the *chasmolithic* occupation of the tiny cracks

Table 3.1 Some chemical characteristics of a range of polar soils.

Type of soil and location	Region	pH	Total organic C (%)	Total N (%)	C/N ratio	Available nutrients (ppm)				
						P	Na	K	Ca	Mg
Cold-polar desert, Inglefield Land, Greenland (1)	High Arctic	8.8	0.2	—	—	—	2.3	11.7	15	3.7
Peat, sedge meadow, Devon I. (2)	Low Arctic	6.2	42.2	2.68	16	0	25	47	—	—
Brown earth, cushion plants, lichens, Devon I. (2)	Low Arctic	7.8	3.7	0.12	31	1	1	12	—	—
Frost-sorted till, LaGorce Mts, Transantarctic Mts (3)	Continental Antarctic	6.82	3.7×10^{-6}	1×10^{-6}	3.7	0.05	0.01	0.3	0.24	—
Dry valley soils, Victoria Land (1)	Continental Antarctic	7.4–8.9	0.02–0.09	0.004–0.085	6–13	0.03–0.27	8–1150	1–245	8–190	1–100
Frost-sorted till, Sky Hi Nunataks, Ellsworth Land (3)	Continental Antarctic	5.42	4.2×10^{-5}	5×10^{-6}	8.4	0.07	0.01	0.11	0.91	—
Frost-sorted till, Coal Nunatak, Alexander I. (3)	Maritime Antarctic (S)	7.67	1.5×10^{-4}	4.7×10^{-6}	31.9	0.044	0.015	0.76	1.4	—
Frost-sorted till, Mars Oasis, Alexander I. (3)	Maritime Antarctic (S)	7.06	3.6×10^{-5}	3.6×10^{-6}	10	0.054	0.021	0.22	1.4	—
Frost-sorted till, Rothera Point, Adelaide I. (3)	Maritime Antarctic (S)	5.47	2.3×10^{-4}	3.1×10^{-5}	7.4	9.2	0.4	0.27	17	—
Frost-sorted till, Signy I. (3)	Maritime Antarctic (N)	6.97	7×10^{-5}	9.3×10^{-6}	7.5	0.1	0.11	0.76	0.89	—
Schist protoranker, Signy I. (2)	Maritime Antarctic (N)	5.4	2.3	0.27	8.5	4	13	7	95	28

(Continued)

Table 3.1 (Continued).

Type of soil and location	Region	pH	Total organic C (%)	Total N (%)	C/N ratio	Available nutrients (ppm)					
						P	Na	K	Ca	Mg	
Marble protoranker, Signy I. (2)	Maritime Antarctic (N)	7.9	2	0.26	7.7	8	24	5	596	17	
Peat under moss turf, Signy I. (2)	Maritime Antarctic (N)	4.7	43.6	1.36	32	3	102	17	213	196	
Brown earth under grass, Signy I. (2)	Maritime Antarctic (N)	5.4	13.8	1.11	12.4	8	47	16	96	58	
Ornithogenic, penguins, Signy I. (2)	Maritime Antarctic (N)	6.1	10	1.8	5.5	460	69	73	106	68	
Elephant seal wallow, Signy I. (2)	Maritime Antarctic (N)	6.7	30	3.58	8.4	66	199	100	220	134	

I, Island; Mts, Mountains. Data from (1) Cameron (1969); (2) Rosswall and Heal (1975); (3) Lawley et al. (2004).

Fig. 3.1   Wind abrasion is a strong weathering force in the deserts of the Maritime and Continental Antarctic, as illustrated by this approximately 2m *ventifact* boulder at Mars Oasis, Alexander Island (photograph by Peter Convey).

and fissures of weathered rocks to colonization of the interstices of porous rocks (*cryptoendolithic*). Chasmolithic and endolithic habitats are not unique to the Antarctic, but here they do represent one of the apparent limits to the survival of life on Earth, as well as being proposed as viable analogues with which to develop and test methodologies for the eventual surveying for the presence of life on other planetary systems. Endolithic microbial ecosystems of the Dry Valleys of Antarctica (Friedmann 1982, Friedmann *et al.* 1993), which have developed in quartz sandstone, are proposed to be among the slowest-growing examples of life, with cells possibly achieving a single binary fission over a year, and carbon-turnover times extending to tens of thousands of years, a unique example of biological and geological timescales overlapping. Prerequisite for the establishment of self-contained endolithic communities is the inclusion of autotrophic (photosynthesizing) members, by definition requiring some light transmission into the rock. Generally, light is absorbed rapidly by rocks, with the exception of types that include light-transmitting elements such as quartz crystals. Endolithic habitats provide a degree of protection from the extreme stresses faced at the surface of the rock (abrasion, radiation, desiccation, temperature extremes and fluctuation; Fig. 3.2). A recently described endolithic habitat in clear gypsum crusts (Hughes and Lawley 2003) also illustrates the requirement to tolerate severe osmotic stress.

Endolithic habitats have the great advantage of conserving water, vital for the maintenance of active life. For example, once wetted, a sandstone matrix can retain moisture for days or even weeks. Within the rock the freezing point of water becomes depressed to as low as −5°C. It is notable that the endolithic communities of frigid and hot deserts are remarkably

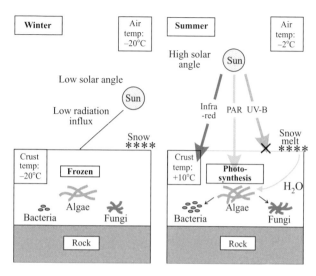

Fig. 3.2   Model figure of an endolithic microbial community. Modified after Hughes and Lawley (2003).

similar in composition and functions, desiccation overriding any effects of temperature. However, in the Antarctic, there is also a suggestion of opportunistic colonization taking place in this most extreme of habitats, with molecular identifications of gypsum-based endolithic communities finding community components that are typical of very different habitats at lower latitudes.

3.2.3 Rock debris

Passing from warm to cold regions of the Earth, chemical weathering of rocks becomes less important and physical processes bringing about decay predominate. These include glacial action, the effects of water (in the forms of vapour, liquid, and ice), salt weathering, insolation, and wind action (Campbell and Claridge 1987). Glacial action is a major agent in the breaking and grinding of rock to rock flour and fragments, large and small. Cold glaciers, frozen to their beds, as found in Greenland and Continental Antarctica, have little erosive power. Warm glaciers, at the pressure-melting temperature of ice, slide over their beds more easily and are more destructive. This type of glacier, which occurs in the less frigid parts of the Arctic and on the sub-Antarctic and Maritime Antarctic islands, Antarctic Peninsula, and parts of the continental coastline produces more glacial *till* (unsorted debris) from which big moraines are built. However, with only just over 0.3% of Antarctica being even seasonally ice-free (British Antarctic Survey 2004), the extent of contemporary till (and soil) formation here is inevitably much more limited than in the north.

Other processes also lead to rock fragmentation: freeze–thaw action and salt crystallization may expand cracks and pores, solar radiation and heating and cooling cycles result in stresses leading to fragmentation, and abrasion by wind-transported ice, sand, or larger particles in effect leads to sand blasting and the formation of spectacular areas of *ventifacts* (as shown in Fig. 3.1).

There are extensive aeolian sands near the Colville River in Alaska (approximately 69°30′N 152°W) and dunes of alluvial sand occur in the lower Victoria Valley in Antarctica (approximately 77°20′S 162°E). Interbedded snow cements sand dunes against high winds. Rock detritus and soils can likewise be stabilized at depth by permafrost, although this is not thought to provide a habitat for biota and its potential (along with that of ice itself) as a reservoir of colonizing material (a *propagule bank*) remains an unproven source of debate. The surface of unconsolidated detritus, under the influence of gravity, water, wind, or freeze–thaw activity, is generally too mobile for permanent colonization by plants. However, it has been demonstrated that primary colonization of glacial forelands on the High Arctic Svalbard archipelago may involve invertebrate detritivores and predators even before plant colonization is seen, with these relying on the rain of aeolian fallout to supply food (Hodkinson *et al.* 2002).

Freeze–thaw activity is particularly important in the formation of well-known and easily recognizable periglacial features, including frost boils, solifluction lobes, and patterned ground (Fig. 3.3). Some of these feature must be at least thousands of years old, with those in parts of the Antarctic Dry Valleys and Trans Antarctic Mountains dating back millions of years to the end of the Miocene glaciations. They can vary in size from less than a metre to 30 m or more in diameter. The sorting process results in the accumulation of fine material (*fines*) in strips or polgons, surrounded by larger stones. The polygon surface is typically domed, as drainage is better among the stones at the edge. However, the stone concentrations tend to accumulate snow, and are more stable, providing water and a substratum for vegetation development (moss and lichen, with vascular plants included in the sub-Antarctic and much of the Arctic).

The periglacial sorting of rock and stones results in a high degree of spatial heterogeneity in the light incident under different parts of the stone/rock patterning and vegetation patterns (Cannone *et al.* 2004). In turn this results in regions of the stone field where light penetrating the stones is sufficient to support photosynthetic carbon assimilation by *hypoliths* inhabiting the underside of the stones. Hypoliths are microorganisms—mostly cyanobacteria and green algae—that grow on the underside of stones and rocks, where they utilize irradiances far less than 0.1% of the incident light for photosynthesis. Known also from hot deserts, where temperatures at stone undersides can reach 65°C, they have recently been described at the

(a)

(b)

Fig. 3.3   Patterned ground in the (a) Arctic, in Ny Ålesund, Svalbard, and the (b) Antarctic, in Davis Valley, Pensacola Mountains (photographs by Peter Convey).

opposite temperature extreme in both Arctic and Antarctic deserts, where they are estimated to be as productive as other primary producers such as the lichens and bryophytes that sparsely inhabit such regions (Cockell and Stokes 2004).

Over time, screes and developing silts and soils may become stable enough for vegetation to establish. The tops of stones become occupied by epilithic lichens whereas deeper in the spaces between them there may be green algae and mosses adapted to lower light intensities. In milder locations, particularly of the sub-Antarctic and Arctic, vascular plants are also an early component of developing communities. Likewise, more stable points in silts and glacial till, for instance around embedded stones, may be colonized and further stabilized by microbial autotrophs and heterotrophs (Wynn-Williams 1996), providing the start of a vegetation succession process.

3.2.4 Permafrost

Permafrost is material which has been below 0°C for several years, so that the components have become ice-cemented (Sage 1986, Fitzpatrick, in Woodin and Marquiss 1997). Continuous permafrost covers some $7.6 \times 10^6 \, km^2$ of the northern hemisphere, with its southern limit, corresponding approximately with a mean annual air temperature of −8°C, reaching 30°N in parts of Eurasia. In places around the Arctic Ocean it extends under the sea. Its depth varies, being mostly between 200 and 700 m. Under the influence of high rates of contemporary climate warming, the thawing of Arctic permafrost is becoming a matter of considerable concern, with ramifications for drainage and water regimes, regional ecology, the balance of global carbon uptake and release, and human engineering and sociological impacts. Less is known of permafrost distribution in Antarctica, where the ice-free surface area available is also much smaller. Ground is frozen to a depth of 1000 m in places in the Victoria Land Dry Valleys (approximately 78°S 162°E) and probably to similar depths in other parts of the continent. Permafrost occurs on the Maritime Antarctic South Shetland and South Orkney archipelagos but not on the sub-Antarctic islands.

While not supporting active life, permafrost has important effects in modifying overlying habitats. It prevents drainage, and cannot be penetrated by plant roots. In the Arctic and where it occurs in the Maritime and coastal regions of the Continental Antarctic, permafrost is normally associated with wet soils, having above it an 'active layer', usually less than 1 m deep, which freezes in winter and thaws in summer. Active layers are not present in the much drier soils of the Dry Valleys and parts of the Trans Antarctic Mountains. Permafrost can drastically affect surface topography and habitats (Fig. 3.4).

The low vapour pressure of unfrozen water in the pore system of the active layer causes diffusion of water vapour from elsewhere and this freezes to form ice lenses. At their greatest development, these give rise to a dramatic feature, the *pingo* (derived from the Inuit for conical hill). This is circular or oval in plan and typically 10–40 m (rarely up to 100 m) in height. The hill contains a massive lump of ice originating from a silted up and shallowed lake, which is forced upwards by its own expansion during freezing and by pressure from the advancing permafrost around it. It can grow by as much as 25 cm per year by capture of moisture from its surroundings.

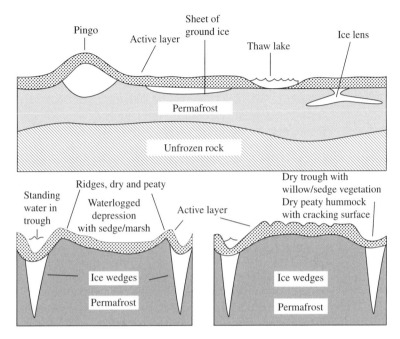

Fig. 3.4   Simplified sections, not drawn to scale, of surface features in Arctic tundra. Modified after Sage (1986).

Pingos are covered with soil and vegetation similar to those around them but the dome may be ruptured to form a crater with a pool in it. Pingos are characteristic of the Arctic, especially in the Mackenzie Delta area in Canada (approximately 68°N 135°W). They are less common and well developed in the High Arctic and in the Antarctic, where ground water is generally less available.

3.2.5 Polar soils

With continued weathering and incorporation of organic matter, rock debris becomes soil. Even in the virtual absence of biological activity, processes of leaching, mineral decay, and segregation continue, and the relations of their structure to time, topography, parent material, and climate give these deposits genetical connection with indisputable soils (Tedrow 1977, Campbell and Claridge 1987). During glacial retreat, the transformation of rock debris into soil is laid out in a chronosequence with distance from the glacier snout. In Low Arctic regions the conversion of till into textured brown soil with developed horizons takes about 150 years (Matthews 1992). Clear soil horizons are rarely seen in the Antarctic.

Cold slows down chemical weathering and the rate of addition of organic matter. Aridity also slows down these two processes and retards segregation of materials by water movements. Modest availability of water permits weathering, some plant growth, and transport of soluble salts with separation into horizons. More water results in leaching of salts and, finally, waterlogging, which, by excluding oxygen, brings about accumulation of organic matter as peat. These various conditions and processes interact to produce from diverse starting materials a variety of different soils. Pedological classification is complex (Campbell and Claridge 1987, Fitzpatrick, in Woodin and Marquiss 1997), but here a simple treatment will suffice. Tedrow (1977) distinguished four soil zones in polar regions (while also not mapping closely on the boundaries of the generally recognized biogeographical zones, see section 3.3; Longton 1988):

1. Tundra,
2. Sub-polar desert,
3. Polar desert,
4. Cold desert.

The representation of these zones is vastly different in Arctic and Antarctic. In the Arctic, tundra and sub-polar desert extend over large areas, the polar desert zone includes all of the regions north of about 80°N, and cold desert is not represented. In the Antarctic, land akin to tundra is confined to the sub-Antarctic islands, sub-polar desert is absent, polar desert is restricted to limited maritime regions, and the rest of the continent is cold desert.

Cold desert soil, distinguished from other polar soils by the absence of organic horizons (but not necessarily by total absence of organic matter) is unique to Antarctica. The Dry Valley soils of Victoria Land are estimated to be several million years old. In the Arctic, polar desert soils are only about 18 000 years old. Antarctic cold desert soils are strongly weathered and, being ahumic, water drains through them rapidly. In aridity, they are on a par with the soils of hot deserts, with which they share features such as granular disintegration and cavernous weathering of surface rocks, oxidation and staining of surface rock, endolithic algae, minimal weathering of buried rock, a crust at the surface and salt horizons within the profile, little organic matter, and being red or dark brown colour in the most weathered soils.

Polar desert soil, under less-dry conditions, has so-called desert pavement—stones and pebbles—at its surface. Organic content remains low but horizons are more prominent, although soluble minerals are unleached. Under more moist but still cold conditions, as in the Maritime Antarctic and northern Arctic coasts, chemical weathering and frost action are more evident, sorted features are formed, horizons are poorly developed, and soluble minerals are leached. The organic content remains low. The mineral soil, having characteristic light grey colour and reducing properties, is of the type known as *gley*. Alongside these ahumic soils, soil with an appreciable organic content may develop in more favoured spots, known as a *protoranker* or dry tundra soil (Table 3.1). In this, pads of humus develop beneath the isolated plant cushions. With more extensive moss or lichen cover the organic content may rise to 10%. The insulation provided by the cover allows surface temperatures to rise, promoting chemical weathering. On poorly drained sites there may be accumulation of peat-like material under moss turf. In the Antarctic this is only seen on the sub-Antarctic islands and at a very few locations in the Maritime Antarctic, in all cases with formation initiated over the few thousand years since retreat from the Last Glacial Maximum. Where present, peat is often cemented by ice crystals and there is only slight penetration of humic material into the mineral soil below.

In the sub-polar zones, where higher plants with root systems penetrating many centimetres deep can grow, organic matter is distributed more deeply and a loamy texture may result. This leads to true organic soils, brown soils, with developed horizons for which, in addition to rooting plants, active soil processes are necessary. Such soils are only rarely found on the Antarctic continent (and then only in the Maritime Antarctic associated with the very limited stands of the two native vascular plants). Soil-mixing organisms, such as earthworms, are absent from the brown soils found on the Maritime Antarctic islands and some more northerly parts of the Arctic, but do occur on the sub-Antarctic islands.

On the fringe of the Arctic in the tundra zone more mature brown soils, although frozen for part of the year, support vigorous root growth with earthworm activity. Brown soils grade into waterlogged peat soils, in which the addition of plant biomass exceeds its decomposition. Waterlogging, resulting in anoxic conditions, low temperatures, and acidity, favours peat accumulation. In parts of the Arctic, peats may be formed from substantial algal growths in pools and streams or from decay-resistant mosses on soils overlying permafrost. Extensive and deep peats in the sub-Arctic become incorporated into the permafrost, the melting of which under current climate warming activity presents a further challenge to global climate stability in the form of release of carbon through reactivated decay processes.

3.3 Communities

Polar biogeographical zones, and within these zones major soil and vegetation types, have been distinguished in various ways (Longton 1988), based largely on a combination of climatic and diversity characteristics, none of which have been entirely satisfactory. Here, the Arctic will be separated simply into High and Low Arctic zones (see Halliday 2002), and the Antarctic into sub-, Maritime, and Continental zones (see Smith, in Laws 1984). These have the merit of simplicity and, while not having directly interchangeable terminology, allow general comparison of Arctic and Antarctic, providing frameworks for accounts of both plant and animal communities (Figs. 3.5 and 3.6).

However, it is increasingly clear that in terms of bipolar comparisons of regional climates and, particularly, biodiversity patterns, the temptation to apply closely similar regional terminology can be counterproductive. Two examples serve to illustrate this point. First, in terms of energy receipt, only the High Arctic (northern fringes of the continental landmasses and Arctic Ocean archipelagoes) lies in the same range as the sub-Antarctic, the least extreme of the Antarctic regions (Table 3.2; Danks 1999, Convey 2007). Second, in terms of terrestrial plant and invertebrate diversity, the High Arctic Svalbard archipelago alone contains far greater species diversity across all major taxonomic groups (by factors of 2 to >10) than is present across the entire Antarctic region (Table 3.3). As the potential of new methods of phylogeographic analyses starts to be realized, it is also

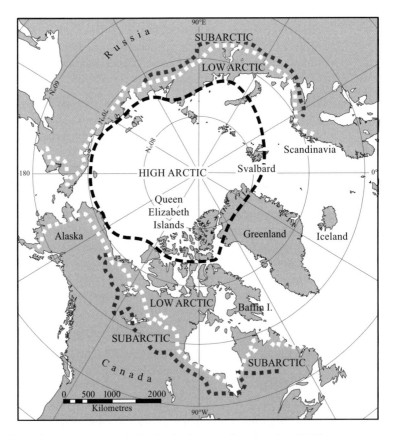

Fig. 3.5   Principal vegetation zones in the Arctic. Redrawn from Longton (1988).

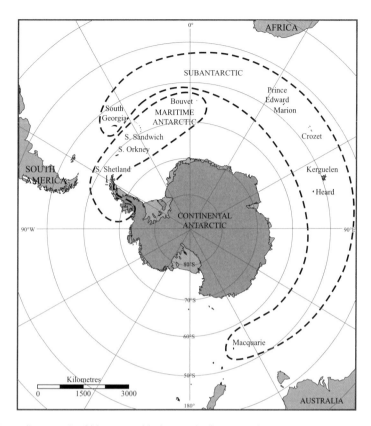

Fig. 3.6   Generally recognized biogeographical zones in the Antarctic.

becoming clear that even the currently accepted regionalization of at least the Antarctic is no longer sufficient (Chown and Convey 2007, Peat *et al.* 2007).

Passing through these zones equatorwards from the poles there is a general increase in productivity and species diversity, but transitions are not smooth nor definite because local features of history, climate, geology, and topography superimpose their effects. In the confusing mosaic of vegetation which is actually encountered, plant sociologists have distinguished various formations, associations, sociations, and assemblages (Smith 1972, Smith, in Laws 1984, Longton 1988). However, this is a level of detail beyond that required here, while there is little evidence for these patterns being passed on to animal communities through specific plant–animal associations.

The successions in time of communities and soils are interwoven. Where conditions are inimical to plant growth, the substratum changes extremely slowly and remains much like its parent material. Microbial

Table 3.2 General climatic comparison between polar zones.

Zone	Months with positive mean air temperatures	Air temperature range (°C)		Day degrees above 0°C
		Mean winter to summer	Extreme range	
Low Arctic	4	−36 to +11		600–900
High Arctic	2–4	−34 to +5	−60 to +20	50–350
Sub-Antarctic	6–12	−2 to +8	−10 to +25	700–1700
Maritime Antarctic	1–4	−12 to +2	−45 to +15	6–100
Continental Antarctic coast	0–1	−30 to −3	−40 to +10	0
Continental inland (deserts and nunataks)	0	<−50 to −10	<−80 to −5	0

Table modified from Convey (2007).

(Wynn-Williams 1996) and subsequent vegetation colonization quickens soil formation and modifies microclimate and hydrology. Under favourable conditions biotic control may be established and a classical succession process then progresses towards a 'climax' community, whatever the starting point may have been. In between these extremes succession is slower and divergent, giving rise to several different types of habitat close together with steep environmental gradients between them. This type of succession does not appear to apply in the so-called fellfield habitats of the Maritime Antarctic, with communities either not developing to a climax state, or following a form of circular succession or autosuccession, with pioneer and climax species occurring together (Müller 1952, Smith 1972, Convey 1996a).

3.3.1 The communities of the Continental Antarctic

Terrestrial (at least seasonally ice free) deserts of the Continental Antarctic occupy some 4000 km² of the Victoria Land Dry Valleys region and lesser areas in the Bunger Hills, Vestfold Hills, and Transantarctic Mountains. Further, smaller, ice-free areas are present associated with the various nunataks and mountain ranges throughout the continent. Climatically, this region has no real comparator in the Arctic, the nearest being the conditions experienced high on the Greenland plateau, about which region little or no biological information is available. The Continental Antarctic presents a range of habitats from those in which extreme aridity makes life

Table 3.3 Overview of species diversity in some of the main representative terrestrial groups in the different Antarctic biogeographical zones, in comparison with three Arctic locations from which recent and partial or comprehensive compilations are available.

Group	Number of species					
	Sub-Antarctic	Maritime Antarctic	Continental Antarctic	High Arctic (Svalbard)	High Arctic (Franz Josef Land)	Greenland
Rotifera	>59	>50	13	154		
Tardigrada	>34	26	19	83		
Nematoda	>22	28	14	111		
Platyhelminthes	4	2	0	10		
Annelida (Oligochaeta)	23	3	0	34		
Mollusca	3/4	0	0	0		2
Crustacea (non-marine)	44	10	14	33		65
Insecta	210	35	49	237		631
Collembola	>30	10	10	60		41
Araneida	20	0	0	19		60
Acarina (free-living)	140	36	29	127		127
Myriapoda	3	0	0	0		1
Mammalia	0	0	0	3	2	8
Aves	0	1	5	17	6	39
Flowering plants	0	2	60	164	57	515
Bryophytes	26	125	335	373	150	612
Lichens	150	250	250	597	>100	950

Svalbard data from Barr (1995), Elvebakk and Hertel (1996), Frisvoll and Elvebakk (1996), Rønning (1996), Coulson and Resfeth (2004) and Coulson (2007); Franz Josef Land from Barr (1995); Greenland from Jensen and Christensen (2003); Antarctic from Convey (2007).

scarcely sustainable to others in which restricted communities containing mosses, lichens, and a variety of invertebrates are able to exist. The biodiversity and biology of terrestrial communities throughout Victoria Land have provided a recent focus of attention (Adams *et al.* 2006, Hogg *et al.* 2006).

The most extreme desert soils of dry valley areas often appear barren, but they have long been known to harbour a range of microbes, while new research indicates that population numbers may until recently have been seriously underestimated (Cowan *et al.* 2002). The numbers of bacteria and protozoa vary with moisture content whereas, even using modern molecular-based survey techniques, it remains difficult to define active presence or community structure. Nematodes, tardigrades, rotifers and arthropods are distributed patchily, presumably both in relation to these food organisms and to their own ecological requirements, particularly for water (Freckman and Virginia 1998, Convey and McInnes 2005, Adams *et al.* 2006).

Progressing from harsh to more favourable environments there is a detection sequence of different kinds of organisms. This is generally:

1. aerobic, heterotrophic, non-pigmented bacteria;
2. microaerophilic (i.e. preferring low oxygen concentrations), heterotrophic, pigmented bacteria;
3. actinomycetes;
4. coccoid green algae and both coccoid and filamentous cyanobacteria;
5. moulds, yeasts, protozoa, and microinvertebrates;
6. lichens and some microarthropods;
7. mosses, with filamentous green algae, diatoms, and nitrogen-fixing cyanobacteria, and the majority of microarthropods.

These various organisms are inevitably dormant for most of the time. Heterotrophic bacteria can find little in the way of substrates for active metabolism in the absence of phototrophic organisms, perhaps suggesting that their presence is not always indicative of forming part of a true community. The assemblage of microbial taxa detected in any one locality seems more a matter of chance colonization than a fundamental characteristic of this zone. Although molecular phylogenetic tools offer a new opportunity, the level of microbial endemicity currently remains uncertain although there is some evidence supporting extended evolutionary isolation and suggesting that the *global ubiquity hypothesis* (Finlay 2002) may not apply equally to the Antarctic continent as proposed elsewhere across the globe (Lawley *et al.* 2004, Boenigk *et al.* 2006).

In some contrast, some Antarctic endolithic communities, particularly those studied in some detail in the Dry Valleys, have evolved to fit a particular niche (Nienow and Friedmann, in Friedmann 1993). Here, among cryptoendolithic communities of various types occurring in sandstone, the most common is lichen-dominated and in fractures normal to the surface shows as distinct, parallel, coloured bands reaching about 10 mm in depth (Fig. 3.7).

Typically, the layer just under the crust is black and next is a white zone. These contain filamentous fungi (the *mycobiont*), the hyphae of which are black near the surface through the accumulation of protective 'sunscreen' pigments and white at greater depth. The green algae accompanying them, the *phycobionts*, include *Trebouxia* spp. Lower down is a green band, usually containing an endemic green alga, *Hemichloris antarctica*, and sometimes below that is a blue-green zone of coccoid cyanobacteria. The presence of *Hemichloris* in the lowest zone, where the photon flux is as low as $0.05–1 \mu mol$ photons $m^{-2} s^{-1}$ when dry, and $2–10 \mu mol$ photons $m^{-2} s^{-1}$

Fig. 3.7   A section at right angles to a sandstone surface from a Dry Valley in Victoria Land, Antarctica, showing layered endolithic growth. Courtesy of D. Wynn-Williams and the British Antarctic Survey (see colour plate).

when wet, may be explained by this alga's sensitivity to high irradiance and its ability to achieve net photosynthesis at extremely low levels.

The dominant organisms sometimes vary between communities, and heterotrophic bacteria and yeasts are also present but only contribute less than 1% of total biomass. Analogous cryptoendolithic communities are found within the clear gypsum crusts that commonly form on boulders at so-called oases on southern Alexander Island in the Maritime Antarctic (Hughes and Lawley 2003; Fig. 3.3b). While having many features in common with those of the Dry Valleys, the dominant fungal genus present has been identified as *Verticillium*, a surprise as elsewhere in the world these fungi are typically insect or other pathogens, and an indication that even the endolithic community in Antarctica may include an important stochastic element.

Estimates of the activity of endolithic communities have been made by measuring rates of carbon dioxide exchange using crushed rock samples (obviously allowing artificially much enhanced levels of gas exchange), and by more realistic measurements of exchange at the rock surface. Net photosynthesis begins at a surface irradiation of about $100\,\mu$mol photons m^{-2} s^{-1} and increases up to $750\,\mu$mol photons m^{-2} s^{-1} without saturation. The lower temperature limit for measurable net carbon dioxide uptake is between -6 and $-8°$C and the optimum between -3 and $+6°$C with the upper compensation point between 8 and 15°C.

These responses are similar to those shown by lichens from sub-polar and alpine situations and do not indicate any special adaptation to frigid desert conditions. The maximum rates of net carbon dioxide uptake lie between 0.22 and 0.78 mg CO_2 m^{-2}h^{-1} as compared with an average value of 500 mg CO_2 m^{-2}h^{-1} for temperate grassland under moist conditions. Estimates of about 3 mg carbon m^{-2}year^{-1} for net ecosystem productivity have been based on the average biomass of the endolithic community in the Dry Valleys (about 30 mg m^{-2}) and its turnover time (10 000 years; Friedmann *et al.* 1993).

However, these estimates are subject to many assumptions and uncertainties, as well as coming nowhere near predictions of gross or net primary production (1215 and 606 mg carbon m^{-2}year^{-1}, respectively) based on very small-scale nanometeorological models. It is suggested that less than 1% of the gross primary production may be incorporated into the standing biomass, with the greater part of the primary production being lost either as extracellular products or specifically through investment in stress-tolerance strategies. Although these are essentially self-contained systems, over geological time through weathering and exfoliation they will contribute a small amount of carbon to neighbouring soil ecosystems, which may eventually be carried by groundwater flows to streams and lakes, thereby supporting the development of other microbial ecosystems.

Away from the more obvious desert ecosystems of the Dry Valleys and other comparable locations, the development of terrestrial communities in the Continental Antarctic is limited to, often small, islands of ice-free ground associated with coastal rock exposures and the summits of inland nunataks. Many specific locations have still never been visited by, or received attention from, biologists (Peat *et al.* 2007). However, those that have, from the continental coastline to the most southern inland rock exposures on the planet, have been found to contain communities comprising the main organism groups already mentioned. In terms of general community structure, though not specific-species composition, these communities are similar to those of the Maritime Antarctic discussed in the following section. It is increasingly being realized that these isolated communities carry important signs of the evolutionary history of the continent, in some cases extending millions of years back in time, and even predating the final separation of the elements of the supercontinent Gondwana (Marshall and Pugh 1996, Maslen and Convey 2006, Stevens *et al.* 2006).

3.3.2 The communities of the Maritime Antarctic and High Arctic

The Maritime Antarctic and High Arctic are often treated as being broadly comparable, although there remain important differences between them. Mean monthly air temperatures rising to 2°C in the Maritime Antarctic, and to 5°C in parts of the High Arctic, allow a greater abundance and variety of life to develop. Vegetation is typically dominated by mosses and liverworts, together with *foliose* (flattened but not firmly attached to the substratum) and *fruticose* (strapshaped or shrubby) lichens, and, in some regions, cushions of flowering plants. The vegetation remains scattered, with bare areas in between. It merges on the one hand with that of the frigid deserts of the Continental Antarctic and on the other with tundra, existing where conditions are not so extremely cold and dry as in the former, but not sufficiently benign to support much growth of flowering plants as in the latter.

The term *desert* is difficult to apply with precision in the polar regions. In temperate regions deserts generally have an annual precipitation of less than 250 mm, but much depends on how this is distributed over the year and on its balance with loss of water by evaporation. Accurate measurement of precipitation is complicated by snow drift, leading to some uncertainty, but around the Arctic basin it is between 70 and 200 mm per year, with an average of about 135 mm. In coastal regions of continental Antarctica, for example, in the Bunger Hills (66°15′S 101°E) and Schirmacher Hills (70°45′S 11°40′E), which have some snow cover in the winter, values are higher, while inland across much of the continent they are much lower. Around the northern Antarctic Peninsula and the Scotia Arc archipelagos annual precipitation may reach 400 mm, but values decrease with southwards progression through the Maritime Antarctic. Low relative humidity is often typical, even in these cold maritime environments, resulting in rapid evaporation/ablation of water input through rain or snow, while the simple mineral soils often have poor water-retention capacity, at least in the surface few centimetres where permafrost does not form a barrier.

At both small and larger scales, topography is an important factor in colonization processes and community development, in particular through protection from prevailing winds, accumulation of scarce water resources (through concentrating snowfall), and optimizing exposure to solar radiation. Because of such effects of small-scale topography, areas in which bryophytes, lichens, or algae find moisture and warmth sufficient for growth in closed stands are to be found alongside what may properly be called desert. Once established, the plant cover provides insulation and promotes the survival and development of both the vegetation and associated microbial and invertebrate communities.

Seasonal snow cover is characteristic of most terrestrial habitats, other than exposed ridges and near-vertical rock faces, in these two regions. During winter it prevents erosion and damage by wind, and provides insulation. Although its insulating capacity varies, snow cover generally allows substratum and vegetation temperature to keep well above air temperature. In principle, snow cover may act as a form of greenhouse insulation: if this is sufficient to provide liquid water at vegetation level, then mosses, lichens, and their dependant invertebrate communities, may be hydrated and active under layers as much as 30 cm deep (Kappen, in Friedmann 1993). On the other hand, it can also retard warming in spring, and biological activity may not recommence until almost the point at which snow melts (Schlensog *et al.* 2004). Late-lying snow provides moisture for summer growth but if it persists too long it reduces the growing season. Separately, even a thin layer of snow is sufficient to absorb the majority of incident short wavelength and damaging UV radiation (UV-B radiation; Cockell *et al.* 2002), a factor important in consideration of the potential biological consequences of the anthropogenically produced Antarctic ozone hole (Convey 2003). The effects of snow on underlying biota are consequently highly variable.

3.3.3 High Arctic biota

The High Arctic mainly forms a fringe around the Arctic basin with its southern limit approximately following the 2°C isotherm. Temperature varies widely over the year, but mean temperatures in July (the warmest month) are below 5°C, and precipitation is low. Parts of this zone, such as Cape Chelyuskin (77°44′N 103°55′E) on the Taymyr Peninsula, are climatically very comparable with much of the Maritime Antarctic, with positive mean air temperatures of less than 2°C only being seen in midsummer. Elsewhere in the High Arctic, such as in parts of the Svalbard archipelago, other Arctic Ocean archipelagoes and the northern Greenland coastline, summer monthly means are several degrees warmer, a factor that highlights the major energetic difference between ecosystems of the Arctic and Antarctic (Convey 1996b, Danks 1999).

During periods of direct insolation, microclimate temperatures may be considerably higher, reaching maxima of 20°C or more above ambient air temperatures. Mosses or lichens predominate in the vegetation, but can cover as little as 2% of the surface area. There is a scattering of flowering plants, although species diversity (e.g. >100 species on Svalbard and 57 on Franz Josef Land alone) greatly exceeds that of the Maritime Antarctic zone (with two species) and even that of the entire sub-Antarctic (Table 3.3, Figs 3.8 and 3.9). They form dense tufts or flattened cushions which may be as much as a hundred years old and have two to six times as much dry weight in their underground parts, which store carbohydrates, as in stems and leaves. High Arctic soils contain a considerable seed bank and can produce a surprising display of flowers under favourable conditions. In some areas vegetation can form extensive patches of closed cover, appearing very lush at the height of summer. This is particularly the case in nutrient-enriched areas associated with vertebrate activity, such as under cliffs used by nesting birds (Fig 3.10).

The species variety of the microfauna varies with the moisture content of the soil, populations typically being highly localized and patchily distributed. They include protozoa, tardigrades, rotifers, nematodes, enchytraeid worms, mites,

Fig. 3.8   High Arctic vegetation: a mosaic of foliose and fruticose lichens with cushions of moss and flowering plants, Svalbard (photograph by David N. Thomas) (see colour plate).

Fig. 3.9   *Dryas octopetalla* in dry Arctic tundra, Ny Ålesund, Svalbard (photograph by Peter Convey).

collembolans (springtails), and insects (best represented by flies and beetles). These invertebrates are generally microbivores and/or detritivores, feeding on bacteria, algae, fungal hyphae, and debris, with very few true herbivores.

Vertebrate herbivores such as the Arctic lemming (*Dicrostonyx groenlandicus*), which feeds on mosses and lichens among other things, and the

Fig. 3.10 Lush vegetation growth is possible through nutrient enrichment under cliffs used by nesting birds; this example in Krossfjord, Svalbard (photograph by Peter Convey) (see colour plate).

Arctic hare (*Lepus arcticus*), the main food of which is Arctic willow (*Salix arctica*), have some impact but do not generally spend much time in this region. Large populations of migratory birds (particularly geese, wading birds, and gulls) utilize the High Arctic for breeding. Reindeer can also be present seasonally (or, in the case of Svalbard reindeer, permanently). Vertebrate predators, represented by the polar bear (*Ursus maritimus*) and Arctic fox (*Alopex lagopus*), are present year-round in this zone, the former relying primarily on the marine environment to provide its main

Table 3.4 Biomass of vegetation and rate of net primary production in terrestrial polar habitats.

Vegetation zone and type	Biomass (g dry weight m^{-2})			Net primary production (g dry weight m^{-2} year^{-1})
	Living	Standing dead	Below-ground living and dead	
Continental Antarctic Dry Valley desert (4)	4.4–7.4[a]	9.5–214[b]		0.01
High Arctic desert (2)	125–185			3
Maritime Antarctic desert (3)	200			100 (?)
Maritime Antarctic moss turf (3)	291–969			321–497
Maritime Antarctic moss carpet (3)	156–204			226–548
Sub-Arctic sedge moss meadow, tundra	959–2083	414–1125		185–280
Sub-Antarctic meadow (1)	937	1598	1642	840
Sub-Antarctic dwarf shrub (1)	1521	517	7536	1605
Sub-Antarctic tussock (1)	7525	5005	5000	6025
Low Arctic moss, lichen, dwarf shrub (2)	1100–2350			100–600
Low Arctic shrub (2)	1400–5850			500–1000
Low Arctic mire (1)	892	450		153

[a]From lipid-phosphate determinations.
[b]From organic-matter determinations.
Data from (1) Rosswall and Heal (1975); (2) Wielgolaski (1975); (3) Laws (1984); (4) Nienow and Friedmann, in Friedmann (1993).

prey, the latter with a widely seasonally varying diet, to which nesting birds and small mammals make a large contribution.

It is difficult to estimate either biomass or primary productivity in such a patchy and heterogenous community. Representative values of these variables are given in Table 3.4. Biomass of living plants is considerably greater than that of the Continental Antarctic, and comparable with (though at the lower end of) several habitats in the Maritime Antarctic. However, primary productivity is very low, reflecting the paucity of opportunities for photosynthesis and growth.

3.3.4 Maritime Antarctic biota

The extensive cryptogamic *fellfields* of the Maritime Antarctic, and also coastal regions of the Continental Antarctic, are characterized by cushion-forming mosses and patches of lichen interspersed with areas of bare ground (Smith 1972, Smith, in Laws 1984). Frost heaving of the soils occurs through substantial freeze–thaw cycling, and the ground is often patterned. Colonization by mosses and lichens on stable substrata can be rapid: on Signy Island (South Orkney Islands), lichen-dominated fellfield can become established in 10–20 years on suitably weathered rock surfaces, while experimental manipulations of soils demonstrate both that an extensive propagule bank is present and that bryophyte (and associated invertebrate) communities can develop on previously bare soil in as little as 3–5 years (Smith 1990, Convey 2003). Some lichen specimens are of a size which, assuming growth rates have not altered significantly, suggests that they are 300–600 years old. Most individual mosses are more transient in presence, but the rare peat banks formed under extensive turves of *Polytrichum* and *Chorisodontium* at a few maritime Antarctic locations have generated radiocarbon dates (5000–6000 years) consistent with a continuous presence since shortly after the commencement of post-Pleistocene glacial retreat at these locations (Fenton and Smith 1982).

Fellfield faunas are relatively poor, although they include representatives of all the invertebrate groups present throughout the Continental and Maritime Antarctic (Block, in Laws 1984, Convey 2007). In this habitat arthropods occur at low population densities of tens to thousands of individuals per square metre. Ecophysiogical studies have focused particularly on these arthropods' abilities to tolerate the cold and desiccation challenges of the Antarctic environment, providing clear demonstrations of a range of these tactics (Cannon and Block 1988, Block 1990, Sømme 1995). These abilities are generally more than sufficient to permit them to survive the stress patterns experienced in their natural habitats (Convey 1996a), although it is the case that some, such as the two resident Maritime Antarctic dipterans, only survive winter extremes though occupation of microhabitats protected by snow or ice cover.

Life-history strategies show little evidence of true specialization and are generally consistent with the predictions of so-called adversity selection (Convey 1996a), typically with slow growth rates, extended life cycles, low rates of reproduction, a lack of dispersal abilities, and considerable investment in stress-tolerance features. Rather than possession of specific features, the absence of some seems to be more important. Thus there are few if any examples of species possessing true (i.e. cued) diapause: overwinter survival is often possible in all life stages, and the possession of multi-year life cycles, and extensive overlap of generations is the norm.

Adjacent to fellfield habitats, and often mixed with them in a mosaic, are various types of vegetation community with more closed cover. The most developed of these is the Antarctic herb tundra formation, which includes the presence of one or both of the native flowering plants, the grass *Deschampsia antarctica* and the pearlwort *Colobanthus quitensis*, together with various mosses (Fig. 3.11). In particularly favourable locations, *Deschampsia* occasionally forms swards of several tens of square metres.

Fig. 3.11   The two Antarctic flowering plants, *Deschampsia antarctica* and *Colobanthus quitensis* (centre), are often found together, as in this example from Signy Island (photograph by Peter Convey).

Both species are capable of producing viable seed, although not every year, and vegetative propagation (by tillers, and wind and bird dispersal of fragments) is important. The two higher plants have formed a focus of research into vegetation responses to the rapid regional climate change being experienced in the Maritime Antarctic, with rapid increases in populations (in terms of both numbers and local extent), and an increased frequency of mature seed production (Convey 2003).

In areas of intense nutrient enrichment around vertebrate colonies, there is no Antarctic analogue for the rich closed vascular plant communities that are seen in the High Arctic. Cryptogamic vegetation has no root system and, instead, is rapidly destroyed by either trampling or over-fertilization. If any vegetation develops in such habitats in the Antarctic, it generally consists of the foliose alga *Prasiola crispa* (a species with a bipolar distribution, but that rarely if ever develops to the same extent in the Arctic). Both closed cryptogamic vegetation, and the vertebrate-associated *Prasiola* vegetation, provide rich habitats for arthropods and other microinvertebrates (particularly nematodes); the former regularly reach densities of tens to hundreds of thousands per square metre (comparable or even greater than found in typical temperate or tropical habitats), whereas the latter may reach millions per square metre (Convey and Smith 1997, Convey and Wynn-Williams 2002).

Investigation of two moss communities on Signy Island over the period 1969–1981 provided information on environmental factors, primary and secondary production, and decomposition processes under cold polar conditions (Davis 1980, 1981). This remains the only such detailed study

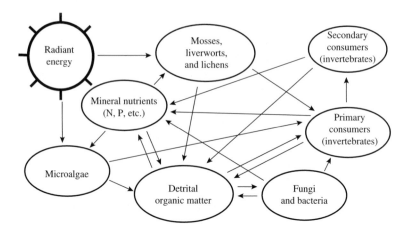

Fig. 3.12   Flows of energy and materials in moss turf on Signy Island. Based on data from Davis (1980, 1981).

to date. The two communities studied were a well-drained *Polytrichum/ Chorisodontium* moss turf and a wet moss carpet consisting of *Calliergidium, Calliergon*, and *Sanionia*. The food web described (Fig. 3.12) possesses a simple trophic structure, lacking true grazers and with few predators, a picture that is still accepted as the best existing description of both Maritime and Continental Antarctic terrestrial food webs.

Studies of one of the two dominant predatory mites (Lister *et al.* 1988) suggest that is has a minimal to undetectable impact on its main prey species. However, few detailed autecological studies have been attempted on these invertebrates, which means that most dietary assignations are in reality based on assumptions drawn from related species or genera (Hogg *et al.* 2006). One study of the common springtail *Cryptopygus antarcticus* has demonstrated considerably greater selectivity in choice of algal food source than expected previously (Worland and Lukesová 2000), indicating that much remains to be learnt of the detail of Antarctic terrestrial food webs. Of the total primary production (Table 3.4; Davis 1980, 1981), less than 0.04% was consumed directly, the food of the invertebrates being algae, other microorganisms, and detritus. Decomposition rates were low in both communities, and lowest in the moss carpet. The occurrence of anaerobic decay processes has been demonstrated in the moss carpet, with experiments *in situ* indicating release of an average 1.24 mg carbon m^{-2} day^{-1} as methane during the summer.

Unfortunately, parallel studies of mineral cycling at these Signy Island sites have not been attempted. Indeed, nutrient availability and cycling are poorly quantified and understood across the Antarctic. Many sources of nutrients are available at coastal locations with large vertebrate populations, such as Signy, and it is often thought that plant nutrients are non-limiting at such sites. However, there is evidence that mineral nutrients can be scarce in some fellfield habitats even here (Davey and Rothery 1992, Arnold *et al.* 2003). At areas more remote from direct or aerosol fertilization from the vertebrates or the sea, and in particular at inland locations, nutrient supply becomes increasingly scarce. Here, available nitrogen and phosphate as well as the entire range of trace elements may approach minimal levels, while soil carbon contents are also extremely low, indicative of very low levels of carbon fixation (Table 3.1; Beyer and Bolter 2002, Lawley *et al.* 2004).

3.3.5 Polar tundra

Polar tundra is a somewhat flexible term. In its strictest interpretation, true tundra is only found in the Arctic, and is a term used to describe habitats dominated by low vascular plants and not including significant tree or shrub growth (with the exception of prostrate forms such as *Salix arctica*), and underlain by permafrost. It can equally well be applied to some habitats already described above at High Arctic locations as it can to large areas of the Low Arctic south of the continental fringes. Sub-Antarctic habitats, while appearing superficially similar and are considered here, develop in a considerably more benign (and lower-latitude) environment, do not have permafrost, and do not include woody plants (Smith, in Laws 1984).

The highest mean monthly air temperatures in the Low Arctic and sub-Antarctic zones vary between roughly 3 and 7°C. However, conditions experienced in winter are strikingly different: Only the coldest sub-Antarctic islands experience mean monthly temperatures slightly below freezing, or long periods of lying snow at low altitudes, and biological activity continues year-round. In contrast, across the Arctic tundra winter temperatures remain well below zero, and biological activity (at least for plants and invertebrates), to all intents ceases. Even with these winter differences, 'tundra' regions both north and south experience longer growing seasons and greater total energy input than those of higher latitudes, and it is this that allows greater abundance of plant and animal life. In both hemispheres, dramatic changes in microclimate take place with the transition from a snow-covered to a snow-free surface which initiates the growing season, as the albedo of bare ground or vegetation is much lower than that of snow. The extent of seasonally snow-free ground is much greater at comparable latitudes in the northern than the southern hemisphere, and it is this difference in energy absorbtion that underlies the fundamental contrasts in climate experienced at the same latitude in both.

3.3.6 Arctic tundra

In this the vegetation varies between upland dry heath-like tundra and wet meadow. Wet tundra with continuous plant cover is the most extensive

of vegetational types in the Arctic, owing its existence more to permafrost than precipitation. There is sharp distinction between these wet and
dry habitats, and the intermediate, *mesic*, condition, in which the soil and
vegetation have developed to buffer extremes in water supply, is generally
absent. A transect from the crest of a hill down into a valley gives a picture
of the general pattern (Fig. 3.13). The crest fellfield has protoranker soil
with frost scars, and supports a varied flora with lichens predominating

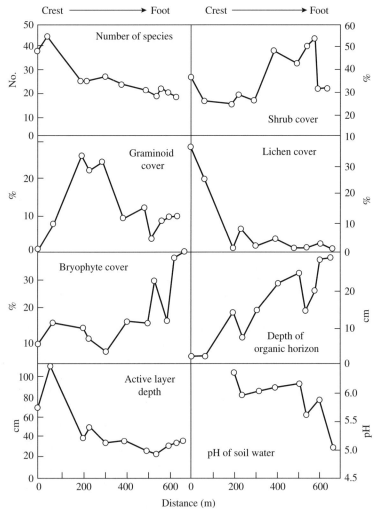

Fig. 3.13 Changes in vegetation types and soil characteristics along a transect from the crest downslope in the Imnavait Creek watershed. Redrawn from Walker and Walker, in Reynolds and Tenhunen (1996).

and abundant prostrate ericaceous species. Going downhill there is a general decrease in species diversity and a steep fall in lichen cover.

The thickness of the organic layer in the soil increases and its pH and concentration of nutrients fall. Shrub cover is less mid-slope, rising to a maximum near the bottom but diminishing again in the mire at the foot. Grasses and sedges are most abundant midway but moss cover, which is fairly uniform over most of the slope, increases steeply near the bottom. This has important effects on the soil environment. The insulation provided keeps the permafrost from thawing and inhibits frost heaving. In spring and summer the moss acts as a sponge and retards drying out. Wherever the bog moss *Sphagnum* appears, its great water-holding capacity promotes the formation of mire. The components of the flora differ but this pattern is much the same throughout the Low Arctic (e.g. Bliss, in Rosswall and Heal 1975, Matveyeva *et al.*, in Rosswall and Heal 1975).

Microorganisms are abundant in tundra soils. Precise description of communities, and estimates of population sizes, remain problematic in all microbial studies, as plate-based counting methods rely on the culturability of the constituent microbes, while approaches measuring molecular diversity cannot differentiate active or viable components. Estimates from Devon Island and Barrow (Bliss, in Rosswall and Heal 1975, Bunnell *et al.*, in Rosswall and Heal 1975), obtained using plate counts, ranged between 7.7×10^6 and $480 \times 10^6 \, g^{-1}$ dry weight of soil in the surface 5 cm. Total biomass estimates to 15 cm ranged between 0.002 and 0.53 g dry weight m^{-2}. At Barrow, direct counts, which include a high proportion of dormant cells, ranged between 3.5×10^9 and $10.4 \times 10^9 \, g^{-1}$. As should be expected in acidic soils, microfungi were also abundant, contributing 4.5–20.3 g m^{-2} on Devon Island. Sterile forms were the major component and they seemed to be cold-tolerant rather than cold-adapted. Other microorganisms recorded from Devon Island included cyanobacteria, at about 1.8×10^5 cells g^{-1}, and protozoa, at 5.3×10^8 cells g^{-1}. Protozoa were estimated to produce two to six generations a year, amounting to a total biomass production of 1.5 g m^{-2} year^{-1}.

The soil fauna, apart from protozoa, includes nematodes, rotifers, enchytraeid worms, tardigrades, and various Crustacea, Collembola, and Diptera, but, notably, not earthworms. Nematodes, producing perhaps 0.16 g dry weight of biomass m^{-2} year^{-1}, are next after protozoa in order of activity in sedge meadow on Devon Island. The total invertebrate production is about 2.6 g m^{-2} year^{-1}, largely consuming bacteria, fungi, and decaying organic matter rather than living plant material.

Among the most conspicuous invertebrates, although they consume only about 0.1% of the above-ground production of plants such as *Dryas* and *Salix*, are the so-called woolly bear caterpillars of the moths *Gynaephora* spp. (Lymantriidae). They persist as larvae for up to 14 years, passing the winters in a desiccated (*anhydrobiotic*) state while their surrounding environment is deeply frozen. They are among the largest invertebrates known to use this stress survival tactic, and are also unique in that as a further means of conserving scarce metabolic resources the larvae dismantle a proportion of their metabolic machinery (mitochondria) in preparation for entering the wintering state (Bennett *et al.* 1999). The short-lived adult moths are present only in summer and do not feed. Flying insects, such as these moths and butterflies, avoid being blown away by keeping low in the calmer air, just above the vegetation, while there is also a generally increased incidence of species that have secondarily lost the ability to fly (through reduction or loss of wings). Diptera are of particular importance in terms of biomass and nutrient turnover in many areas of Arctic tundra.

One of the most striking differences between northern and southern polar regions lies in the former having a well-developed terrestrial vertebrate fauna, including large herbivores and carnivores, and birds (Sage 1986, Berta *et al.* 2005). The muskox (*Ovibos moschatus*, Fig. 3.14) is one of the most specialized of Arctic terrestrial mammals, well adapted to the severe conditions of the polar desert and tundra. Its remarkably thick coat provides insulation sufficient for it to spend much of its time resting rather than expending energy in foraging. In this way, year-long existence in areas of low productivity is possible. On Devon Island it has been estimated

Fig. 3.14   Muskoxen grazing on summer Arctic pastures at Kangerlussuaq, Greenland (photograph by Christian Haas).

that they have little impact on the vegetation, removing less than 1% of the available herbage, considerably less than other herbivores.

In the Barrow area the principal herbivore is the brown lemming (*Lemmus sibiricus*). Its numbers vary cyclically with a periodicity of 3–5 years, grazing pressure altering correspondingly from less than 0.1 up to 25% of the above-ground net primary production. At a population peak, with a density of per-haps $200\,ha^{-1}$, there is widespread destruction of the habitat. Their grazing changes the floristic composition of the tundra, encouraging monocotyledons with relatively protected growing points and a capacity for vegetative repro-duction. Elsewhere across the Arctic tundra, other herbivores are important, including the Arctic hare, reindeer (caribou), and avian herbivores such as snow geese. Larger mammals and birds are generally migratory, leaving this region for the forests to the south, or further afield, during the winter.

The main insectivores of tundra ecosystems are Lapland and snow bunt-ings (*Calcarius lapponicus* and *Plectrophenax nivalis*), which are also con-siderable grazers of seeds and berries. These are joined in summer by a range of shore or wading birds that include insects in their diet. Current climatic changes being experienced in Arctic regions are already placing considerable pressure on some tundra ecosystems and species, pressures that are both shared by and contributed to by use and exploitation, both by indigenous and other human populations (Convey *et al.* 2003).

The primary production upon which these ecosystems depend varies widely (more than two orders of magnitude) between habitats (Bliss, in Rosswall and Heal 1975). However, again, little of the living plant material is con-sumed directly, the main flow of material being through the decomposition cycle. The input of material is greater than the decomposers can assimilate, limited as they are for most of the time by low temperature and the short active season, and the surplus accumulates as peat. Decomposition results in mineralization but mineral nutrients are in short supply in the tundra. This is shown by the more luxuriant growth of plants around old bones and demonstrated more exactly by increases of growth of between 0.5- and 15-fold after treatment with fertilizer. Phosphorus is usually the principal limiting nutrient. Some plants (e.g. the heath, *Cassiope tetragona*), take up phosphorus and other nutrients directly from organic matter decomposed by their associated mycorrhizal fungi. The low concentration of soluble phosphorus in the soil results in minimal loss by leaching and run-off. Loss by leaching of combined nitrogen is greater, although fixation of nitro-gen by bacteria and cyanobacteria in the soil offsets this to some extent. Mineralized nitrogen and phosphorus are quickly released during spring thaws but are rapidly taken up and immobilized, mainly in soil microor-ganisms (Bliss, in Rosswall and Heal 1975). Grazing vertebrates can alter this pattern, returning about 70% of the material eaten to the tundra as fae-ces; these accumulate over winter and again release their nutrients during spring melt. The complete absence of grazing vertebrates, and of significant invertebrate grazing, in Antarctic terrestrial ecosystems results in a bottle-neck in nutrient cycling through decomposition, highlighting further the differences between ecosystems of the two polar regions.

3.3.7 Antarctic tundra

Terrestrial ecosystems of the sub-Antarctic islands are often compared with the tundra of the Arctic, although there are fundamental differences between the two regions as described above. Compared with the corresponding Arctic zone, the strong oceanic influence experienced by these isolated islands leads to markedly reduced environmental variation, particularly in terms of thermal seasonality, making for cooler summers and milder winters (Convey 1996b, Danks 1999). Their lower-latitude location (the major sub-Antarctic islands lie between approximately 47 and 54°S, well north of the Antarctic Circle), despite the extreme cloudiness, also results in a longer growing season.

Flowering plants assume dominance in grass heath, herbfield, and tussock grassland but the presence or absence of dwarf shrub heath hinges some-what on the fine distinction between a woody shrub and a *suffruticose* herb (like a wallflower). Ferns and large rosette-forming flowering plants are a significant part of the flora. There are no native land mammals and very few terrestrial birds. The latter are limited to three species of duck (one endemic, two islands), a single passerine endemic on South Georgia (*Anthus antarcticus*, the South Georgia pipit), and two scavenging sheath-bills. As with all the zones considered, there are mosaics of communities, but here these are often distinctively different between islands.

Short grassland on Macquarie Island is dominated by the grass *Festuca contracta* and occurs on coastal terraces. However, the most striking grass-land both here and on South Georgia (and previously existing on the cold temperate Falkland Islands to the north, but long since largely destroyed by farming activity) is the tussock, that grows at lower levels on disturbed and well-drained peaty soils. These large grasses can be as much as 2 m high with individual plants developing a pedestal of roots and foliage. The species dif-fer between islands: on South Georgia it is *Parodiochloa flabellata* (Fig. 3.15), while on Macquarie *Poa foliosa* forms tall tussocks and *Poa cookii* and *Poa litorosa* form smaller ones. *P. foliosa* forms dense stands with *Stilbocarpa polaris*, a large-leaved plant with a fleshy rhizome which regenerates readily.

Plants with large growth forms such as *Stilbocarpa* and *Pringlea antiscor-butica* (known as the Kerguelen cabbage) are known as megaherbs (Fig. 3.16), and are another characteristic vegetation type of many sub-Antarctic islands and other more northern islands such as the New Zealand shelf islands (Meurk *et al.* 1994). A common factor between all of these islands is the lack of vertebrate herbivores, which is proposed to have encouraged their evolution in these locations (Mitchell *et al.* 1999). The accidental and

Fig. 3.15   Tussock grass in January, South Georgia (photograph by David N. Thomas).

Fig. 3.16   Megaherbs dominate native vegetation on Iles Kerguelen (photograph by M.R. Worland) (see colour plate).

deliberate introduction of grazing vertebrates to most sub-Antarctic islands has had drastic and damaging effects on many large tracts of native vegetation (Frenot *et al.* 2005). In its undamaged form, this community has high standing biomass, primary productivity, and reproductive investment and output (Convey *et al.*, in Bergstrom *et al.* 2006; see Table 3.4).

Studies of the microbiota and invertebrates of most sub-Antarctic islands have generally taken a back seat in comparison with those of the more charismatic vertebrates, for which the importance of these islands as centres of world populations and conservation importance is well recognized (for instance, Macquarie Island has been defined as a World Heritage Site, and all or extensive portions of most other islands are variously defined under national legislation as National Parks, reserves, etc.). It is known that decay processes are slow and humus accumulates in the soil. Although detritivorous invertebrates, including earthworms, insects, microarthropods, and nematodes, are present and often abundant, detailed studies on Marion Island have indicated that these are insufficient to overcome a bottleneck in the decomposition cycle (Slabber and Chown 2002).

Tussock provides an attractive habitat for many mammals and birds, affording shelter and a potentially rich source of food in the carbohydrate reserves laid down in shoots and leaf bases. However, few vertebrates other than the endemic South Georgia pipit and the aforementioned introduced species take advantage of these food resources. Elephant seals (*Mirounga leonina*) and fur seals (*Arctocephalus* spp.) find the gulleys between and tops of the tussocks comfortable places to rest and moult. Penguins, petrels, albatrosses, and other birds nest among tussock. Many of these birds, even those as large as the wandering albatross (*Diomedia exulans*), as well as indigenous invertebrates, have been impacted by predation from introduced rats and mice, to the extent that some are now only found where there are no rodents (Frenot *et al.* 2005).

South Georgia is one of two major sub-Antarctic islands (the other being Heard Island) that lie south of the oceanic Polar Frontal Zone, and thus experience a markedly colder and more seasonal climate than some of the other sub-Antarctic islands. Both these islands are also extensively glaciated today, and would have been more so at Pleistocene glacial maximum. While the other islands also experienced glaciation, for none of them was it complete. Nevertheless, local isolation of populations by glacial extension has been an important evolutionary force driving differentiation processes for sub-Antarctic species (illustrated for invertebrates by Marshall and Convey 2004, Mortimer and Jansen van Vuuren 2006).

Unlike the more northern sub-Antarctic islands, mean monthly temperatures near sea level on South Georgia do become negative (over the year they range between −1.5 and 5.3°C). As would be expected of an oceanic island lying in the cyclonic belt of the Southern Ocean, mean wind velocity is high (15.8 km h^{-1}), as is mean annual precipitation (1405 mm). Conditions in the mountains, which rise to 2934 m, are much more severe (Headland 1984). Some sub-Antarctic islands (notably Marion and Macquarie), although not lying in the Antarctic Peninsula region that is experiencing very rapid regional warming (Turner *et al.* 2005), are experiencing significant climatic changes in the form of warming and much reduced precipitation (Convey, in Bergstrom *et al.* 2006).

The nearest approach to dwarf shrub heath that develops in the sub-Antarctic is the community on South Georgia dominated by the greater burnet (*Acaena magellanica*). *Acaena* is a perennial prostrate herb with stems, woody at the base, arising from long, intertwined, woody rhizomes. It is often one of the first colonists on bare screes, being well adapted to stabilize them, and is almost invariably accompanied by *Syntrichia robusta*, a turf-forming moss, which provides moist, warm, conditions, and humus for the rhizomes and roots. On stabilized ground where a peaty loam has developed, the grass *Festuca erecta* becomes codominant with *Acaena* and on well-drained stable ground replaces it as dominant. On wet seepage slopes the rush *Rostkovia magellanica* accompanies *Acaena*. Biomass and primary productivity in the *Acaena–Festuca–Syntrichia* heath (Table 3.4) are higher than corresponding estimates from the Maritime Antarctic, but less than those for tussock.

The most abundant invertebrates in sub-Antarctic vegetation communities are mites and springtails. As in the Maritime Antarctic, population densities are typically numbered in tens, and sometimes hundreds, of thousands of individuals per square metre. There are also enchytraeids, earthworms, tardigrades, nematodes, spiders, beetles, flies, and moths, with smaller representation of some other insect groups (Gressitt 1970, Convey 2007). As in the other zones discussed, few are herbivores, with the exception of some beetles and the moths. Spiders, and a diving beetle indigenous to South Georgia, are carnivorous but, other than these, large invertebrate predators are absent. The introduction of carabids to parts of South Georgia and Kerguelen is leading to rapid and spatially extensive changes to local community structure which threatens the continued existence of some indigenous and/or endemic invertebrates (Ernsting *et al.* 1995, Frenot *et al.* 2005).

Moving equatorwards from the zones described, before the development of true woodland is possible, there is a region dominated by dwarf shrubs (stems ascending up to 50 cm) such as *Salix arctica*, as distinct from prostrate woody perennials. This forms an extensive circumpolar band in the north but is restricted in the south to a small area including the Falkland Islands and a narrow coastal strip of southern Chile. Highest mean monthly air temperatures are between 6 and 12°C but soils may be frozen up to as much as 3–6 months each winter, although some soils may not freeze at all. Although a logical continuation towards lower latitudes of the polar zones already described, this zone will not be a subject of further discussion in this chapter, except to note that polewards movement of the treeline is a predicted and in some northern areas an already detected consequence of global climate warming (Walther *et al.* 2002), hence placing pressure on the continued existence of parts of the tundra biome. Furthermore, extensive peat deposits, in part contained within permafrost, are characteristic of this zone, and with the zones already discussed comprise over 75% (or 370×10^9 t carbon, roughly half that present in the atmosphere) of the peat existing worldwide. Changes in relative rates of carbon uptake and release from this peat provide a potentially important feedback into global warming processes, but as yet predicting the magnitude or even direction of these changes remains uncertain, and appears to depend on the balance between warming and wetting or drying processes (Callaghan 2004, Arctic Climate Impact Assessment 2005).

3.4 The physiological ecology of polar plants and invertebrates

It is appropriate to consider the effects of polar conditions, as distinct from freezing specifically, on biota. These effects are complex and, perhaps surprisingly, the success of organisms in the polar environment depends more on their responses to stresses experienced during the summer than on those of winter. The many different processes involved in growth, maintenance, and metabolism have different temperature coefficients and optimum temperatures so that the responses of the organism as a whole depend on how well these are orchestrated in harmony with environmental conditions. One general physiological characeristic of polar plants is that they are less sensitive to low temperature than is seen in their temperate counterparts. Photosynthesis, which may be perceptible at –10°C, and show a considerable proportion of the maximum rate at temperatures near to 0°C, usually has an optimum temperature of around 15°C as compared with 25°C in warmer climes (Longton 1988). This low temperature optimum seems to be related to high Rubisco content (see Chapter 2). The oxygenase activity of this enzyme can result in extra respiration in light, thereby consuming the products of photosynthesis. This light respiration is not significant at low temperatures but when Arctic plants are transplanted south high Rubisco oxygenase activity makes survival impossible.

Chlorophyll contents are not any higher in polar plants than in temperate or tropical species. This might be expected since the function of this pigment is light absorption, which is temperature-independent. The annual amount of carbon assimilation is limited mainly by the length of the growing season. The *in situ* growth rate of Arctic plants is comparable with, or even higher than, that of similar temperate plants growing in ambient temperatures 15–20°C higher, and production levels of maritime Antarctic mosses are likewise within the range of temperate values (Davis 1980, 1981). A range of enzyme and metabolic adaptations have been proposed to operate in various polar terrestrial invertebrates and plants, which act to have the net effect of increasing stress tolerance and metabolic activity in these organisms at low temperatures relative to those seen in comparable temperate species (Block 1990, Convey 1996a, Hennion *et al.*, in Bergstrom *et al.* 2006, Peck *et al.* 2006).

With the increasing focus in the last two decades on understanding and predicting the consequences of global climate change and ozone depletion, workers in both polar regions have deployed a wide variety of 'greenhouse' field manipulations (including cloches, screens of various types, ground heating, lamp augmentation; Molau and Molgaard 1996, Convey 2003) to examine the abilities of various elements of the biota to respond to changing environmental conditions. Almost universally these are found to have dramatic effects in increasing growth. However, while conceived with the idea of experimentally imitating the trends predicted in climatic models, these manipulations are difficult to interpret in detail. Apart from raising air and soil temperatures by varying degrees, other important environmental variables and their interactions are affected, including protection from wind, changes in relative humidity, precipitation and water flow regimes, radiation climate, and changes in the duration of the growing season (Kennedy 1995). Some more recent studies have attempted to take a more formally planned multivariate approach to such manipulations, and also included the multiple trophic levels of both plant and animal communities in their analyses (Day et al. 1999, Convey et al. 2002).

It should be noted that, as with the life-history flexibility, highlighted above, that is typical of Antarctic invertebrates (Convey 1996a), the success of an Arctic plant species depends greatly on its plasticity, both phenotypic and genotypic. For example, the purple saxifrage (*Saxifraga oppositifolia*), which grows further north than any other flowering plant and is present on the northernmost Arctic islands north of Greenland, has specialized forms with a range of alternative strategies for survival. These are adapted to either cold and wet, or warm and dry, microhabitats and are genetically distinct but interfertile, providing a mutual support system for survival (Crawford 1995). In evolutionary terms this can be seen as a form of bet hedging. *Polyploidy*—duplication of chromosomes, believed to confer increased adaptability to extreme conditions—increases in frequency at higher latitudes (55% of flowering plants in Iceland, 77% on Svalbard; Halliday 2002), and is also seen in some animals, including freshwater planktonic crustacea.

Completion of the life cycle by the production of viable propagules is complicated for polar organisms by the short growing season. Species with obligate annual life cycles are, therefore, rare in both Arctic and Antarctic. Biennials, requiring two seasons, are more common, but most polar plants are perennials and most animals have multi-year development. Some species that are annuals at lower latitudes (e.g. the grass *Poa annua*) show the flexibility of multi-year development in the polar regions, a factor proposed to underlie their success as invasive species when inadvertently introduced to these regions by human activity (Frenot et al. 2005, Convey et al., in Bergstrom et al. 2006). By drawing on reserves accumulated during the previous season, having wintergreen leaves that can begin photosynthesis without delay in spring, and producing in late summer flower buds which can overwinter to bloom the next season, many polar plants sucessfully flower except in the most unfavourable summers.

Many Arctic vascular plants, and all Antarctic ones, are wind-pollinated. The large diversity of bryophytes also rely entirely on mechanical (wind, water) transfer mechanisms for spore formation (where it occurs) (Longton 1988). However, over 50% of Arctic plant species, including some of the commonest, are insect-pollinated. In contrast, there are no species of pollinating insect indigenous to Antarctic (including sub-Antarctic) locations. In terms of future community trends, this illustrates both a limitation (in terms of the establishment and expansion potential of pollinator-requiring plant species) and a danger (in terms of the risk of parallel introductions of non-indigenous pollinator insects to locations where suitable plant species are already established but unable to spread; Frenot *et al.* 2005).

For many species cross-pollination is not essential and self-pollination can result in viable seed. Some Arctic plants (e.g. *Poa* spp. and *Potentilla* spp.) reproduce by *apomixis*, in which seed is produced without fertilization. A parallel feature that is more widespread in all the common groups represented in polar invertebrate communities, is that of inclusion of species that can reproduce asexually, either facultatively or obligately (Convey *et al.*, in Bergstrom *et al.* 2006). In the event of an unfavourable summer preventing successful production of propagules, possession of perennial or multi-year life cycles provides several other opportunities, while long-term survival in propagule banks also ensures the survival of a species in a given habitat. Finally, as with asexual reproduction in animals, the various mechanisms of vegetative reproduction employed by vascular plants and crytpogams provide an important, in some cases a sole, means of propagation.

3.5 Specialized communities

The communities described so far fall into a series generally corresponding to climatic zones and soil types, extending from extreme to moderate polar conditions. However, there are other distinctive but spatially restricted communities which, being produced by some locally operating factor, are not distributed zonally. Among these are communities on evaporite soils, on biogenic soils, and in fumarolic areas.

3.5.1 Communities on evaporite soils

All polar deserts tend to be saline, but soils formed in wet hollows where salts are concentrated to particularly high levels need special consideration. These evaporite soils are usually associated with saline lakes and the salts may originate from wind-blown sea spray or by leaching from rocks. Accumulations of salt vary from 0.1 to $100\,kg\,m^{-2}$ in the oldest and driest soils. The kind of salt varies with origin but chlorides, nitrates, and sulphates of sodium, potassium, calcium, and magnesium are most common (Campbell and Claridge 1987). Such salinity levels strongly restrict colonization and are sufficient to prevent the growth of most organisms, and evaporite soils often appear sterile. On close examination, numbers of psychrophilic yeast cells in Antarctic Dry Valley soils decrease with increasing salinity, to the point that no sites with more than $60.5\,M$ equivalents g^{-1} contain yeasts. The bacterium *Planococcus*, isolated from a Dry Valley soil, is able to grow in $2.0\,M$ sodium chloride at a temperature of $0°C$, aided by accumulating amino acids as compatible solutes. *Planococcus* has no specific requirement for sodium chloride and its growth rate decreases with increasing concentrations of this salt (Vishniac, in Friedmann 1993). A variety of other bacteria have also been isolated from these soils. One biological advantage of high salinity is that, by lowering freezing point, it extends the period of availability of liquid water and reduces the damaging effects of freeze–thaw cycles. At the extreme this results in the formation of liquid hypersaline layers or slimes at the bottom of some otherwise permanently frozen lakes, with the best-known examples existing in the Victoria Land Dry Valleys and some other specific locations in the Transantarctic Mountains (see Chapter 5). Only a small number of cyanobacteria are known to be able to exist within these exceptional and osmotically extreme habitats.

3.5.2 Communities on highly polluted (ornithogenic and seal-wallow) soils

Substantial and locally concentrated inputs of organic matter are associated with breeding colonies and haul-out locations used by birds and marine vertebrates in both polar regions, generally at coastal sites but also affecting some inland nunataks. The soils which result have nutrient concentrations which may be so high as to be toxic (Table 3.1). In bird colonies organic matter is contributed in the form of guano, feathers, and dead birds. The organic content of these deposits is initially around 30%, but the uric acid from guano is rapidly degraded into ammonia, which may reach a concentration of 140 ppm and then be volatilized and dispersed, and the phosphate content is several hundred times that of neighbouring mineral soils. Elephant seal wallows contain even higher organic and ammonium nitrogen contents (Table 3.1; Campbell and Claridge 1987, Vishniac, in Friedmann 1993).

Through the mechanical disturbance (trampling) of penguins (Fig 3.17) and seals and the anaerobic, toxic, nature of their excrement, no cryptogam or vascular plant is likely to survive in intensively colonized areas (e.g. Smith 1988) unless, as with tussock, the plants are massive and well established before occupation by the vertebrates. Some microorganisms can take

Fig. 3.17   Penguin colonies, because of the disturbance from trampling, are difficult habitats for crytogram or vascular plants to establish despite the high fertilization from the guano (photograph by David N. Thomas)

advantage of the high nutrient levels, and numbers of bacteria may exceed $2 \times 10^{10}\,g^{-1}$. Despite this, carbon dioxide production by these soils is low, and comparable with temperate peat bogs unless it is warm and moist. Annual loss of carbon is less than 0.5%. Loss of nitrogen from ornithogenic soils is much more rapid, occurring mainly by volatilization of ammonia: for instance, of the 275 kg of ammonium nitrogen introduced each day into a penguin rookery on Marion Island, some 220 kg is volatilized. Wind and rain are the main agents removing and redistributing solid material.

Nutrient transfer from bird and seal colonies can benefit adjacent vegetation. This has already been illustrated in the High Arctic using the example of exceptional vegetation development beneath bird nesting cliffs (Fig. 3.11). A further and equally striking example from the Continental Antarctic is found at Robertskollen, a group of inland nunataks at 71°28′S 3°15′E, which have breeding colonies of snow petrels (*Pagodroma nivea*). Although, as across the Antarctic continent, water availability is the major limiting factor for biological activity here, vegetation cover (in this instance mostly lichens, with a few mosses, and the alga *Prasiola crispa*), is almost doubled on sites adjacent to the colonies, as compared with the immediate vicinity of nests, and about seven times as much as that on nunataks without birds (Ryan and Watkins 1989). Certain arthropods are specifically associated with this enriched vegetation.

3.5.3 Communities in geothermal areas

Geothermal activity (e.g. fumaroles, warmed ground, heated pools or springs) occur in some places, such as the shores of Bockfjord (Svalbard),

Disko Island (west Greenland), on Mounts Erebus, Melbourne, and Rittman on the Antarctic continent, and on the South Shetland and South Sandwich islands and Bouvetøya in the Maritime Antarctic. Three sub-Antarctic islands are also the sites of significant volcanic activity, these being Heard and Macdonald Islands, which are currently continuously active, and Marion Island, which has shown bouts of activity throughout at least the Pleistocene, although is not currently thought to host active fumaroles. The latter have not been the subject of studies in the context of identifying any geothermal influence on biota. Collectively, these locations provide a unique set of conditions for biota otherwise exposed to the various environmental challenges of polar habitats. They maintain warm conditions which can both promote biological activity and lead to intense chemical weathering producing clay minerals. There is often a fairly constant supply of liquid water from steam condensation while there are also chemical challenges in the form of high mineral and sulphur levels. In Iceland, the effect on growth of vegetation is clear around alkaline springs, where dense stands of dwarf willows, grasses, herbs, and heaths develop. The warm springs of Greenland support some species which are unrecorded elsewhere in the region and others which reach their northern limits in the favourable temperatures which are available (Heide-Jørgensen and Kristensen 1999, Jensen and Christensen 2003), and the same is seen in the exceptional vegetation and faunal communities that develop around fumaroles on the South Sandwich and South Shetland Islands (Fig. 3.18; Convey *et al.* 2000a, 2000b, Smith 2005, Convey and Smith 2006) and the Continental Antarctic volcanoes (Broady, in Friedmann 1993, Bargagli *et al.* 2004).

Habitats heated by volcanism in parts of Antarctica present a particularly unusual biological situation since they are isolated by long distances over sea or ice from potential sources of colonizing organisms. Nevertheless, on the previously inactive Deception Island in the Maritime Antarctic South Shetland Islands, 2 years after a violent series of eruptions between 1967 and 1970 which destroyed most existing vegetation stands and covered much of the island deep in ash, warmed surfaces were rapidly colonized. Mosses and species of protozoa were among the first colonists. Within 11 years these areas supported closed stands of bryophytes covering several square metres. These areas, then and now, included both species found in local non-thermal habitats and others associated only with fumaroles (Smith 2005). The same pattern is a feature of the vegetation and also the invertebrate fauna of the South Sandwich Islands (Convey *et al.* 2000a, 2000b, Convey and Smith 2006). In the Continental Antarctic, heated ground and fumaroles at high altitude (>2500 m) support growths of bryophytes and algae, some of which are unknown elsewhere on the entire continent (Bargagli *et al.* 2004).

The degree of heating experienced in such habitats is very variable, both between sites and at a single site over varying timescales. In terms of geological timescales, these geothermal features are almost by definition ephemeral,

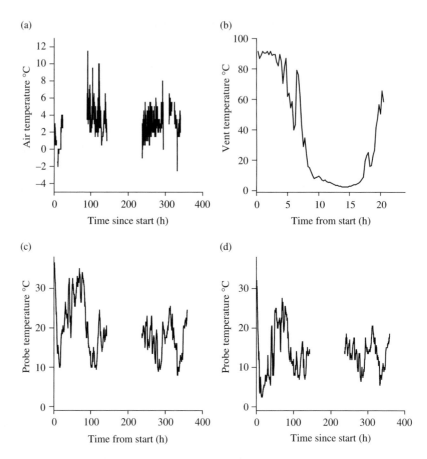

Fig. 3.18   Large changes in the level of geothermal activity occur over short timescales (hours) at a fumarole on Candelmas Island, South Sandwich Islands, and these changes are reflected in patterns of temperature variation within adjacent vegetation. (a) Air temperature recorded at 1 m above ground level; (b) emission temperature of gases within a fumarole vent (note shorter timescale on the *x* axis); (c, d) example (c) sub-surface and (d) surface temperature records from pairs of probes inserted at 2.5–5 cm below surface level and at the vegetation surface of a carpet of *Campylopus introflexus*. Taken from Convey and Smith (2006), with the permission of Opulus Press.

meaning that their biota faces the perennial challenges of local extinction of populations and colonization of newly active areas. Although many fumaroles produce a relatively steady level of warming, in some cases, such as seen on Candlemas Island (South Sandwich Islands), vent temperatures, and nearby vegetation and soil temperatures, may show both cyclicity and rapid changes of up to tens of degrees over only a few hours (Fig. 3.18; Convey and Smith 2006). The level of warming may be marginal (i.e. only a few degrees), as seen in the warm springs of Svalbard and Greenland, or much more extreme.

Some areas of fumarolic activity on Bellingshausen Island (South Sandwich Islands) emit superheated steam at temperatures over 100°C. These are

Fig. 3.19   Exceptional moss and liverwort communities growing around sources of geothermal heat within the crater of Bellingshausen Island, South Sandwich Islands (photograph by Peter Convey).

often surrounded by progressively cooler zones, which may be successively colonized by thermophilic/thermotolerant bacteria and algae, and bryophytes. This leads to a distinctive zoning pattern developing in the macroscopic vegetation (Fig. 3.19), as species have different tolerances to the thermal and other environmental stresses (Convey and Smith 2006). Most tolerant of heat extremes among the mosses are those of the genus *Campylopus*, that are able to tolerate temperatures within the upper surface layers of living shoots of more than 40°C, and sub-surface temperatures of more than 60°C as little as 2.5 cm below the surface. Liverworts (*Cryptochila, Marchantia*) typically dominate the next ring of vegetation at temperatures of 25 to more than 30°C. *Campylopus* is also the only moss identified from the high-altitude fumaroles in Victoria Land, where the soil varies in temperature between 14 and 31°C and has a regular moisture supply from condensed steam. These fumaroles remain free of ice, but after blizzards may become coated with spectacular crusts or canopies of ice raised a few centimetres above the surface.

3.6 Comparison of Arctic and Antarctic terrestrial habitats and communities

Some generalizations can be applied across both polar regions. Beyond the boundaries of the zones considered above in this chapter, dwarf shrub and shrub communities have annual primary production rates (see Table 3.4) which are not greatly different from those of birch forest in England but only about a tenth of that of tropical rain forest. Proceeding polewards one finds continuous stands of low-growing vegetation becoming patchier with a mosaic of communities related to local geology, topography, and hyrdology. In the Low Arctic, sedge moss meadow on Devon Island is about as efficient (0.79%) in utilization of photosynthetically available radiation as is temperate grassland. On the basis both of unit area of ground and of plant cover, biomass and primary production diminish to about the same levels in Arctic and Antarctic with progression polewards; however, this process happens at roughly 10–15° of latitude higher in the Arctic than in the Antarctic (much of the continental area of which has no environmental or biological comparison in the Arctic). Liquid water rather than low temperature is the major limiting factor (Kennedy 1993, Block 1996) and the moister Antarctic 'tundra' communities exceed those of the Arctic both in biomass and primary production. The deserts of the High Arctic and Maritime and Continental Antarctic have extremely low biomass and primary productivity.

Both in the Arctic and Antarctic, mosses, lichens, algae, and cyanobacteria among phototrophic organisms, and mites, springtails, and microfauna (particularly nematodes, tardigrades, and rotifers) among invertebrates are the most able to tolerate the various extreme environmental stresses presented. Endolithic communities of phototrophic and heterotrophic microorganisms, largely self-contained, are the forms of life which can achieve some activity and growth under the severest terrestrial conditions which either pole can inflict, and represent one end of the spectrum of life-supporting habitats on the planet.

Notwithstanding these similarities, the differences between Arctic and Antarctic terrestrial communities are striking. This is evident both in species diversity and in the species themselves. Thus, against about 900 species of vascular plants in the Arctic there are only two on the Antarctic continent and, whereas the Arctic has 48 species of native land mammals, Antarctica has none. Nevertheless, it is far from straightforward to make any valid detailed numerical comparisons, and it is difficult to match communities north and south. In particular, inventories of species remain incomplete with many gaps in different groups for different localities, and the potential (and complications associated with) of molecular taxonomy is only now starting to be applied. In Table 3.3 a comparative overview is attempted for some broadly similar regions from the Arctic and Antarctic.

It is evident that the oceanic and atmospheric isolation of Antarctica, in constrast with the continuous southwards continental connection of much of the Arctic, has been an important driver of the differences seen. The same can also be argued from the observation that a relatively low number of established alien species are known from Arctic locations such as Svalbard (either vascular plants or invertebrates; Rønning 1996, Coulson 2007), in comparison with the large numbers known from the sub-Antarctic (Frenot *et al.* 2005): as many of the latter alien species are often cosmopolitan northern hemisphere and boreal 'weeds', this may indicate that they have had greater opportunity to reach polar latitudes by natural means in the north than the south. Among the bryophytes and lichens of both poles there are a number of cosmopolitan or bipolar species (Longton 1988). Whereas their evolutionary relationships have yet to be addressed by molecular analyses, this does suggest a degree of connection between these two most remote regions through long-distance aerial transport of propagules.

Seasonality is generally very pronounced at high latitudes, and certainly beyond the polar circles. Short but relatively favourable seasons across the Arctic, even at the highest latitude terrestrial sites, allow for rapid growth spurts while restricting development and reproduction overall. In essence the same is seen in the Maritime Antarctic and at some, particularly coastal, locations in the Continental Antarctic. The difference seen in these regions relates to the continuously greater risk of facing desiccating and freezing conditions, factors that drive high investment in protective strategies and the loss of certain characteristics such as true diapause that are widespread in the Arctic. The sub-Antarctic is distinctively different again, with much longer, but cool windy and wet, growing seasons promoting continuous biological activity and vegetative production.

Terrestrial food webs are of a similar general pattern with the main pathway of energy flow flowing through plants to decomposers to organic matter stored in the soil. Herbivores and carnivores mediate a small flow of biomass and energy in the Arctic (largely insignificant in invertebrate communities), while this component is negligible in the Antarctic. The fewer species involved in the Antarctic make for greater simplicity and suitability for meaningful numerical modelling of energy-flow systems (Block 1994).

In the absence of large terrestrial carnivores, Antarctica provides a habitat for a distinctive marine fauna, quite different from that of the Arctic. Seven species of penguin, all flightless and nesting ashore in large colonies extremely vulnerable to land-based predators, do not and could not exist in the Arctic. Likewise, whereas some Antarctic seals breed on sea ice three species breed ashore whereas none do in the Arctic. The repercussions of this for the supply of nutrients to the terrestrial biota of the Maritime and sub-Antarctic are significant.

Finally, there is the major difference that the extensive area of moderately productive wetlands in the Arctic provides for large and diverse populations of migrant animals. Only 11 species of birds are winter residents in the Arctic; the others, around 90 species, must make long migratory flights to other areas where conditions are milder and food is available. Enormous numbers of these migrants assemble in traditional breeding grounds in the Arctic each spring, where, given a good season, ample amounts of plant or insect food are available. Birds, such as geese, arrive with energy reserves, the amount of which determines their reproductive success. The food found in the Arctic must suffice to fuel juvenile maturation and replenish fat reserves for the return journey. The majority of migrations are relatively short, usually to more southern destinations on the same continental landmass. Exceptions are provided by the Arctic tern (*Sterna paradisaea*), which, on fledging migrates the length of the Atlantic to the Southern Ocean, and is often seen around the Antarctic Peninsula, and Baird's sandpiper (*Calidris bairdii*), which makes a journey over 100 degrees of latitude from the High Arctic to the Andes of South America. Most bird movements from Antarctic locations are somewhat different, as the majority involve marine species that simply spend the non-breeding season at sea in the Southern Ocean, although some of these undertake journeys to other oceans (for example the south polar skua, *Catharacta maccormicki*, has been recorded in the North Pacific during the southern winter). An exception is provided by the snowy sheathbill (*Chionis alba*), a scavenging species which, somewhat implausibly given its apparently poor flying ability, migrates between the Antarctic Peninsula and southern South America.

The reindeer is one of the most numerous large wild mammals now existing, with a total population of over 2 million. In addition to the feral herds there is more than an equal number of domesticated animals. Those in the north-eastern Canadian Arctic overwinter on the tundra, while the small Svalbard population of perhaps 2000 individuals (which is thought to have originated by travel across sea ice from northern Russia) remains on the High Arctic archipelago. Otherwise, most herds carry out extended seasonal migrations, focusing on restricted calving areas, which may involve twice-yearly movements of many hundreds of kilometres.

A final, but fundamental, difference between the two polar regions is that, along the Eurasian and North American Arctic coasts there are indigenous peoples. Those of the Eurasian region (Lapps, Samoyeds, Yakuts), and others also traditionally migrate, following the reindeer into the tundra in summer and retreating south into forest in winter. This way of life has developed in the last 2000 years and has not extended to Arctic America, where the Inuit remained hunters and gatherers until recently. Part of the reason for this is that the Eurasian Arctic is a thin strip across which there is relatively easy passage, whereas the American Arctic presents longer distances interrupted by straits, severely hampering a nomadic lifestyle. Additionally, herding had a long tradition, dating to neolithic times, in the Eurasian Arctic whereas, on the large scale, it was unknown in the American Arctic. These indigenous peoples, and their traditional lifestyles, face many contemporary pressures, ranging from climate change to so-called development, both economic and social, and an uncertain future.

4 Glacial habitats in polar regions

4.1 Introduction

The lack of water in its liquid form in the glacial ice of the polar regions on Earth makes these desert-like regions seemingly inhospitable for life (Fig. 4.1). However, the ice environments in the glaciers and ice sheets of the Arctic and Antarctic have a surprising number of ice-bound habitats where liquid water is available. Early exploration of the poles combined with studies over the past several decades have shown considerable microbial presence and diversity in these environments, such that our view of the ice sheets as expansive wastelands devoid of life are changing.

Snow fields in the maritime Arctic and Antarctica contain large regions where snow algal community development colours the snowfields red, green, and sometimes yellow (Kol 1972, Kol and Eurola 1974, Ling and Seppelt 1990). Multicoloured cyanobacterial mats can be found in ephemeral pools and streams of glacial meltwater along the surface of ice at the margins of the Antarctic ice sheet (Vincent *et al.* 2000). Small pockets of liquid water in glaciers (formed through the localized melting induced by solar radiation) have productive algal, cyanobacterial, and bacterial consortia (Gerdel and Drouet 1960, Wharton *et al.* 1981, Mueller 2001).

The interior of the ice sheets as well as the snow fields are now also recognized as not only having microbial propagules as spores but also having active cells being sustained in crystal grain boundaries (Christner *et al.* 2000, Price and Sowers 2004). The basal ice and subglacial lakes also offer additional habitats of liquid water where polar microbial life survives and grows (Siegert *et al.* 1996, Priscu *et al.* 1999, Skidmore *et al.* 2000). Thus, the glacial ice of the poles is not viewed by polar biologists as a sterile environment, although there is still a lively debate about how much biology is really present in these habitats and how to sample them correctly without contamination (Inman 2007). Rather, the glacial ice sheets and

Fig. 4.1   The wind-swept surfaces of ice sheets and glaciers can appear to be barren wastelands devoid of life (photograph by David N. Thomas) (see colour plate).

glaciers of the Arctic and Antarctic are now viewed as expansive areas whereby a range of oases exist for microbial life survival and growth: akin to the hot deserts on Earth and their many and varied oases.

4.2 Life in the interior of polar ice sheets and glaciers

4.2.1 Snow and glacial ice

Freshly fallen snow is crystalline and loosely packed. The air volume in freshly deposited snow is high and the overall density is generally 0.1–0.3 g cm^{-3}. Compaction and bonding between ice granules leads firn formation with a density of approximately 0.55 g cm^{-3}. Depending on the local temperatures, the depth of firn development on ice sheets and glaciers is approximately 10 m. The further densification of firn results in recrystallization of the ice grains and at a density of 0.83 g cm^{-3} the ice is impermeable to air. The conversion of firn to glacial ice occurs at roughly 70 m. Thus, glacial ice is produced through snow deposition and compaction. The accumulation of snow and ice in the accumulation areas of ice sheets and glaciers leads to ice flowing towards the periphery of the ice sheet where sublimation, melting, ablation, and shelving processes lead to the net loss of mass at the margins. Within all of these areas microbes and habitats are found (Fig. 4.2).

Snow impurities accompany the ice crystals during this glacial-ice formation process. Bacteria, yeasts, fungi, algal remains, and cysts have

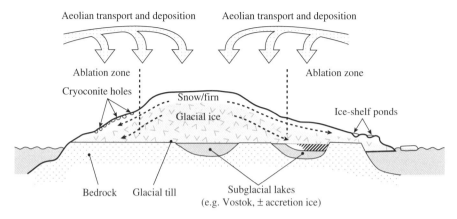

Aeolian transport and deposition   Aeolian transport and deposition

Ablation zone   Ablation zone
Cryoconite holes   Snow/firn
                        Glacial ice   Ice-shelf ponds

Bedrock   Glacial till   Subglacial lakes
                            (e.g. Vostok, ± accretion ice)

Fig. 4.2   Schematic representation of an ice sheet's dynamics and habitats.

been recovered from the dry snow regions of the ice caps. Early on it was presumed that the bacteria and other cultured biota from these dry and cold snow regions were revitalized from inactive spores. However, more recent experiments suggest that some heterotrophic activity may occur *in situ* (Carpenter *et al.* 2000). Amplification of bacterial DNA in South Pole snow showed a preponderance of sequences that clustered with *Deinococcus* species. *Deinococcus* is a known radiation-resistant bacterium that has the ability to repair its DNA from a fragmented condition (Battista 1997). This adaptation is one that is presumably useful in repairing damage in conditions where radiation, desiccation, and freezing are all enocuntered.

Bacteria, algae, yeasts, and other fungi have also been recovered from greater depths in glacial ice sheets of the Antarctic and Arctic (Abyzov, in Friedman 1993, Willerslev *et al.* 1999, Christner *et al.* 2000). DNA and amino acid biomarkers can also be recovered from the silty impurity-rich basal sections of deep ice cores, enabling the reconstructions of past flora and fauna. This enabled Willerslev *et al.* (2007) to show that conifer trees and a diverse insect fauna was present in southern Greenland in the past 1 million years. This is the most ancient DNA to be retrieved and analysed so far, although Siberian permafrost is also a source of very ancient DNA (Willerslev *et al.* 2003).

The biota surviving the aeolian transport and deposition to these regions must survive harsh conditions such as freezing, desiccation, and high radiation. Thus, some have pointed out that it is not surprising that a many of the biota recovered are pigmented; the pigmentation conferring a resistance to radiation damage and maintaining membrane fluidity. Moreover, Gram-positive spore-forming bacteria and spore-forming fungi are among the microbiota that are desiccation-resistant and their presence and survival during aeolian transport and burial in the polar ice sheets is thought to be enhanced over that of less-tolerant biota. Molecular surveys, however, also confirm a diversity of biotic 16S rDNA beyond the spore-forming microbes (Christner et al. 2000, Sheridan et al. 2003, Priscu and Christner 2004). Actinobacterium (in particular Arthrobacter spp.) sequences are notable in ice and they become dominant sequences in old (between 300 000 and 600 000 years) permafrost. This suggests that some species in this particular genera may have adaptive strategies or characteristics that allow their selection or survival in the extended icy habitats (Willerslev et al. 2004).

Until the mid-to-late 1990s it was largely assumed that liquid water in the vast polar ice sheets was absent and recovered microbiota from ice cores were in a dormant condition. However, there have been several reports since that time arguing that liquid water exists at ice-crystal boundaries and crystal triple junctions (Mader 1992, Price 2000). The liquid-water milieu in these micrometre-scale inclusions exists as a result of pressure, freezing-point depressions by impurities (salts and acids), and surface-tension and water-molecule ordering processes at surfaces. The low temperatures and chemical environments in these crystal boundaries is not thought to be overly conducive to high productivity and growth (bacterial cell concentrations typically are low and range between 10^3 and 10^4 cells per milliliter of ice core meltwater); yet, it is an environment in which microbial metabolism is tenable (Gilichinsky and Wagener 1995, Christner 2002, Miteva et al. 2004, Price and Sowers 2004). Slow metabolism may explain the long-term survival of microbes deep within the polar ice sheets, and activity in situ in deep and older ice has been argued as being a necessity due to the radiogenic damage to cells that would accumulate and be lethal in completely dormant cells without repair.

Even low-level repair processes require energy and therefore some level of active metabolism. The energy that may allow metabolism within deep glacial ice may come from chemolithoautotrophy in situ or reduced matter transported and deposited on the ice surface and its subsequent burial. Direct evidence for chemolithoautotrophy in situ in ice sheets is highly limited. The supply and stocks of reduced organic matter in ice sheets is generally low, yet reduced compounds, such as methanesulphonate, and reduced particulate and dissolved organic carbon deposited at the ice surface and then buried has the distinct potential to fuel the energetics of ice-crystal boundary ecosystems (Price 2000). Although there is evidence for ice cores from low-latitude glaciers having products that could be the result of microbial metabolism and transformations in situ (Campen et al. 2003), the extent and nature of the utilization of this energy and the extent of the cycling of matter within the polar ice sheet's presumptive microbial habitats remains undocumented.

4.2.2 Subglacial ice and sediments

The subglacial basal ice is typically sediment-rich and the contact pressure and basal ice sheet friction and flow can create liquid water at the base of ice sheets and glaciers (Skidmore *et al.* 2000). This subglacial sediment-ice environment is one that has received attention recently as another area to look for active microbes and unique microbial consortia. Microbes recovered from the debris-rich ice beneath the John Evans Glacier on Ellsmere Island contained nitrate reducers, sulphate reducers, and methanogens, as well as aerobic heterotrophs that were predominantly psychrophilic (Skidmore *et al.* 2000) and exhibited measurable activity near 0°C. Analysis of DNA recovered from the basal ice compared with supraglacial ice and proglacial sediments showed distinct differences between there microbial consortia, with over 100 unique DNA fragments being found only in the subglacial ice (Bhatia *et al.* 2006). Similar findings of culturable bacteria and evidence of activity *in situ* from beneath glaciers in the Southern Alps of New Zealand showed that the subglacial microbial communities are perhaps spread throughout the subglacial bed environments throughout the world and that these communities are likely to develop *in situ*.

Although the environments are typically regarded as oligotrophic, it is recognized that the organic matter at the base of glaciers and ice sheets is likely to be derived through pre-glacial sediment and organic matter and hence the geochemical processing of the subglacial matter, and has been argued as being important in the cycling of sequestered carbon on the glacial/interglacial timescale (Skidmore *et al.* 2000).

4.2.3 Subglacial lakes

Through enhanced mapping of the Antarctic continent, over 100 subglacial lakes have been identified since the 1970s (Fig. 4.3; Siegert *et al.* 2005). Inventories and characterizations of these lakes have shown them to vary in size from Lake Vostok (which is the approximate size and depth of Lake Ontario) to much smaller and shallower bodies of water that would be akin to subglacial swamps (Priscu 2003). It is thought that these lakes may have been isolated for thousands of years, although recent evidence has shown that they may in fact not all be discrete lakes but rather an interconnected network representing a drainage system that may be flushed entirely on a periodic basis (Wingham *et al.* 2006, Fricker *et al.* 2007).

None of these lakes have been sampled to date, yet the number and size of liquid-water bodies beneath thousands of square kilometres of ice being isolated for millions of years begs the question as to whether or not these lakes harbor contemporary microbial ecosystems. Lake Vostok is the most studied of these lakes and its general dynamics are such that the glacial ice at its northern end melts and provides freshwater input into the lake

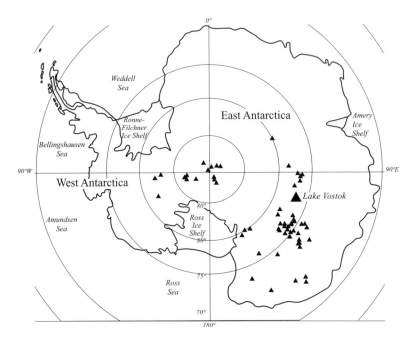

Fig. 4.3   Distribution of subglacial lakes in Antarctica, shown by black triangles.

while freezing of water in the centre and at the southern end adds ice back to the base of the ice sheets. This accretion ice near the southern part of the lake has been sampled during ice-coring operations at Vostok Station. At this drilling site the accretion ice starts at 3538 m and different depths in the accretion ice are believed to have originated from different areas in the lake due to the movement of the ice sheet (Bell *et al.*, in Castello and Rogers 2005).

A range of bacterial cells and 16 S rDNAs have been recovered and sequenced from this accretion ice at various depths (Karl *et al.* 1999, Priscu *et al.* 1999, Abyzov *et al.* and Christner, in Castello and Rogers 2005). Sequences aligned with those from *Acidoverax, Actinomyces, Afipia, Caulobacter, Commomonas, Cytophaga, Bryachybacterium, Hydrogenophilus, Methylbacterium, Friedmanniella, Serratia,* and *Sphingomonas* genera are among those that have been reported to date. There is still no apparent pattern in the distribution of different sequences and bacteria at different depths, although there may be some correspondence between some genera (e.g. *Cytophaga* spp. and *Caulobacter* spp.), with ice believed to have originated from different areas from within Vostok's shallow embayments or deep waters. Despite cleaning of ice samples during analysis there is some speculation that several of these recovered sequences may be due to contamination during core recovery, handling, or analysis (Bell *et al.*, in Castello and Rogers 2005).

If the microbial cells and DNA are indicative of a larger population of microbes in the water from which the ice originated, it is worth considering that the water column of this lake may be supporting an ecosystem that has been buried for millions of years with little energy. Energy for this presumptive subglacial ice ecosystem will be nominally derived by the input of small amounts of matter through melting of glacial ice. Yet, this small amount of energy may be augmented or perhaps superseded by that derived through possible geothermal fields or vents in the lake (Petit *et al.* 2005). It seems the questions regarding the inhabitants and the energetics of this ecosystem will remain until the water column of the lake is characterized and sampled directly. To do so there are considerable ethical and logistical considerations for gaining access to one of Earth's most extreme and most pristine environments.

As we consider the environments at the center of the polar and glacial ice sheets it is apparent that snow, glacial ice, and subglacial ice habitats contain microbial life and there is evidence for active metabolism and growth. These glacial environments are probably best characterized as being near the limits of life being able to survive and not overly productive in these regions, where energy and liquid water are extremely limited. However, due to the huge scales of the systems they canot be ignored in terms of global production: Priscu and Christner (2004) estimate that the Antarctic ice contains approximately 8.8×10^{25} prokaryotic cells and that Antarctic subglacial lakes a further 1.2×10^{25} prokaryotes. In terms of organic carbon they calculate that this biomass is comparable with the total organic carbon in all of the Earth's combined rivers and lakes.

4.3 Life at the margin

As ice sheets and glaciers flow from regions of accrual to areas of net sublimation and melting there is a transition in the properties of the glacial ice, which becomes more hospitable to life through the occurrence and the seasonal persistence of liquid water at or near its freezing point. Near the margins, the proximity of these habitats to the new aeolian transport of propagules, organic matter, sea salts, and nutrients presumably contributes to the habitability, diversity, and productivity of microbial life closer to the margins.

4.3.1 Cryoconites

Cryoconite holes are examples of a type of icy habitat where more hospitable conditions favour the establishment of an active and productive icy ecosystem. Cryoconite holes form when aeolian-deposited dark material accumulates in depressions on a glacier's surface and contributes to localized solar-induced melting during the summer. This localized melting creates a process of positive feedback whereby the dark material (sediments as well as live and dead organic matter) helps create deeper surface depressions that can lead to the trapping of more material, which in turn contributes to more localized melting and the formation of a hole in the glacier.

Cryoconite holes are found on glaciers at both poles (Podgorny and Grenfell 1996, Mueller 2001) and typically range from a few centimetres (Fig. 4.4a) to 50 cm in diameter. The depth of cryoconite holes is generally shallower than 50 cm due to the lessening of the solar radiant energy penetrating the glacial ice as the sediments melt deeper into the ice surface. The melting process creates a miniature vertical ice-bound water column with a benthic sedimentary layer (Fig. 4.4b). Holes can be sealed with an ice lid or be open at the glacier surface depending on the local air and ice temperatures. The holes in glaciers of the McMurdo Dry Valleys of the Antarctic mostly occur as ice-sealed systems except during notably warm summers (e.g. January 2001). In regions of the Arctic where summer air temperatures are typically higher than in the McMurdo Dry Valleys the holes typically do not have an ice lid (Mueller and Pollard 2004).

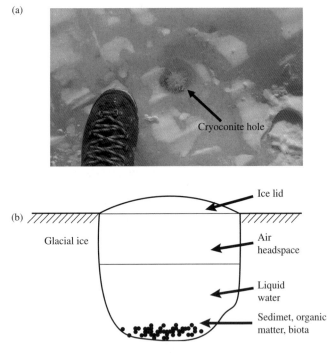

Fig. 4.4   (a) Surface view of a cryoconite hole of the Canada Glacier, Antarctica. The hole has an ice lid yet is discernable by the dark appearance of the sediments beneath the lid, with a distinct bubble produced from the re-freezing of the glacial ice meltwater. (b) Schematic of the vertical structure of cryoconites during their warm liquid-water phase.

The biotic communities in cryoconite holes are surprisingly diverse with a range of algae (desmids, diatoms), phytoflagellates, cyanobacteria, protests, fungi, and bacteria often comprising a cryoconite's microbial consortia. Filamentous cyanobacteria (e.g. *Phormidium* and *Lyngbya* spp.) are the autotrophs most commonly reported in the cryoconite holes of glaciers in the McMurdo Dry Valleys (Wharton *et al.* 1981), the Canadian High Arctic Axel Heiberg Island (Mueller and Pollard 2004), and in Svalbard. These genera of cyanobacteria are often mat-forming and as they grow within the cryoconite environment the filaments serve to aggregate the sediment debris. *Gloecapsa* and *Chroococcous* are also commonly found within the sediment-filamentous cyanobacterial aggregates (Wharton *et al.* 1981, Mueller and Pollard 2004). The cryoconite hole of the White Glacier in the High Arctic is notable for having desmids as community dominants on occasion. Pennate diatoms are also regularly observed in samples from both the Arctic and Antarctic glaciers, yet are usually not found in any significant numbers in most cryoconite systems.

Rotifers and tartigrades occur in cryoconites of some of the glaciers in the McMurdo Dry Valleys (Commonwealth, Canada, and Taylor glaciers), yet abundances are low or absent in others (e.g. Howard and Hughes glaciers). Significant relationships were found between their abundances, physical parameters (depth and sediment), and the chemical constituents (cations, pH, dissolved organic matter) in cryoconite holes during the austral summer of 2001 (Porazinska *et al.* 2004). These relationships generally coincided with the productivity gradients in the Dry Valleys that may allow more productive systems to support microfauna only in some areas. Heterotrophic bacteria and fungi in cryoconite holes undoubtedly cycle carbon and nutrients in these systems.

Cryoconites and their associated microbiota have often been viewed as interesting oddities of glacier surfaces from both poles and alpine regions. Yet, it is becoming more evident that they impart a greater dynamic influence on the glacier's overall dynamics and geochemistry than perhaps previously recognized. The dark holes on the surface of glaciers can provide a substantial amount of the annual meltwater in some systems (Fountain *et al.* 2004, 2006) and this meltwater is not always isolated and retained in the holes. Rather, cryoconite holes can melt to the degree that they become connected and form part of the near-surface hydrologic system of glaciers. The geochemistry of glacier's runoff has the potential of being altered by the productivity cycles in cryoconite holes due to productivity processes that enhance and deplete nutrients, change pH, and change the dissolution kinetics of certain minerals such as those containing silica, calcium, and phosphorous (Tranter *et al.* 2004, Säwström *et al.* 2007). The alteration of glacier runoff and runoff geochemistry has downstream implications for connected aquatic systems (streams, lakes). Reciprocal interactions are also believed to exist between the local landscape productivity and soil geochemistry surrounding the glaciers in determining a glacier's chemistry, productivity, and ecosystem richness (Porazinska *et al.* 2004). The specific reciprocal interactions among these landscape features and the icy biosphere are surprisingly intricate.

4.3.2 Supraglacial pools

Cryoconite holes may enlarge or meltwater may be impounded to form substantial pools on an ice surface. In ablation zones of glaciers, especially near the terminus of some glaciers, these ponds and streams can become large and interconnected. The McMurdo Ice Shelf at 78°00′S 165°35′E, an area of gently undulating ice covered with a sediment, has numerous pools and temporary streams (Vincent 1988, James *et al.* 1995, Hawes *et al.* 1999). While these may persist for decades, the habitat is essentially ephemeral, changing as the ice shelves and glaciers move and seasonal water flows. Usually the ice-shelf pools contain thick mats of cyanobacteria, pink or grey in colour, and generally are similar to those in cryoconite holes (Fig. 4.5). Some have *Phormidium* spp. as the dominant forms, others *Nostoc punctiforme*, and diatoms are present. Planktonic algae form only a small proportion of the total biomass in these pools, a varied assemblage of species being found with heterotrophic cryptophytes tending to dominate. Rotifers (e.g. *Philodina gregaria*) are also present. These ice-shelf pools are known for their relatively high biomass (reaching up to 400 mg chlorophyll $a\,m^{-2}$) and productivity during the austral summers (Howard-Williams *et al.* 1989, Hawes 1993). Moreover, these systems are also known to have

Fig. 4.5   Sampling a pond on the Markham Ice Shelf in the High Arctic in summer. See Vincent *et al.* (2004) (photograph by Warwick Vincent) (see colour plate).

a range of physicochemical environments, from freshwater, to brackish and hypersaline conditions. The productivity and range in biogeochemical conditions undoubtedly enhance the biodiversity of the ice-shelf ecosystems (Suren 1990); however the extreme ephemeral conditions—yearly freezing and thawing—are likely to be a countering effect.

Although these ponds' liquid-water environments are ephemeral, their productivity and overall physical stability over several years allows the cyanobacterial mats to mature and become geochemically stratified, much like cyanobacterial mats in other polar and non-polar environments (Whitton and Potts 2000). The geochemical stratification has zones of oxygen abundance and oxygen depletion, and various zones of oxidation–reduction potential that allows microbial processing of organic matter through alternative oxidative pathways such as sulphate reduction and methanogenesis (Hawes *et al.* 1999, Mountfort *et al.* 1999). These microzones also aid in structuring and enhancing the bacterial autotroph and heterotroph diversity within these systems (Sjoling and Cowan 2003). The overall productivity, relative physical stability, biomass accumulation, and vertical structuring of the mat communities in the supraglacial ponds of the McMurdo Ice Shelf sets these environments apart from the ice-sediment environments and communities found in the perennial lake ice (Priscu *et al.* 1998) and cryoconite holes (see above) and makes them in some respects more similar to the terrestrial ponds of the McMurdo Dry Valleys region, especially those in the moraine soils in the region.

Marine-derived ice shelves along the northern shore of Ellesmere Island (e.g. Ward-Hunt Ice Shelf, 83°N 74°W; Markham Ice Shelf, 83°N 70°W) share several physical, geochemical, and biological similarities with the McMurdo Ice Shelf of the Antarctic. The surface of these shelves also undulates and contains cryoconites, meltwater ponds, and streams where productive mat communities are abundant in areas of sediment accumulation and exposure. These communities also appear to sustain productivity to the degree that biomass can accumulate to extremely high levels (in excess of $400 \, \mathrm{mg \, m^{-2}}$) yet the mat composition and architecture appears to differ from those in the McMurdo Ice Shelf and Antarctic cryoconites. Specifically, the mats of these shelves have a preponderance of chlorophytes (*Palmellopsis*, *Pleurastrum*, *Chlorosarcinopsis*, and *Bracteocossus*) in addition to the mat-forming cyanobacteria (*Phormidium*, *Nostoc*, *Gleocapsa*, and *Leptolyngbya*) that are also common in the Antarctic streams, lake ice, and cryoconite hole ecosystems. Although viruses, bacteria, fungi, ciliates, nematodes, and tubellarian worms have also been found in these cryoecosytems, their activities and functional roles in the ice-shelf dynamic systems have only begun to be appreciated and studied (Vincent *et al.* 2004).

Although glaciers of the polar regions are not generally reported to have insects and ice worms (cf. Shain *et al.* 2000) associated with them, it should be noted that glacial copepods (*Glaciealla yalensis*), worms (*Mesenchytraeus solifugus*), and stoneflies (*Eocapnia nivalis* and *Scopura longa*) are found on the surface of some alpine glaciers in tropical and subpolar regions of the world. These biota form an expanded view of the cryoecosystems of the world yet the factors defining their biogeographic range have yet to be determined.

4.4 The snow alga community

Snow fields in all parts of the world, in alpine regions as well as in the Maritime Arctic and Antarctic, often develop patches of colour where melting and ablation are occurring. The patches, usually red, but sometimes green, yellow, or grey, are striking. A cartoon by George Cruikshank, depicting John Ross's return from his Northwest Passage expedition, even shows a barrel labelled 'Red snow 4 BM' (meaning for British Museum) as its centrepiece (Fig. 4.6).

Snow algal spores are often bright red (e.g. those of *Chlamydomonas nivalis*, *Chlorosphaera antarctica*, and *Chloromonas*) due to the production and accumulation of carotenoids (Hoham and Duval 2001). Thus, when snow banks begin to melt or ablate and these spores are present, the snow can become pink or red, thus yielding the name watermelon snow. Green snow (which occurs where and when snow becomes waterlogged and the general availability of nitrogen is higher) contains other green algae, such as *Hormidium subtile* and *Raphidonema nivale*. Yellow snow (due to algal coloration) may contain chlorotic green algae or chrysophytes. Often grey snow is that colour because of rock dust but, sometimes, it contains a desmid, *Mesotaenium berggrenii*, a green alga with a reddish-brown pigment. Cyanobacterial trichomes are found in some snow banks in the Arctic and Antarctic yet do not appear to grow to the extent that they are known to colour snowfields.

Fig. 4.6   George Cruikshank cartoon depicting John Ross's return from his Northwest Passage expedition, entitled 'Landing the Treasures, or Results of the Polar Expedition at Whitehall 17 Dec. 1818'. There is a barrel labelled 'Red snow 4 BM' (for British Museum).

The primary requirement for growth and accumulation of a snow algal community is liquid water in the snow for a duration that allows time for algal populations to grow. Hence algal-coloured snow is usually found where the mean air temperature reaches 0°C or more and the liquid water in the snow remains for a period of about a month or more. This constraint generally limits the distribution of significant snow algal populations to alpine regions, terrestrial polar regions, and the margins of the polar ice sheets. Snow algal patches have been occasionally seen in southern Victoria Land (approximately 77°S 160°E), where even in summer the mean air temperatures lie several degrees below zero. Being pigmented, the algae absorb radiant energy and can contribute to the localized trapping of energy in the snow that helps creates higher temperatures than the surrounding snow and liquid water for growth.

Various processes, both active and passive, affect the distribution of algae through the depth of snow. Some snow algal species have motile stages and disperse by swimming in water films around snow crystals. These stages in green algal flagellates generally occur after zygospore germination that is triggered when liquid water and new nutrients become available during the initial snow melt (Hoham 1980). These motile swimming stages (zoospores) migrate through the snow pack to find favourable irradiance and nutrient regimes. Within the water environment of the snow pack the algae utilize the available nutrients, grow, and undergo gamete formation and fusion which creates the resting zygospores that persist through the summer and winter until the cycle is repeated.

Resting spore formation seems to coincide with nutrient depletion in several algae and snow systems (Czygan 1970, Hoham *et al.* 1998). During summer and autumn snow algae have to acclimate to extreme temperature regimes, high irradiance and UV radiation, and low nutrient levels. For instance, in high altitudes (above 2500 m) UV radiation can be very high and spherical integrated photosynthetic active radiation can often reach 4500 µmol photons m^{-2} s^{-1} and occasionally 6000 µmol photons m^{-2} s^{-1}. If less liquid water becomes available, and therefore nutrients become limited, most flagellated stages turn into immotile hypnoblast stages, which is the form most resistant to environmental changes. The transformation into hypnoblasts is characterized by a massive incorporation of reserve material, including sugars and lipids, and by formation of esterified extraplastidal secondary carotenoids. Studies have shown that the cells mainly form oxycarotenoids and in particular astaxanthin, which has a red colour (Müller *et al.* 1998). Hypnocygotes and other resting cells have thick cell walls and sometimes mucilaginous envelopes (Müller *et al* 1998). They can survive dry and warm periods in a dormant state, and tolerate high pressure such as under thick snow. They also tolerate freezing in ice blocks at temperatures down to −35°C during winter. However, some of these resting stages can remain photosynthetically active even under very high photon flux densities because of well-protected photosystems by secondary carotenoids (Remias *et al.* 2005).

Physical dynamics during the snow melt process also affect the distribution of algae in the snow. During periods of ablation patches of algae may appear within a few hours, giving an impression of rapid growth. This is often illusory, as the doubling time of snow algae is generally on the scale of a few weeks. The vertical distribution of algae after a snow fall has the peak concentrations remaining at the top of the firn. Therefore, as the fresh snow ablates, the old surface can be exposed and with further ablation the algae remain at the surface and any cells deeper in the snow become part of the surface population. The seemingly rapid development of a snow algal patch can be more a matter of the concentration of cells already present than multiplication to produce new ones.

Numbers of bacteria in algal-coloured snow are generally greater than in white snow, up to $6 \times 10^5 \, ml^{-1}$, and bacterial production is higher. The production and growth of bacteria are undoubtedly stimulated by the production and release of organic matter through extracellular release of photosynthate or the mortality of algal cells. The taxa of bacteria that are being documented in snow environments resemble the broad taxanomic groups that commonly occur in glacial ice, cryoconite holes, lake ice, and sea ice. Namely, the *Proteobacteria* (α, β, γ), the *Cytophaga–Flavobacteriium–Bacteriodes* group and the Gram-positive genera are common in these environments, suggesting that the combined capabilities of being able to maintain membrane fluidity, form spores, and survive freezing, thawing, and desiccation (characteristics that confer low-temperature and freezing tolerance) are common among these groups.

The production of carbon and the allocthanous deposition of organic matter on snow banks can support other forms of biota. 'Snow fleas', mainly small wingless insects of the genus *Podura*, appearing on the top of snow or ice, are well known in the Arctic. These, however, are soil-dwelling and work their way up to the snow surface when conditions get warmer in the spring. As mentioned above, ice worms (*Mesenchytraeus solifogus*) are known from snow banks and glaciers in Alaska and are found in association with snow algae which are believed to form a substantial portion of their food base (Shain *et al.* 2001).

Like the communities of cryoconite holes and supraglacial pools, snow algae patches are ephemeral, liable to be destroyed abruptly by ice movements or a change in weather pattern. If the snow ablates completely, their cells, which tend to agglutinate, are left on the substratum to desiccate and be distributed by wind. Because of their wide distribution, snow algae have potential importance on the global scale. The patches of low albedo which they produce are capable of growing and multiplying and have appreciable local effect in accelerating the disappearance of snow, although this does not often extend over a whole snow field (Thomas and Duval 1995).

4.5 Wider perspectives

Despite the activity of scientists keen to sample ancient organisms and DNA from ice cores and subglacial lakes discussed above, some of the most recent interest in the microbiology of the cryosphere comes from astrobiologists who scrutinize the ice-covered seas of Jupiter's moons Europa, Ganymede, and Mars (Lunine 2004, Schulze-Makuch and Irwin 2004). There are also researchers speculating about the life on *snowball Earths*, the last of which is thought to have ended 635 million years ago (Vincent and Howard-Williams 2000, Walker 2003, Corsetti *et al.* 2006). If life forms do exist, or have existed, in these systems, it does seem likely that they will be quite different from those that dominate the sea ice found on Earth today (Cavicchioli 2002, Chyba and Phillips 2002, Marion *et al.* 2003). However, since we have no other proxies for life in such systems, the cryosphere, and in particular the coldest ice sheets, will continue to be a major source of inspiration for predictions of how life may survive in extraterrestirial systems and previous major ice events in the Earth's history. Probably one of the champion organisms in this respect is the already-mentioned bacterium, *Deinococcus radiodurans*, found in Antarctic ice sheets. As well as surviving extremely low temperatures this organism is also capable of withstanding bursts of up to 1000 times the dose of ionizing radiation that is lethal to human cells. It is thought that it survives such extremes because of very efficient DNA-repair mechanisms (Daly 2006).

5 Inland waters in polar regions

5.1 Introduction

In the past decade the study of polar lakes has gained considerable momentum. These investigations include the Long-Term Ecosystem Research (LTER) programmes in the McMurdo Dry Valleys (Antarctica) and Toolik Lake in Alaska, as well as various other detailed studies, for example the Vestfold Hills, Larsemann Hills, and Bunger Hills in Antarctica and the Canadian Arctic. Together these data have provided a much more detailed picture of the history, chemistry, trophic structure, and dynamics of these extreme ecosystems, allowing us, for example, to produce models of carbon cycling (Hobbie *et al.* 1999, McKenna *et al.* 2006). Moreover, as the global warming progresses Antarctic lakes are increasingly being recognized as barometers of climate change (Doran *et al.* 2002, Quayle *et al.* 2002).

Polar lakes are usually in inaccessible regions, but where there are adequate means of transport (e.g. all-terrain vehicles, helicopters) and access to laboratories, detailed investigations are possible, although most studies are confined to the summer. The Vestfold Hills in Antarctica is one of the few locations where studies spanning an entire year have been conducted. There are major differences in the species diversity and complexity of food webs in Arctic and Antarctic lakes. Whereas the Arctic is largely made up of the high-latitude regions of continental land masses, the Antarctic is an isolated continent. This fact has an impact on biodiversity, as species can move north or south in the Arctic in relation to climatic variability, whereas propagules can only reach Antarctica on the wind or on the feathers and feet of birds. Consequently Antarctica is species-poor compared with the Arctic.

There is a great variety of inland waters in the polar regions. Lakes range from small freshwater ponds to large lakes, among which are epishelf lakes

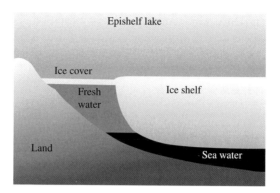

Fig. 5.1 A cross-section of an epishelf lake. These are tidal freshwater lakes where fresh water has come to overlie colder, denser sea water.

that are almost unique to Antarctica (Fig. 5.1). These are freshwater lakes that lie between the land and the edge of an ice-shelf where fresh water has drained from glacial melt or from another inland lake to overlie colder, denser sea water. Consequently they are tidal freshwater lakes. As well as freshwater lakes there are many saline lakes, especially in Antarctica, that range from brackish to hypersaline (around 10 times the salinity of sea water). Some of the saline lakes are meromictic; that is, permanently stratified with strong chemical and physical gradients. Their lower waters (monimolimnion) are permanently anoxic and the domain of anaerobic prokaryotes. It is worth noting that although 98% of the Antarctic continent is covered by a polar ice cap, the continent has a large variety of lakes on its coastal ice-free regions. As indicated in Chapter 4 there are also many subglacial lakes underlying the polar ice caps and in summer glaciers develop mini-lakes called cryoconites where biological activity can be high. The difference in flowing waters between the two polar regions is extreme. Whereas Antarctica has only a few minor streams that flow during the brief austral summer, the Arctic regions have some of the world's major rivers, creating large-scale drainage systems.

The character of any lake is determined largely by its heat budget. In polar regions an equivalent of the energy given out and dissipated when ice forms in winter must be reabsorbed to melt it; consequently water temperatures have little opportunity to rise much more than a few degrees in the brief summers. Ice protects underlying water from wind so that the water column can stabilize and any mixing is restricted to the ice-free period. Polar lakes are extreme environments and on a global scale are unproductive or oligotrophic, unless they are adjacent to seal wallows or bird colonies where they receive inputs of carbon and the essential nutrients such as nitrogen and phosphorus needed for photosynthesis. Many Arctic lakes lie in vegetated catchments and receive inputs of allochthonous carbon and nutrients, while Antarctica lacks vegetation and all the carbon within a lake is derived from photosynthesis within the system (autochthonous).

5.2 Arctic lakes

Lakes are abundant in the Arctic but distributed irregularly, with most being on the flat coastal plains. Of the many agencies which produce lakes, glacier ice and permafrost have been most active. The excavation of basins in the bedrock and blocking of drainage systems by glacial deposits have produced a large number of small and shallow lakes in northern North America. One, however, Great Slave Lake (approximately 61°N 114°W), has an area of 30 000 km² and a maximum depth of 614 m. Other large lakes of this type are Nettilling (66°30′N 70°30′W, 5525 km²) and Amadjicak (65°00′N 71°00′W, 3105 km²) on Baffin Island. In the mountainous areas valley glaciers have produced lakes of various sorts, including some dammed by ice. These are prone to sudden draining in the warm season, either by an overflow cutting a gorge through the ice or, if the water deepens, by the main body of ice floating up and allowing subglacial drainage.

5.2.1 Glacial lakes

Toolik Lake in Alaska (68°38′N, 140°36′W) is the site of a LTER programme funded by the US National Science Foundation. The lake is a multiple-basin kettle lake that has an area of 1.5 km² and a maximum depth of 25 m and was formed some 12 600 years ago. It has ice cover (1.5–2 m) from late September until late June. Water temperatures may reach 12°C in summer (Table 5.1). The lake is dimictic; that is, it is thermally stratified in both summer and winter with phases of mixis in between. The hypolimnion is close to full oxygen saturation during summer stratification, but during winter stratification it decreases to less than 50% in the deepest part of the lake. Toolik Lake has been subject to detailed investigation that includes artificial fertilization of water contained in what are now described as limno-corrals, although such tubes that extend from the surface of the water column to the sediment were used many years ago in the English Lake District by John Lund and were known as Lund tubes.

Toolik Lake is oligotrophic; however, it is more productive and has greater species diversity than Char Lake that lies 6° further north. The concentration of total dissolved ions in Toolik Lake is four times that of Char Lake, resulting in higher phytoplankton biomass. Typical chlorophyll *a* concentrations and primary production in Toolik Lake (Table 5.1) can be enhanced by the addition of nutrients in limno-corrals to over 1600 µg Cl⁻¹ day⁻¹ (O'Brien *et al.* 1992). By comparison with other Arctic lakes this

Table 5.1 Typical chlorophyll a, light-climate, and productivity values for polar lakes. Unless otherwise stated values pertain to summer. Temperatures are summer maxima. M, meromictic; PAR, photosynthetically active radiation. Data from Alexander et al. (1980), Lizotte and Priscu (1992), O'Brien et al. (1992), Vincent et al. (1998), Markager et al. (1999), Henshaw and Laybourn-Parry (2002), Laybourn-Parry and Marshall (2003), and Laybourn-Parry et al. (2006).

Lake	Depth (m)	Temperature (°C)	PAR ($\mu mol\,m^{-2}\,s^{-1}$)	Chlorophyll a ($\mu g\,l^{-1}$)	Net primary productivity ($\mu g\,C\,l^{-1}\,day^{-1}$)
Arctic					
Char Lake, Canada	0.5	4.7	85	0.46	4.41
	15	4.5	62	0.78	2.97
Meretta, Canada	0.5	5.5	163	1.11	20.4
	6	7.2	162	0.96	14.1
Eleanor, Canada	0.5	2.0	52	0.97	5.52
	20	1.7	48	1.02	10.32
Sophia, Canada (M)	0.5	6.0	167	0.31	11.28
	12	6.8	50	1.24	4.56
Garrow, Canada (M)	0.5	5.2	196	1.12	10.75
Toolik, Alaska	0.25	15.0	91–281	1.8	Approx. 100
Barrow tundra ponds, Alaska	–	15.0	–	0.1–2.0	4.8–432
Ossian Sars, Svalbard	2	7.0	151–256	0.2–0.3	3.1–29.5
Antarctic					
Crooked, Vestfold Hills	1.0	4	320	0.2–1.0	0.3–38.6
	10	4	56	–	Mean over year
Ace, Vestfold Hills (M)	4.5	1.2–6	566	1.3–5.0	22.8–195.6
	12	6–7			Mean over year
Highway, Vestfold Hills	2	0.8–1.2	216	0.6–2.4	8.2
	8	2.1–4.4	52	1.2–2.3	0.6
Beaver, MacRobertson Land	6–8	0.22–0.55	105–205	0.05–0.4	1.5–6.9
	90–100	0.47–2.7	0–15*	0.04–2.75	1.3–13.9
Vanda, Dry Valleys (M)	10	3	80	0.05	0.13
	55	13.4	25	0.25	0.27
Fryxell, Dry Valleys (M)	5	2–5	6–13	6	Up to 30
	8.5	9–15	2.5–10	1	
Bonney, Dry Valleys (M)	4	1		1.21	0–3.2
West Lobe	8	3		1.39	

*Extrapolated.

a relatively high level of productivity (Table 5.1). Nitrogen and phosphorus co-limit primary production during the open-water phase. Moreover, dissolved organic nitrogen may provide as much as 35% of phytoplankton nitrogen requirement, but recycling of ammonium within the water column was the predominant source of inorganic nitrogen (O'Brien *et al.* 1997 and references therein).

The phytoplankton has high species diversity, with over 135 recorded species. The assemblage includes cyanobacteria, for example *Anabaena* spp.and *Microcystis incerta*, diatoms including *Asterionella formosa*, *Cyclotella* spp., and *Tabellaria flocculosa*, desmids such as *Cosmarium*, dinoflagellates such as *Ceratium hirundinella*, and a range of chlorophytes, chrysophytes, cryptophytes (Fig. 5.3), and euglenophytes. The phytoplankton is dominated by chrysophytes, as are many other nutrient-poor Arctic lakes, with cryptophytes and dinoflagellates of second and third importance. Among the chryosophytes the genera *Ochromonas*, *Dinobryon* (Fig. 5.2), and *Chromulina* are common (O'Brien *et al.* 1997). These species are well known mixotrophs; that is, they are both photosynthetic and heterotrophic, feeding on bacteria to supplement their carbon budgets and possibly also to gain phosphorus and nitrogen for photosynthesis. As we will see section 5.3, mixotrophy is a common and important phenomenon in Antarctic lakes, but it may also play a role in Arctic lakes. Lakes in Spitzbergen (Svalbard Archipelago, 78°N) have lower diversity than Toolik Lake, but they have a dominance of cryptophytes and chryosphytes. Among the latter *Dinobryon* and other genera of the family Dinobryacées are common and mixotrophic during early summer, possibly providing a competitive advantage over non-mixotrophs (Laybourn-Parry and Marshall 2003).

The zooplankton of Toolik Lake is diverse and in 1975 contained two species of *Daphnia* (*Daphnia middendorffiana* and *Daphnia longiremis*), *Holopedium gibberum*, *Bosmina longirotris*, *Heterocope septentrionalis*,

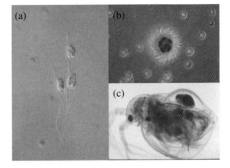

Fig. 5.2   (a) *Dinobryon* from a High Arctic Canadian lake; each lorica contains a single flagellate cell (courtesy of Warwick Vincent). (b) *Mesodinium rubrum*, a ciliate from Highway Lake in the Vestfold Hills; note the darkly stained region inside the cell: this is the endosymbiotic cryptophyte. (c) *Daphniopsis studeri*, a cladoceran from the freshwater and slightly brackish lakes of the Vestfold Hills; note the single juvenile in the dorsal brood pouch (see colour plate).

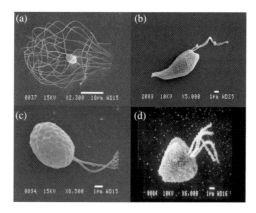

Fig. 5.3   Scanning electron micrographs of some planktonic organisms. (a) *Diaphanoeca grandis*, a heterotrophic flagellate from Highway Lake; this is a collared flagellate surrounded by a scaffold of silica that aids in floatation. (b) Cryptophyte from Pendant Lake. (c) A cryptophyte from Ace Lake; cryptophytes are common in both Arctic and Antarctic lakes. (d) *Pyramimonas gelidicola* from saline lakes in the Vestfold Hills; the genus is also found in the Dry Valley lakes. Scanning electron micrographs courtesy of Gerry Nash, Australian Antarctic Division.

Diaptomus pribiliofensis, and *Cyclops scutifer*. Two of the larger species, *D. middendorffiana* and *H. gibberum*, had virtually disappeared by the mid-1990s. The demise of these species is difficult to explain, but there is evidence that changes in the dominant fish species, possibly resulting from bird predation, may have imposed greater predation on these larger cladoceran species. Although the zooplankton species diversity is relatively high, the numbers of individuals in the water column is low, for example *Daphnia* and *Bosmina* are usually seen at around $1 l^{-1}$ and the copepods *Diaptomus* and *Cyclops* may reach maxima of $10 l^{-1}$ in late June and early July. It is suggested that the crustacean zooplankton does not impose a significant top-down control on the phytoplankton. Eight species of rotifer also contribute to zooplankton biomass, reaching densities of between 13 and $53 l^{-1}$. The fish community comprises five species—lake trout, burbot, Arctic grayling, round whitefish, and slimy sculpin—that all undergo ontogenetic feeding shifts. The trout feed mainly on the benthos molluscs, and in their first young stages feed on zooplankton, incorporating caddis larvae and chironomids as they grow. All of the fish in these unproductive waters are slow-growing and long-lived (O'Brien et al. 1997). Thus the food web of Toolik Lake is complex, with linkages between the plankton and benthos (Fig. 5.4).

The microbial loop plays a role in Arctic lakes as it does worldwide. Bacterial concentrations in Toolik lake ranged between 0.1×10^6 and $3.1 \times 10^6 \, ml^{-1}$ over a year. These values are at the lower end of the reported

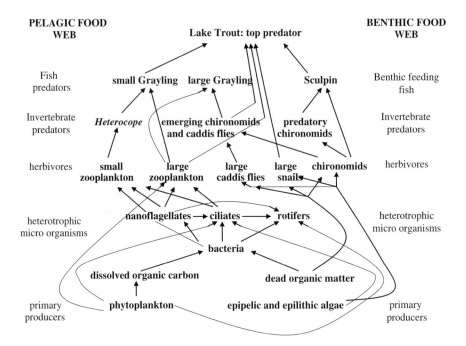

Fig. 5.4 The food web of Toolik Lake, Alaska. Information on some fish species is omitted as their feeding biology is poorly known. Based on O'Brien *et al.* (1997) and other Toolik Lake LTER publications.

spectrum for lakes. The bacterial community can achieve a production of $22\,\mu g\,Cl^{-1}\,day^{-1}$ in May. This is a high level of productivity, but some of the dissolved organic carbon used by the bacteria may be of allochthonous origin. This production supports a heterotrophic flagellate community up to $1 \times 10^3 - 3 \times 10^3\,ml^{-1}$ that includes the species *Monas*, *Oikomonas*, and *Bodo*. In turn heterotrophic flagellates, bacteria, and elements of the phytoplankton support a ciliate community that contains common planktonic species such as *Halteria*, *Strombidium*, *Strobilidium*, and *Vorticella* (O'Brien *et al.* 1997, Vincent and Hobbie 1999).

Many Arctic lakes have well-developed benthic communities. In Toolik Lake chironomid larvae and molluscs predominate, with caddis larvae, some mites, and benthic microcrustacea. At least 25 species of chironomid have been collected from the sediments. They form an important component of fish diet, as do the gastropod molluscs *Limnea elodes*, *Valvata lewisi*, *Gyraulus* spp., and *Physa* spp., and two bivalves *Pisidium* and *Sphaerium*. In shallow regions the lake supports macrophytes that provide another type of benthic habitat.

As one progresses northwards lakes become more species-poor and unproductive. The lakes of Cornwallis Island and Little Cornwallis Island in High Arctic Canada (73–75°N, 92–95°W) are situated in a polar desert climate. The catchment is sparsely vegetated so there is little allochthonous input of carbon or nutrients to the lakes. They are ice-free for only 1–10 weeks each year depending on the local climate and summer temperatures. Low water temperatures result in the lakes remaining unstratified during the summer ice-free period (Table 5.1). Among these lakes are two meromictic lakes, Lake Sophia and Lake Garrow.

Char Lake on Cornwallis Island has an area of $0.52\,km^2$ and a maximum depth of 27.5 m. The surface is only ice-free in August. The main primary producer in Char Lake is not plankton (Table 5.1) but benthic vegetation, which contributes about 80% of the total. Benthic algae in the rocky shore and deep silty zones are most active, while the luxuriant-looking mosses, as much as 40 cm in length, which grow in beds at depths of between 3 and 15 m, are only about half as productive. Phytoplankton primary production in the High Arctic Canadian Lakes is low compared with the more southerly Toolik Lake and tundra ponds in the Barrow region of Alaska (Table 5.1). This is undoubtedly related to temperature and higher nutrient inputs. It is worth noting that the Arctic can be cloudy in summer and this affects the levels of photosynthetically active radiation that drives photosynthesis. Antarctic lakes that are much colder have more sunshine in the short austral summer, and this results in levels of primary production that are comparable, or higher, than those of many Arctic lakes (Table 5.1).

The benthic and planktonic animal communities in Char Lake are less diverse than in Toolik Lake. One copepod, *Limnocalanus macrurus*, dominates the plankton. Char Lake contains one species of fish, the Arctic char *Salvelinus alpinus*, as major predator. The biomass of this fish, at $1.56\,g\,C\,m^{-2}$, is almost three times as large as that of all the other animals put together. However, its growth rate is extremely slow, around $0.03\,g\,C\,m^{-2}\,year^{-1}$, and the production of invertebrate prey seems sufficient to maintain this.

5.2.2 Permafrost lakes

Permafrost is most effective in lake formation in extensive deposits of unconsolidated, fine-grained material, where it produces *thermokarst* or *thaw lakes*, unique to the Arctic. They occur in profusion on the north coasts of Siberia, Alaska, and Canada in flat lowland areas where permafrost prevents underground drainage. Soil water is not uniformly distributed so discrete lenses and wedges of ice form and grow by accretion. If such masses melt then a pool or lake is formed. In Alaska thaw lakes tend to be elongated perpendicular to the prevailing wind, being from several hundred metres to several kilometres in length, and arranged in a parallel fashion. They are rarely deeper than 3 m. The mechanism producing this peculiar configuration may perhaps be subsurface currents, set up by a constant wind, eroding the shore into elliptical an shape (Livingstone 1963). There is a cycle as thaw lakes advance across the terrain, merge, drain, and reform in the course of thousands of years. After capture and drainage by small streams, empty basins are subject to frost action and low-centred polygons with ponds in the middle are established. These may coalesce and the cycle starts again. Some lakes in permafrost areas are larger than thaw lakes. Ozero Taymyr at about 74°N 102°30'E in Siberia, with an area of 4650 km^2 and a maximum depth of 26 m, is one of the largest. It is a relict lake in a recently uplifted land surface, including a basin that was once covered by sea.

Aquatic flowering plants and mosses grow in the shallow pond margins but there is little phytoplankton. Detritus supports large populations of chironomids and other aquatic insect larvae, which provide a rich source of food for diving ducks and other insectivorous vertebrates.

Where thaw lakes are near the sea they may become saline. The effects of salinity are seen in meromictic Garrow Lake on Little Cornwallis Island, with a maximum depth of 50 m (Dickman and Ouellet 1987). Isotope dating showed that the anoxic bottom water (monimolimnion) had been unmixed for nearly 2500 years. The top 10 m remained free to circulate in the ice-free period. This situation seems to have come about by an advancing permafrost wedge forcing brine from rock strata below through an unfrozen chimney in the centre of the lake basin (Fig. 5.5). This had a striking effect on water temperatures. The ice was clear and not usually covered with snow, so it transmitted solar energy and this heated up the water at 20 m to around 9°C, which was maintained throughout the year, although the mean annual air temperature was −16°C and that for July only 4.3°C.

These physical circumstances provided a suite of different habitats. The surface water had flora and fauna rather similar to that of Char Lake. Centric diatoms grow on the undersurface of the ice and are replaced on thaw by phytoflagellates including cryptophytes. The only crustacean zooplankter was *Limnocalanus macrurus* and the only fish was the four-horned sculpin. A dense layer of *Chromatium*-type photosynthetic bacteria, with between 0.1×10^5 and 5×10^6 cells ml^{-1}, occurred around the chemocline (the boundary between the upper mixed water (mixolimnion) and the lower unmixed layer (monimolimnion)). Purple sulphur bacteria such as *Chromatium* require anoxic conditions and the presence of hydrogen sulphide to act as a hydrogen donor to carry out photosynthesis. These photosynthetic bacteria, besides providing a significant source of food for the *Limnocalanus*, were the main agents absorbing the radiant energy which warmed the water (Fig. 5.5). In November 1981 a mining company began discharging wastes into the lake, amounting to 100 t of

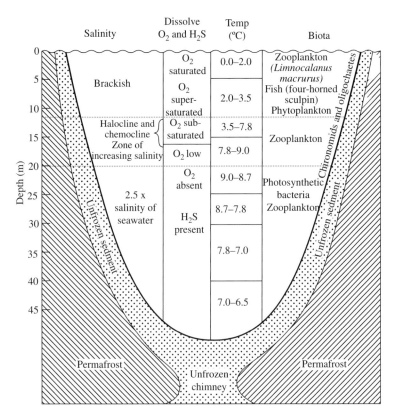

Fig. 5.5 Schematic section of Garrow Lake, Little Cornwallis Island, Northwest Territories, showing physical features and biota. Modified from Dickman and Quellet (1987).

zinc and lead mine tailings per hour. Whereas little immediate effect on phytoplankton was observed, the abundance of photosynthetic bacteria declined within 2 years to negligible levels. Since these bacteria played key roles, both in determining the physical conditions within the lake and as primary producers, it must be concluded that this remarkable ecosystem has been damaged irreversibly (Dickman and Ouellet 1987). However, recent studies on phytoplankton production showed levels comparable to meromictic Lake Sophia (Table 5.1) (Markager *et al.* 1999).

5.3 Antarctic lakes

The lakes of Antarctica were formed by glacial processes and isostatic rebound, superimposed on which were phases of climatic cooling and warming and eustatic changes in sea level. The inland lakes of the McMurdo Dry Valleys have a complex history. Palaeoclimatology indicates that there was a period of severe cooling between 1000 and 1200 years ago. During this time Lakes Vanda, Bonney, and Fryxell appear to have lost their ice covers and dried down to small hypersaline ponds. Lake Hoare either did not exist then or dried down completely. In the following warmer period the lakes refilled (Lyons *et al.* 1998). Lake Bonney is an old lake system and has undergone several phases of filling and drawing down (Lyons *et al.* 1999). The coastal lakes of the Vestfold Hills are examples of much younger lakes formed after the last major glaciation some 10 000 years ago by isostatic rebound. As the ice retreated the land rose, trapping sea water in closed basins and cutting off fjords. During this time eustatic sea-level changes occurred more rapidly than isostatic rebound, resulting in marine incursions to some lakes (Hodgson *et al.* 2004). There are a number of other coastal ice-free areas carrying suites of lakes, including the Larsemann Hills, the Bunger Hills, and the Schirmacher Oasis, but they have not been subject to long-term investigations.

All continental lakes in the Antarctic investigated to date are dominated by the microbial loop. There are few zooplankton and no fish (Fig. 5.6). The

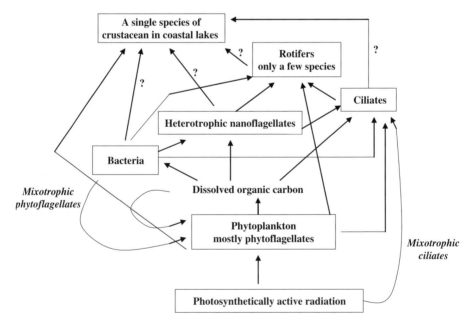

Fig. 5.6   The typical food web in the plankton of Continental Antarctic lakes. Note that the Dry Valley lakes lack planktonic crustacea. Mixotrophy plays an important role. There are mixotrophic ciliates with endosymbiotic algae and mixotrophic flagellates that take up both bacteria and dissolved organic carbon. The mixotrophic ciliate *Mesodinium rubrum* and some heterotrophic flagellates also take up dissolved organic carbon.

phytoplankton is usually dominated a small number of protozoan phyto-flagellate species, heterotrophic flagellate grazers of bacteria and ciliated protozoa. Typically species diversity is much lower than in Arctic lakes, but some of the players are the same, for example cryptophytes (Fig. 5.3) are common in both polar regions. Although we have a reasonably good picture of protozoan diversity, we are only beginning to unravel the molecular biodiversity of the bacterial communities in the plankton and benthos. What is emerging is a marked degree of endemicity (Laybourn-Parry and Pearce, 2007).

5.3.1 The lakes of Signy Island

Signy Island lies at 60°S and is a maritime polar environment with a relatively mild climate compared to Continental Antarctica. It has a number of small lakes, the largest of which is 0.04 km^2 in area and the deepest with a maximum depth of 15 m, lying in valleys and depressions in the narrow coastal lowland. Typically, the lake basin is a steep-sided trough surrounded by a shelf, usually at about 1 m depth, formed by moraine damming. Snow, which varies greatly from year to year, has an important effect on the lake environment since it determines the depth to which the water freezes (1–2 m), the duration of the ice cover (8–12 months), and the irradiance immediately under the ice (0.1–20% of incident visible light). Maximum temperatures in the summer range between 1 and 6°C when the water column is isothermal due to stirring by strong winds. The lakes are inversely thermally stratified in winter.

Moss Lake, occupying a cirque basin of maximum depth 10.4 m in a small catchment area of rock, scree, and small but permanent patches of snow and ice, is one of the most oligotrophic of these lakes, precipitation being its major source of mineral nutrients. The water is clear and phytoplankton sparse, chlorophyll *a* concentrations varying from 0.5 to 8.0 μg l^{-1} (see Table 5.1 for comparison). Zooplankton is sparse, but more abundant than in Continental Antarctic lakes. One of the more abundant cladocerans, *Alona rectangula*, although an active swimmer, is mainly benthic. The bottom of the lake is dominated by the aquatic mosses, *Calliergon sarmentosum* and *Drepanocladus* spp. These mosses support a complex community of epiphytic algae and invertebrates, which differ according to the moss species. The epiphytes include cyanobacteria, *Oedogonium*, and diatoms. The fauna, mainly of opportunistic grazers, has sessile and swimming rotifers and cladocerans as its most numerous members, together with ostracods, tardigrades, nematodes, and gastrotrichs. There are no fish in Moss Lake. The water remains oxygenated throughout its depth during the period of ice cover. It seems that organic production in Moss Lake is largely consumed by decomposers and there is little storage of organic material in the sediment (Heywood, in Laws 1984).

Heywood Lake, is different principally because its catchment, largely moss-covered, is accessible to seals which contribute considerable amounts of organic matter to the water. The activities of these animals have transformed the lake from an oligotrophic to a mesotrophic condition in the last 30 years. The water is turbid in summer, horizontal visibility being reduced to about 20 cm, but it becomes clear in winter under ice. The relatively abundant phytoplankton, which includes small species of *Ochromonas*, *Cryptomonas*, *Chlamydomonas*, and *Ankistrodesmus*, reaches a peak in the spring. A cryptophyte, *Rhodomonas minuta*, may become dominant later but diatoms are rare. A chlorophyll *a* concentration of around 48 μg l^{-1} has been recorded in spring and primary production sometimes reaches 40 mg m^{-2} (Hawes 1985). The copepod *Boeckella poppei* occurs in the plankton along with a range of ciliate genera including *Uronema*, *Vorticella*, *Halteria*, and *Monodinum*. Representatives of these genera are also found in the lakes of the Dry Valleys. Heterotrophic flagellated protozoa, the major grazers of bacterioplankton, reach their peak in December at 174 $\times$ 10^5 l^{-1}. This is four times higher than in neighbouring Sombre Lake, which is not enriched by seal faeces, and it reflects differences in bacterioplankton concentrations. In Heywood Lake bacterial concentrations reached a maximum of 80 $\times$ 10^8 l^{-1} in summer while in Sombre Lake they peaked at 31 $\times$ 10^8 l^{-1}.

5.3.2 The lakes and ponds of the McMurdo Sound area

Cape Evans, a low ice-free area of black basaltic lava on Ross Island at 77°38'S 166°25'E, has a number of small lakes with diverse biological characters. Skua Lake is a favourite haunt of the local birds and Algal Lake was so called because of a conspicuous mat of cyanobacterial remains on its leeward side. Coastal pools receive sea spray and become saline to varying degrees. Pony Lake, near the Cape Royds Adélie penguin rookery (77°33'S 166°08'E), has concentrations of ammonium which are sometimes more than 263 μM and soluble reactive phosphorus greater than 65 μM. Its phytoplankton biomass may rise to as high as 347 μg chlorophyll *a* l^{-1}. All these lakes freeze almost solid during the winter but a residuum of saline water at the bottom becomes concentrated more than 15-fold so that it remains liquid even at –13°C. Thus, the physical and chemical characteristics of these water bodies are highly unstable and the duration of conditions suitable for plant and animal growth is brief. Nonetheless plankton with low species diversity develops in summer including the phototrophs *Ochromonas*, *Chroomonas*, cryptophytes, and the picocyanobacteriaum *Synechococcus* and ciliates such as *Vorticella*, *Euplotes*, and *Cinetochilum*. The abundances and occurrence of species was related to a range of factors, of which conductivity was a prime parameter. All of these species must have the ability to form resting stages, for example cysts, so that they can withstand the freezing of the water body and the highly concentrated bottom brine in winter. An endemic rotifer is well adapted to such conditions, being reported as able to survive in its dormant state salinities up to 250 and temperatures down to −78°C.

There is strong inhibition of phytoplankton photosynthesis by high light intensities in summer. Algal and Skua Lakes both show periodicity in carbon fixation out of phase with the diurnal variation in radiation, with maximum rates of photosynthesis being found at midnight (Fig. 5.7). This effect is more pronounced near the water surface (5 cm) than at a depth of 50 cm, and in the more productive Skua Lake, in which phytoplankton is self-shading, than in the clearer Algal Lake. When samples are shaded by neutral filters maximum photosynthesis occurs at 20% of the incident solar radiation at noon. Rise in temperature increases the rate of

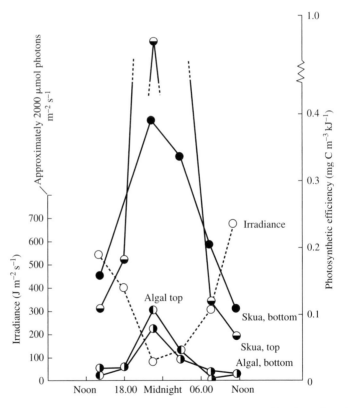

Fig. 5.7   Variation over 24 h in efficiency of photosynthesis, as determined by the radiocarbon method, in two small lakes, Skua (turbid with phytoplankton) and Algal (clear water) at 77°38′S 166°24′E. Marked reduction was shown at noon and maximum efficiency, both at the surface and at the bottom, occurred at midnight. Irradiance is given in joules but the approximate equivalent in photon flux density is indicated. It should be noted that reduced efficiency is caused by a combination of photoinhibition and irradiance above the saturating level. Modified from Goldman *et al.* (1963).

carbon dioxide fixation in photoinhibited samples with a Q_{10} of about 7 whereas similar but non-inhibited samples have an average Q_{10} of about 2. Photoinhibited algae recover if kept in dim light for a few hours. Since experiments have been conducted in glass bottles, opaque to ultraviolet radiation, these effects must be produced by visible, photosynthetically active, radiation (and these experiments were done before the advent of the ozone hole). Although the phytoplankton of the two lakes, differing markedly in transparency, standing crop, and rates of production, show large differences in extent of inhibition, their responses in terms of overall ecological efficiency (energy fixation by phytoplankton as a fraction of incident radiant energy) are similar. Higher biomass-related production in Skua Lake is offset by its greater sensitivity to high light intensity and by the greater depth of water column available for photosynthesis in the clearer Algal Lake (Goldman *et al.* 1963).

5.3.3 The McMurdo Dry Valley lakes

The lakes in the Dry Valleys have perennial ice covers up to 3–4 m thick. The ice contains wind-blown debris from the surrounding hills that reduces light transmission (Fig. 5.8). The levels of photosynthetically active radiation immediately under the ice are very low, for example in Lake Vanda that has relatively transparent ice, between 5.2 and 20% of surface irradiation, while in Lake Fryxell it is between 0.5 and 3.2% and in Lake Hoare between 0.5 and 2.8% (Howard-Williams *et al.* 1998). Consequently

Fig. 5.8   Lake Frxyell, Dry Valleys. Note the perennial ice surface with its undulating surface and load of grit. The polar haven covers the sampling hole in the ice and helps prevent it from re-freezing during the field season. The Canada Glacier that feeds the lake with melt streams in summer is visible in the background (see colour plate).

the phytoplankton and benthic algal mats exist in extreme shade. Water temperatures in summer are given in Table 5.1. Among polar lakes these are some of the most extreme. Their communities live in a dark, cold, nutrient-poor environment. Some of the lakes are meromictic, for example Lakes Bonney, Fryxell, and Vanda.

One of the most remarkable and largest Dry Valley lakes is Lake Vida (area of 6.8 km²). It was thought that this was frozen to its base at 10 m, but recent work has shown that it is an ice-sealed lake with a 19-m-thick ice cover below which is a brine layer with a salinity seven times that of sea water and a temperature of –10°C. It has been sealed for 2800 years. The ice cover is stabilized by a negative feedback between the ice growth and the freezing-point depression of the brine. In contrast, the other lakes are not ice-sealed and receive glacier meltwater below floating ice covers in summer. Trapped in the ice of Lake Vida are cyanobacteria and heterotrophic cells that are viable when melted. Radiocarbon (^{14}C) dating showed that at 12 m below the surface of the ice cover these organisms were 2800 years old (Doran *et al.* 2003).

The lakes of the Dry Valleys typically lack any crustacean plankton. The only metazoans are a few species of rotifer belonging to the genus *Philodina*. However, the protozoan plankton is rather more diverse than seen in coastal lakes like those of the Vestfold Hills. The species diversity of the ciliates is relatively high, with around 11 genera, including *Plagiocampa*, *Askenasia*, *Cyclidium*, *Monodinium*, *Euplotes*, and *Vorticella*. Moreover, a species of predatory ciliate, the suctorian *Sphaerophrya*, that feeds on other ciliates occurs. These various genera show clear stratification in the stable ice-covered water column. For example in Lake Fryxell *Plagiocampa*, the most common species, occupies a position on and just above the chemocline (Roberts and Laybourn-Parry 1999). The phototrophic plankton is dominated by phytoflagellates, in particular species of cryptophyte, *Chlamydomonas*, and *Pyramimonas* (Fig. 5.3). The dominant species varies between lakes (Spaulding *et al.* 1994, Kepner *et al.* 1999, Roberts *et al.* 2004a, 2004b). All of these genera also occur in the coastal lakes like those of the Vestfold Hills. In both locations the cryptophytes and *Pyramimonas* are mixotrophic. They feed on bacteria and dissolved organic carbon (Marshall and Laybourn-Parry 2002, Laybourn-Parry *et al.* 2005). This enables them to supplement the carbon they gain from photosynthesis and operate when the light climate is particularly poor.

The benthic and littoral regions of the lakes are covered by extensive well-developed algal mats made up largely of cyanobacteria with some diatoms. In Lake Hoare, for example, the microbial mats vary in structure and species composition. These differences arise from the quantity and quality of incident down-welling irradiation. There are three distinct morphologies: smooth moat mats that occur in the littoral regions, columnar lift-off mats and prostrate mats that can have pinnacles of up to 3 cm. Measurements *in situ* of photosynthesis in Lake Hoare showed that mats down to a depth of 16.6 m were net producers of oxygen during the summer period. Moreover, photosynthesis occurred at, or close to, maximum efficiency. The light climate on the lake bottom is particular poor under the ice and ranged between 1.0 and 4.6 μmol m^{-2} s^{-1}, enabling net oxygen production between 100 and 500 μmol m^{-2} h^{-1} (Vopel and Hawes 2006). These extensive mats make a significant contribution to carbon fixation in the lakes. Stable isotope (δ^{13}C) analysis of organic matter shows that Lakes Fryxell and Hoare are dominated by benthic productivity, while in contrast Lake Bonney has a carbon cycle dominated by the pelagic component (Lawson *et al.* 2004).

With such low mean air temperatures one may wonder how any lakes are able to retain any liquid water at all. Partly, this is because lakes lying over permafrost act as heat sinks, collecting 'hot' run-off water which is then protected from wind disturbance by 3-m-thick permanent ice cover. For a given air-temperature regime this thickness is fairly uniform from lake to lake because each winter a new layer of ice is added to the bottom of the cover and the latent heat released is about equivalent to the heat loss from the ablating ice surface. However, in meromictic Lake Vanda (77°32′S 161°33′E), 5.2 km^2 in area and maximum depth 66.1 m, another factor comes into play. There is only one inflow, the Onyx River, and the heat input from this is too limited in duration to affect the temperature of the lake significantly. There is no outflow from the lake. The main source of thermal energy is solar radiation penetrating the ice and clear water during the 24-h days of summer. The vertical turnover of the ice constrains its crystals to grow in the vertical direction, forming optical pipes transmitting radiation into the water. Meromixis ensures that heat is not dissipated by convection and the bottom water reaches about 25–46°C above the mean temperature in the valley. The origins of the meromictic condition are not clear. Possibly there have been large fluctuations in lake area and depth during the last few thousand years and, after a period of salt concentration by ablation of ice cover during a colder period, the Onyx River resumed its flow and overlaid the brine with fresh water.

Logistics restrict access to the Dry Valleys except in a brief November-to-January field season but observations starting in mid-September have been made in adjacent Lake Bonney. Even in summer light penetration of the ice cover is poor (1.7–3.3% of surface irradiation). Nevertheless, photosynthesis begins in early September and the maximum for phytoplankton biomass and production moves progressively down the water column, following the seasonal increase in irradiance (Lizotte *et al.* 1996). Whereas the Dry Valley lakes usually have relatively high concentrations of inorganic nitrogen, they are poor in phosphate and Lake Vanda is one of the most phosphorus-deficient and oligotrophic freshwaters known. The phosphorus deficiency may result from its removal by abiotic formation of the mineral hydroxyapatite in the bottom waters. Chlorophyll *a* is low (Table 5.1) but increases on the chemocline at 57 m. The planktonic primary producers are mainly flagellates, which, being able to swim, can accumulate at the depth most suitable for their growth, or small non-motile forms which sink very slowly. This results in stratification of the communities, each adapted to the light, temperature, and chemical conditions in their respective layers. Three floristically distinct communities have been found. Microflagellates, including *Ochromonas miniscula* and *Polytomella* spp., occur just under the ice. They benefit from the rather higher nutrient levels resulting from solute exclusion as fresh ice is formed. Nitrate, for example, is between 50 and 100% higher at 5 m than it is at 30 m. This community seems to be adapted to the low irradiances of spring, when most of its growth evidently takes place, rather than to relatively higher summer values. A second community, with *Phormidium* spp. as well as flagellates, exists in convection cells between 15 and 38 m. The third community, responsible for a deep chlorophyll maximum at 58 m, is dominated by two cyanobacteria, *Phormidium* spp. and a *Synechocystis*-like form, adapted to warmth and dim light. It must suffer severely from phosphorus deficiency since the nitrogen/phosphorus ratio is greater than 5000:1 (in sea water it is usually 16:1).

In the oxygenated water the distribution of heterotrophic bacteria follows approximately that of the algae but their numbers rise steeply in the anoxic bottom water. Different biochemical types of bacteria show a well-defined layering, reflected in the vertical distribution of nitrous oxide, an intermediate in nitrification (oxidation of ammonia to nitrate). In the upper community, where nitrate is reduced to ammonia and assimilated into protein, its concentration is in equilibrium with that in air (Fig. 5.9). In the deep chlorophyll maximum, where ammonia diffusing up from the anoxic layer is nitrified, it increases to over 20 000% of the air equilibrium concentration. Below this, sulphate reduction is the dominant process, nitrate is removed by denitrification, ammonia is produced by degradation of protein, and the concentration of nitrous oxide falls to less than the air equilibrium value (Vincent and Vincent 1982).

5.3.4 Vestfold Hills lakes

The Vestfold Hills lie between 68°25'S 77°50'E and 68°40'S 78°35'E. They are an area of low-lying hills of about 500 km² with several hundred saline and freshwater lakes and ponds (Fig. 5.10). The limnology of this area has been studied in detail in the last 10 years or so, in many cases over annual cycles, because the Australians maintain a permanently staffed station, Davis, on the coast. In common with the Dry Valley lakes the plankton is dominated by microbial forms (Fig. 5.6); however, these lakes also support a sparse crustacean component. In the saline

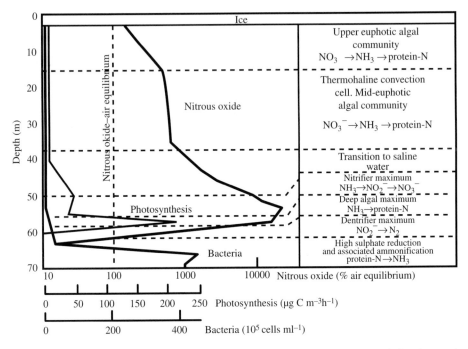

Fig. 5.9 Schematic section of Lake Vanda, Wright Valley, southern Victoria Land, showing stratification of different processes in nitrogen cycling. The various microbial components are located in specific depth zones. The transition from oxygenated to anoxic water at around 58 m is critical. Data from Vincent (1987), courtesy of the author and the manager, SIR Publishing, Wellington, New Zealand.

Fig. 5.10 The Vestfold Hills. Bare rock, punctuated by lakes and ponds. Watts Lake is in the foreground and Lake Druzhby is to the left.

lakes with salinities between around 15 and sea water (salinity of 35) the marine copepod *Paralabidocera antarctica* is found, whereas in the freshwater lakes the endemic Antarctic cladoceran *Daphniopsis studeri* (Fig. 5.2c) occurs. This species has also invaded the slightly brackish lakes such as Highway Lake. The saline lakes were formed by isostatic uplift and subsequent marine incursions, so they possess marine species. Through time the isolated marine communities underwent simplification with only the most robust and adapted species surviving. There are no fish and only a single crustacean species and a few rotifer species. In the hypersaline lakes species diversity is severely reduced with a loss of the crustacean and many of the ciliate and the phytoplankton species and a predominance of prokaryotes. Deep Lake is one of the most saline lakes in the Vestfold Hills with a salt concentration around 10 times that of sea water. It is so saline it never develops an ice cover and has a large annual temperature range as a result: winter temperatures plummet to –17°C, while in summer the upper waters (epilimnion) may reach between 7 and 11°C. Deep lake is thermally stratified in summer and mixed in winter.

A number of the saline lakes have been studied in considerable detail, these include meromictic Ace Lake, that has an upper water column (mixolimnion) with a salinity around 18 and lower, permanently anoxic waters (monimolimnion) with a salinity close to that of sea water, Pendant Lake (salinity 18–19) and Highway Lake (salinity around 5). The saline lakes possess a phytoplankton similar in many respects to that seen in the Dry Valley lakes, with a dominance of cryptophytes, the prasinophyte *Pyramimonas gelidicola* (Fig. 5.3d) and lesser numbers of *Chlamydomonas*. Both the cryptophytes and *Pyramimonas* practise mixotrophy. Some of the marine dinoflagellates have also survived in the lacustrine environment including both phototrophic and heterotrophic species. Among the heterotrophic flagellates the marine choanoflagellate *Diaphanoeca grandis* (Fig. 5.3a) dominates in Highway Lake. The ciliate community of the brackish and saline lakes with salinities between approximately 5–60 is usually dominated by the remarkable mixotrophic marine ciliate *Mesodinium rubrum* (Fig. 5.2b). It is found worldwide in the sea and estuaries where it can form red tides. *M. rubrum* contains an endosymbiotic cryptophycean. As well as gaining photosynthate from its symbiont the ciliate is capable of taking up dissolved organic carbon from the surrounding water. On occasion it can reach concentrations of $100\,000\,l^{-1}$ and can contribute between 15 and 40% of primary production in the plankton of Ace Lake and Highway Lake.

The Vestfold Hills also have a number of large freshwater lakes, the largest being Crooked Lake and Lake Druzhby. Crooked Lake has an area of $9\,km^2$ and a depth greater than $100\,m$, while Lake Druzhby has an area of 7 km² and a maximum depth of 40 m. Water temperatures are typically below 4°C (Table 5.1). These large lakes are extremely unproductive (ultra-oligotrophic) with chlorophyll *a* concentrations below 1 μg l⁻¹ and low levels of nutrients, particularly soluble reactive phosphorus. Like the saline lakes their phytoplankton is dominated by phytoflagellates including *Ochromonas*, *Chlamydomonas*, some chlorococcales *Chorella* and *Gloetila*, as well as very small numbers of a small species of the diatom *Fragillaria*. The heterotrophic microbial community contains the ciliates *Askenasia*, *Strombidium*, and some scuticociliates, heliozoans, and heterotrophic nanoflagellates including *Paraphysomonas*, *Heteromita*, and *Monosiga consociata* (Tong *et al.* 1997). Unlike the Dry Valley lakes the algal mats of the Vestfold Hills lakes are poorly developed. However, the lakes of the Larsemann Hills, some 80 km from the Vestfold Hills, have well-developed cyanobacterial mats.

The lakes of the Vestfold Hills have clear annual ice covers up to 2 m thick (Fig. 5.11). Unlike the Dry Valley lakes there is good light transmission to the underlying water column (Table 5.1). Snow cover, which can attenuate light severely, does not accumulate because it is quickly blown off by the katabatic winds that flow down off the continental ice cap. The phytoplankton of these lakes functions for most of the year. As soon as the light returns there is measurable photosynthesis (Table 5.1). Chlorophyll-specific rates of photosynthesis or assimilation numbers varied between 0.05 and 44.9 μg C μg chlorophyll $a^{-1}h^{-1}$ in Crooked Lake and Lake Druzhby and photosynthetic efficiency between 0.02 and 5.19 μg C μg chlorophyll $a^{-1}h^{-1}μmol\,m^{-2}s^{-1}$ suggesting that the phytoplankton is adapted to low irradiance levels (Henshaw and Laybourn-Parry 2002). The saline lakes

Fig. 5.11 (a) Drilling the ice surface with a motorized Jiffy Drill prior to sampling on Crooked Lake; note the smooth glassy surface of the annual ice-cover. (b) Close up of the ice surface in early summer; note the transparency of the ice: the auger is clearly visible through the ice (see colour plate).

are more productive with higher chlorophyll *a* concentrations (Table 5.1). In Ace Lake and Highway Lake carbon fixation started to increase in late July and peaked in January (Laybourn-Parry *et al.* 2005).

The most important finding from annual studies in the Vestfold Hills lakes is that processes, both photosynthetic and heterotrophic, continue during the winter. Bacterial production can reach levels comparable with summer during the winter months, and consequently many bacterial grazers remain active. The crustaceans *Paralabidocera antarctica* and *Daphniopsis studeri* also remain active throughout the year. The evidence suggests that the communities of Arctic lakes effectively become dormant during the winter. The summer in the Arctic is longer and the water temperatures are higher (Table 5.1), thereby allowing greater scope for growth. In contrast the austral summer is very short, and consequently organisms have to be physiologically capable of switching from a dormant stage to full activity in a short time period (Laybourn-Parry 2002).

Viruses are now recognized as important elements in plankton dynamics. They infect bacteria and elements of the phytoplankton. When they lyse bacterial and other cells the carbon cycle is short-circuited, because carbon is returned to the organic carbon pool before it can be transferred up the food chain. Viruses also play a role in transferring genetic material between hosts and may also act as food to heterotrophic flagellates. Investigations in the saline lakes of Vestfold Hills and Dry Valleys have revealed high numbers of viruses in the water column, for example over an annual cycle 1.2×10^7–$12.0 \times 10^7\,\mathrm{ml}^{-1}$ in Pendant Lake and 0.9×10^7–$6.1 \times 10^7\,\mathrm{ml}^{-1}$ in Ace Lake, with a maximum of $3.4 \times 10^7\,\mathrm{ml}^{-1}$ in Lake Fryxell in summer (Laybourn-Parry and Pearce 2007). In all of these lakes the virus/bacteria ratios were high, up to 141. It is likely that in these microbially dominated lakes viruses may play an important role in biogeochemical cycling. Interestingly the ultra-oligotrophic freshwater lakes have viral abundances below the range reported for lower-latitude freshwater lakes.

5.3.5 Epishelf lakes

As indicated in section 5.1 epishelf lakes are almost unique to Antarctica. Only a few have been described from the Arctic. Until recently our knowledge of their geochemistry and biology was extremely limited, because they are remote and therefore difficult to study. Some are thought to be old; for example Beaver Lake in MacRobertson Land predates the last major glaciation, and White Smoke Lake in the Bunger Hills has an east basin over 3000 years old and a west basin that is only a century old. Beaver Lake, situated at 70°48′S 68°15′E, is the largest epishelf lake in Antarctica. It is 15 km wide and 30 km long with a freshwater/seawater interface at 220–260 m. This interface lies below the floating Charybdis Glacier that covers the outer reaches of the lake. The inner reaches are covered by blue

Fig. 5.12   A Twin Otter aeroplane on the surface of Beaver Lake, an epishelf lake in MacRobertson Land in eastern Antarctica. The ice cover is perennial and has the appearance of cobbles. The Dragon's Teeth hills are in the background.

ice up to 4.2 m thick (Fig. 5.12). During late summer some moating occurs at the edges where there are significant regions of rafted ice caused by tidal action.

A recent summer study has provided a reasonably good picture of the community structure and plankton dynamics of Beaver Lake (Laybourn-Parry *et al.* 2006). Water temperatures within the 110-m water column at the sampling site were between 0.5 and 1.9°C (Table 5.1). Photosynthetically active radiation profiles show that light reaches the lake bottom. Surprisingly this is where highest levels of chlorophyll *a* and primary production occurred during December and January (Table 5.1). Low temperatures and phosphorus limitation undoubtedly contribute to low levels of carbon fixation. However, at the lake bottom soluble reactive phosphorus may be recycled from the sediment, allowing higher photosynthesis. Chlorophytes occurred throughout the water column, euglenoids being common in the upper layers, and replaced by colonial forms in the lower waters. Cryptophytes, the ubiquitous, common component of Antarctic lake phytoplankton, occurred throughout the water column, and prasinophytes, particularly *Mantoniella* spp., and chrysophytes were also present.

The ciliate community was sparse, with fewer than 100 cells l^{-1}, contributing only 7 ng C l^{-1}. The major species were *Askenasia* and *Mondinium*, but it is unlikely that they play any significant role in carbon cycling. Heterotrophic flagellates that contributed between 6.9 and 22.4 µg C l^{-1} removed more than 100% of bacterial production in early December; following the decline, bacterial production increased in late December and

January to a maximum of 12.0 ng $Cl^{-1}h^{-1}$. The lake has a single crustacean, a dwarf form *Boeckella poppei*, whose larger relatives are also found on Signy Island and western Antarctica, but not elsewhere in eastern Antarctica, except in lakes neighbouring Beaver Lake in the Amery oasis. The evidence suggests that the populations in eastern Antarctica have been isolated from a time predating the current interglacial period. The population in Beaver Lake has low fecundity, females carrying only a few eggs in their egg sacs. This and their dwarfism undoubtedly is related to the ultra-oligotrophic nature of the lake. Their faecal pellets were well colonized by bacteria and flagellates and it is possible that they re-ingest them. At times Beaver Lake appears to become carbon-limited; on occasion the concentrations of dissolved organic carbon, which provides a substrate for bacterial growth, fell below $100\,\mu g\,l^{-1}$. If Beaver Lake is typical of epishelf lakes, it represents an end-member system of productivity. It really is life on the edge.

5.4 Streams and rivers

As in other parts of the world, the streams and rivers of Polar regions have taken second place to lakes with limnologists.

5.4.1 The Arctic

No major rivers have their sources in the Arctic but about 14%, more than 10 million km^2, of the land area of the world drains into the Arctic Ocean via five of its largest rivers – Yenisey/Angara, Ob/Irtysh, Lena, Kolyma, and Mackenzie. These discharge through deltas which usually present a maze of channels and shallow lakes. These river systems naturally input high amounts of suspended solids, dissolved organic matter, and inorganic nutrients (Dittmar and Kattner 2003). Naturally these rivers also discharge pollutants such as polychlorinated biphenyls, heavy metals, and biological contaminants that will be absent from the Antarctic. As a result of such large amounts of freshwater input, much of the surface waters of the Kara, Laptev, East Siberian, and Beaufort Seas have low salinities. Arctic ice formed in coastal areas can be heavily laden with sediments, which are then transported large distances in the moving ice fields carried by the transpolar current. Even material such as tree trunks and soil turfs become encased in Arctic ice floes and are eventually released many thousands of kilometres from the place they were initially caught up in the ice (Johansen 1998).

The Mackenzie River, draining nearly a fifth of the total land area of Canada, has a flow which is more evenly distributed through the year than that of smaller rivers with catchment areas, mostly tundra, with run-offs having a peak at the spring melt. The Colville River in Alaska had 43% of its discharge in 1962 concentrated in a 3-week period. Such spates make the substratum highly unstable. Where high-energy streams flow through non-cohesive deposits they carry heavy loads and develop braided multiple channels, constantly shifting in position. The fine sediments are often colonized by tundra vegetation, which is particularly rich and productive in such a situation. For some 6 months in the year Arctic rivers and streams are ice covered and their huge deltas are frozen to depths of several metres. The cataclysmic break-up of the ice in the spring gives overwhelming erosive power (Sage 1986). In all, these rivers do not provide habitats favouring the establishment of either plankton or stable submerged communities. In any case, most of these rivers have their sources in latitudes well south of the tree line and flow through regions which are non-Arctic in character for much of their lengths. Water temperatures are higher than those of the surrounding terrain and both benthic and pelagic species would tend to be those of temperate latitudes.

The classification of Alaskan streams and rivers by Craig and McCart (1975) is applicable generally in the Arctic. The three major types distinguished are:

- Mountain streams: fed by springs and surface run-off; waters rarely exceeding 10°C. They flow for about 5 months in the year and the density of benthic invertebrates is low, of the order of 100 organisms m^{-2}. Arctic char (*Salvelinus alpinus*) is the common fish species.
- Spring streams: small, spring-fed, tributaries of mountain streams providing a more stable habitat; mean temperatures are 2.5°C in winter and 7°C in summer. The banks are often overgrown with vegetation and beds largely covered with moss or algae. There are high densities of benthic invetebrates, around 10 000 organisms m^{-2}, and Arctic char are the main fish species.
- Tundra streams: draining the peat of foothills and coastal plains; tend to be small and meandering, flowing erratically for 3.5–4.5 months in the year. They are usually 'beaded' with alternation of pools and riffles, and the pools are liable to be isolated in dry periods and grading into the static pools discussed above. The waters are more acid, with less calcium in solution than in mountain and spring streams, and they are often discoloured with humic materials. There is little organic matter in suspension but much, up to 14 mg l^{-1}, in solution, exporting around 3 g C m^{-2} annually from the drainage basin (Oswood et al., in Reynolds and Tenhunen 1996). In summer temperatures rise to over 16°C, and densities of benthic invertebrates are intermediate, around 1000 m^{-2}. Conforming to the pattern for all three types of stream, densities are inversely proportional to stream discharge. Grayling (*Thymallus arcticus*) use these streams for spawning and are the characteristic fish.

5.4.2 The Antarctic

In comparison to the Arctic, the Antarctic has few streams and rivers. However, in the McMurdo Dry Valleys glacial meltwater streams are a critical linkage between the glaciers and the closed basin lakes on the valley floors. As part of the McMurdo LTER study the streams feeding Lake Fryxell have been studied in detail. There are some 14 streams flowing into the Fryxell basin. Lake Fryxell is also fed by direct glacial melt from the Canada Glacier that dams one of its ends. Such streams typically flow for only 6–8 weeks each year and exhibit very considerable inter-annual and diurnal variations in flow. For example in the summer of 1990–1991 total flow to the Fryxell basin was $3440 \times 10^3 \, m^3$ while in the 1994–1995 summer it was only $160 \times 10^3 \, m^3$ (Conovitz *et al.* 1998). The dominant controls on fluctuations in diurnal flow are solar position and melt from glacier faces.

The stream beds are covered by algal mats and these are most abundant at sites with moderate gradients that have stream beds covered by large cobbles arranged in a flat stone pavement resulting from periglacial processes. Four types of mat have been distinguished in streams flowing into the Fryxell basin in the Taylor Valley. These are black mats dominated by *Nostoc* species, orange mats dominated by species of *Oscillatoria* and *Phormidium*, green mats largely composed of *Prasiola calophylla* or *Prasiola crispa*, and lastly red mats that, like orange mats, are made up of both *Oscillatoria* and *Phormidium* (McKnight *et al.* 1998). These mats undoubtedly support communities of bacteria and protozoa, as well as rotifers and nematodes, just like the mats on the lake bottom. Gross primary production in mats typically approached an upper limit of $4 \, \mu g$ $C \, cm^2 \, h^{-1}$. Net and gross photosynthesis increased with temperature (ambient temperature range 0–8°C) indicating that temperature is a prime controlling factor in carbon fixation. The accumulation of new growth on exposed surfaces was slow so that communities were at least 3–4 years old (Hawes and Howard-Williams 1998). During the winter the mats desiccate and withstand temperatures down to –60°C but they possess remarkable powers of recovery. For example, desiccated *Nostoc* mat recovered to pre-desiccation rates of photosynthesis and respiration within 10 min of being wetted (Hawes *et al.* 1992).

One unusual stream in the Taylor Valley is Blood Falls. This is an iron-rich, saline subglacial discharge from the terminus of the Taylor glacier that provides episodic discharges into meromictic Lake Bonney. It is thought to be a remnant marine feature originating below the glacier and is probably the oldest liquid-water feature in Taylor Valley, dating back to the time when the valley network was fjord-like. Biological analysis of the outflow revealed a viable actively growing microbial assemblage that included *Schwenella frigidamarina*, a facultative iron reducer that has also been isolated from the sea ice and marine-derived lakes in the Vestfold

Hills. The episodic inflow of water to Lake Bonney can be significant. The density of the water from Blood Falls indicates that it would sink below the lake's chemocline at 14 m. It has been estimated that in the summer of 1994–1995 Blood Falls contributed up to 95% of the chloride flux to the west lobe of Lake Bonney (Mikucki *et al.* 2004).

Rivers are rare in Antarctica, and the largest river (40 km long) is the Onyx River that flows into Lake Vanda in the Wright Valley in southern Victoria Land. Its annual discharge varies from 15 million m^3 to zero. The river only flows in summer and in some years the flow is so low that it does not actually reach the lake. Cloud cover immediately stops its flow and consequently day-to-day variations in discharge are enormous (Vincent 1987). Like the streams described above cyanobacterial mats are the most conspicuous vegetation. There are of course many other localized glacier run-offs in summer all around the Antarctic, but not enough to actually form rivers, and they are at best described as small temporary creeks.

The algal mats found in streams are organized as photosynthetic tissue like those in pools and lakes (Vincent *et al.* 1993). The surface layer is carotenoid-rich and provides protection against high levels of radiation and, by absorbing radiant energy, warms the mat. This layer has only low photosynthetic activity. Below is a stratum in which concentrations of chlorophyll and the accessory photosynthetic pigment phycocyanin reach a maximum. Here, the algae are in an orange-red shade environment, optimal for their requirements. If the mat is shaded, the cyanobacteria, which are motile, may in some cases move up to the surface in 2 h or so, bringing about a change in colour of the mat. The strong self-shading within the mat makes it behave on the whole as a shade-adapted photo-synthetic system.

5.5 Conclusions

The polar regions contain some of the most extreme, unproductive aquatic environments on the planet. Although the Arctic and Antarctic have their high-latitude locations in common, it is clear that there are some very significant differences between their lakes and running waters in terms of trophic structure, diversity, and productivity. Perhaps one of the most striking differences is that the communities of most Antarctic lakes con-tinue to function in winter, while the current evidence suggests that is not the case in Arctic lakes, but as yet data are limited for Arctic systems. Equally surprising is the wide diversity of lake types, especially in the Antarctic. The study of polar limnology is gaining momentum, but we still have much to learn about biogeography from molecular analysis and the biochemical mechanisms that allow survival at such low temperatures.

6 Open oceans in polar regions

6.1 Introduction

The basic structure of the pelagic and benthic realms of the Arctic and Antarctic oceans is similar to those of the adjacent, more temperature seas. However, they are unique in the adaptations and acclimations of their inhabitants to cold temperatures, seasonal or permanent ice cover and, maybe most specifically, their enormous seasonality of light and consequently newly formed organic matter. Polar seas interact with ice and the atmosphere, thereby playing important roles in polar thermal regimes, and providing the organic production which sustains the animal life of both Arctic and Antarctic oceanic and coastal regions (Smetacek and Nicol 2005).

The study of polar marine biology has lagged behind that in other sea areas mainly due to high costs and logistical difficulties of operating in these hostile waters. However, it is evident that the Arctic seas and the Southern Ocean play crucial parts in global processes and that knowledge of their biological characteristics—in conjunction with chemical and physical oceanography—is essential if we are to understand and manage the environment. The recent increased interest in the ecology and functioning of the polar pelagic realms was mainly driven by questions related to potential impacts of environmental changes, ranging from increased ultraviolet radiation to the global carbon cycle. The hydrographical and physical background has already been outlined and this chapter is concerned with the ecology of this environment.

6.2 Gradients in waters

If a water body is not stirred it will tend to stratify, because the surface layer heats up and becomes less dense, or because of the introduction of water of different salinity, or for other reasons. The pycnocline marking the boundary between water masses of different densities in polar waters has already been mentioned. Melting ice produces a surface layer of less-saline water above a pycnocline of only a metre or so thickness, a situation which is particularly important for phytoplankton production. Stratification may be well developed in sheltered lakes and especially in those protected from wind by ice cover (see Chapter 5). The water column inevitably has a gradient of irradiance intensity and quality from top to bottom. As a result of biological activity, sedimentation, and upward transport from bottom deposits, gradients may develop in concentrations of oxygen, nutrients, or other substances. If the water column is stabilized, different species come to dominate at different depths, each proliferating where it finds the conditions which suit it best. This can sometimes be seen by divers as water layers, perhaps only a centimetre in thickness, picked out by a dense growth of some pigmented microorganism. Interstitial water in a sediment or ice is more protected from mixing and here chemical gradients may develop in which biologically significant changes may occur in a matter of millimetres. Clearly, to understand the ecology of such situations one needs to be able to make measurement of physical and chemical factors within small distances across these gradients.

6.2.1 Fronts in the sea

Fronts are narrow zones of demarcation appearing at the surface between different water masses (Fogg and Thake 1987). The Antarctic Polar Front is the grandest example, but smaller ones are frequent, especially in shallow coastal waters where tidal stirring may differentiate a water mass from adjacent stratified waters. Fronts between meltwater from glaciers, which appears milky or pale blue because of the content of rock flour, and sea water, a clear green or indigo, are particularly striking. While the Polar Front is a permanent feature, these others may have only a seasonal existence, although they appear regularly in the same positions each year. Fronts have steep gradients in properties such as temperature, salinity, and nutrient concentrations across them and are of considerable biological significance. In studying fronts, satellite images or air photography give the best idea of their form but measurements *in situ* over short distances are needed to give the physical background necessary for biologists. Packages of sensors which give continuous readings as they are lowered to depth or towed horizontally to obtain transects are most useful here.

6.2.2 Scales of turbulence

Turbulence affects all aquatic organisms. For plankton its effects are immediate; although irradiance and nutrients are the actual determinants of phytoplankton growth, it is water movement which positions the cells in the intensity or concentration gradients and enables access to these essentials (Reynolds 1992). For larger organisms turbulence usually oper-ates indirectly, but nonetheless is important, even for whales and seabirds, in determining the distribution of their food. In polar waters, which range from the roughest seas in the world, stirred by great ocean currents, to the quietest ice-bound lakes, turbulence is of particular interest (Fig 6.1).

Water movement is a continuum, varying in space and time, from circu-lations filling ocean basins, through gyres and eddies, down to molecu-lar motion in which gradients of concentration occupy fractions of a millimetre and dissipate in milliseconds. The motive power for all these movements comes ultimately from the global changes which have been discussed in earlier chapters. Depending on wind speed, current shear, and stratification, timescales for cycling of plankton by turbulent eddies and mixing in the sea have been found to vary from about 0.5 to 100 h for vertical displacements of 10 m. The mixing response in the surface layer to the onset of strong winds is rapid, taking a few hours during which cycling times are reduced to less than an hour. Large eddies transfer their energy to smaller ones and those around 100 km in diameter may be char-acterized by the time taken to transfer half their kinetic energy to eddies of half the original size. Comparing this time with the doubling time of the plankton one comes to the conclusion that such eddies can maintain within themselves populations of the more rapidly multiplying plankton species. Rings of 200–300 km can retain their characteristic phytoplank-ton flora for 2–3 years although the larger zooplankton, with long response times and migratory habits, tend to disperse. At around a critical diameter of 1 km, patches of phytoplankton disperse. Eddies thus play an import-ant part in determining patchiness of phytoplankton in the sea and their

Fig. 6.1   The turbulent Southern Ocean effectively mix the surface layers of the oceans and coastal waters (photograph by David N. Thomas).

internal circulation probably plays a part in holding it near the surface. Unfortunately for biological oceanographers, direct measurement of turbulence is extremely difficult and only approximate estimates of its intensity can be obtained from time-averaged fluctuations in the horizontal and vertical velocities observed at a particular point. Turbulence remains one of the least understood areas of classical physics and the variety of complicated non-linear behaviour is rich in situations which lead to chaos.

At the lower end of the scale turbulent energy is finally dissipated in molecular motion; that is, thermal energy. This is a crucial frontier for microorganisms. The transition of turbulent flow into the realm in which molecular motion predominates and flow becomes laminar is dependent on shear rate and viscosity. The ratio between these two is expressed in the dimensionless Reynolds number (R_e). The critical value for R_e in water lies between 500 and 2000, flow being laminar on the lower side, turbulent on the upper.

The domain of low Reynolds numbers is one in which molecular diffusion is the dominant agency for transport of materials and in which inertia is irrelevant. Molecular diffusion over distances of around $100\,\mu m$, such as we are dealing with in this domain, is rapid and the circumstances that microorganisms have high surface area/volume ratios and that their small radii of curvature have the effect of steepening concentration gradients, allows rapid exchange of materials between cell and water (Riebesell and Wolf Gladrow, in Williams *et al.* 2002). For a cell of $1\,\mu m$ in diameter—about the size of a bacterial cell—the flux of phosphate from a low concentration such as is usually found in natural waters is several hundred times what is needed to maintain maximum growth rate. Small organisms have a tremendous advantage over larger ones in competing for dissolved nutrients (e.g. macronutrients such as nitrate, ammonium, nitrite, phosphate, and micronutrients such as iron and zinc). Because the inertia of the cell becomes unimportant, its motility is of a different kind and serves different functions to those for larger forms.

There is, in fact, a distinct break in form and function at a size of around $8\,\mu m$ between smaller organisms, such as bacteria, picoplankton, and microflagellates, and the larger plankton diatoms and crustacea. Life in the domain of low Reynolds numbers is different from and to some extent separate from that in the world of turbulent flow but is nevertheless of basic importance in aquatic ecology. It would seem that since the viscosity of water increases at low temperatures—being at 0°C twice what it is at 25°C—that the extent of domains of laminar flow should be greater in polar as compared with tropical waters, but there is no information as to whether this is of significance.

6.3 The plankton

Plankton includes all those organisms, from viruses, bacteria, and microalgae to animals, which live freely in the water column and whose

Table 6.1 Classification of polar plankton based on size and taxonomy.

Size fraction	Femto-	Pico-	Nano-	Micro-	Meso-	Macro-	Mega-
Taxonomic group	<0.2 µm	0.2–2 µm	2–20 µm	20–200 µm	0.2–20 mm	2–20 cm	>20 cm
Viruses	*						
Heterotrophic bacteria	*	*					
Cyanobacteria		*					
Dinoflagellates			F	*	F		
Diatoms (including colonies)			F	*	F		
Prymensiophytes			*	*	*		
Prasinophytes		F	*				
Heterotrophic flagellates, amoeba		*	*	F			
Ciliates			*	*	F		
Copepods				* Juveniles	* Adults		
Euphausiids				*	* Juveniles	* Adults	
Amphipods					*	*	
Jellyfish					*	*	*
Salps						*	*
Chaetognaths					*		

* Indicates that the species are only or mostly in those size class ranges; F indicates that a few species are found in those size classes.

movements are determined more by currents and eddies than by their own motive powers. The smallest size fractions of the plankton (from viruses to algae) are typically sampled with water samplers, while larger animals are collected with plankton nets of various mesh sizes. As the sampling approach plays a major role in analysing plankton communities, plankton has been devided into certain classes based on size (Table 6.1). A different and overlapping classification deals with its taxonomic groups, distinguishing virio-, bacterio-, phyto-, and zooplankton. Note that these definitions are not always as clear as it appears. For example, heterotrophic ciliates, which at times harvest intact chloroplasts from their prey, store those in their cytoplasm and function as either mixotrophic or phototrophic members of the polar food webs.

6.3.1 The femto-, pico-, and nanoplankton

The smallest size fraction is called *ultraplankton* (Table 6.1), a group of microorganisms with linear dimensions less than 20 µm: *nanoplankton* (2–20 µm), *picoplankton* (0.2–2 µm), and *femtoplankton* (<0.2 µm). These dimensions are small in comparison with the smallest water eddies and the only motion to which the cells are exposed directly is that of molecular diffusion. At these low Reynolds numbers, the viscosity of sea water is a major force and laminar flow and molecular diffusion dominate over turbulent processes (see the previous section). Only over the last 30 years, largely driven by the development of new technologies, has research started to understand the functional role of the smallest size fraction in the oceans, including the polar seas, leading to the development of the new paradigm of the microbial loop or, even more precise, the microbial network.

Most information about the large-scale distribution of plankton organisms of the polar seas has been obtained by sampling with a traditional plankton net with a mesh size of 20–63 µm. This retains perhaps only 10% of the total phytoplankton and will miss all the smallest size fractions. To obtain correct representative samples including femto- and picoplankton it is necessary to collect water and centrifuge it or filter cells from it using a micro-pore membrane. The smallest size fraction (femto- and parts of the picoplankton) are even too small to be visible under normal light microscopy and the discovery of their role in aquatic ecosystems was linked to the development of adequate techniques, mainly epifluorescence and electron microscopy (for examples see Fig. 6.2).

In addition, determinations of plankton productivity involving filtration have sometimes been ambiguous because the pore size of the filters used have not been specified so that one is not sure whether picoplankton was included or not. In the northern Foxe Basin (approximately 68°N 80°W) between 10 and 70% of the photosynthetically active flora passed through a 1-µm filter (i.e. it was largely picoplanktonic). Many recent studies using proper techniques have highlighted the role of small size fractions in both the Arctic and Antarctic. The application of molecular techniques has opened the door to discovering the hidden diversity in these size fractions (e.g. Diez *et al.* 2001, Lovejoy *et al.* 2006): in Antarctic and Arctic waters ultraplankton contributes frequently more than 50% of the total phytoplankton biomass and productivity (Gosselin *et al.* 1997, Lee and Whitledge 2005). Such observations show that polar seas resemble other oceanic areas, in which ultraplankton is at times ubiquitous and can be the major primary producer specifically under oligotrophic conditions.

Only very recently have viruses, as major contributor to the femtoplankton size class, been identified as major components in marine ecosystems (Suttle 2005). It is estimated that there are about 3×10^9 viruses l^{-1} of seawater. Assuming the volume of the oceans is approximately 1.3×10^{21} l, then there are about 4×10^{30} viruses in the oceans or about 200 Mt of carbon (equivalent to the carbon in 75 million blue whales, which are about 10% carbon). However, our knowledge about their impact on the mortality of bacteria and algal species in polar waters is very little (Le Romancer

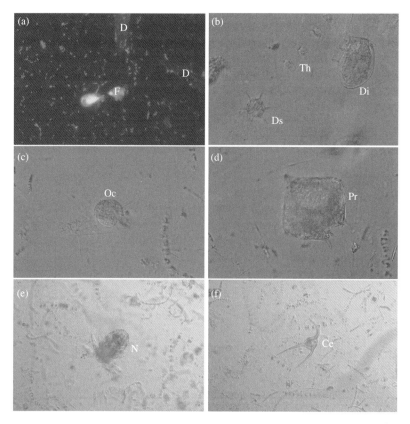

Fig. 6.2   Images of Arctic plankton collected in the Chukchi Sea in May 2004. (a) Epifluorescence image of DAPI (4,6-diamidino-2-phenylindole; a DNA stain)-stained bacteria (abundant small particles), diatoms (D), and phototrophic flagellates (F) (magnificiation 100×); (b) examples of a chain-forming diatom (Th; *Thalassiosira* sp.), dinoflagellate (Di; *Dinophysis* sp.), and silicoflagellate (Ds; *Distephanus speculum*) (magnificiation 40×); (c) oligotrich ciliate (Oc) (magnificiation 40×); (d) hetertrophic dinoflagellate (Pr) (magnificiation 40×); (e) copepod nauplius (N) (magnificiation 10×); (f) phototrophic dinoflagellate (*Ceratium* sp.; Ce) (magnificiation 10×). Note the high abundance of chain-forming diatoms in e and f (see colour plate).

et al. 2007). The few existing studies in polar waters have showed that viruses do occur at the same concentrations in Arctic and Antarctic waters as elsewhere, and even in concentrations one or two orders of magnitude higher in the sea-ice ecosystem. It seems likely that they have high multiplication and high decay rates with correspondingly high rates of lysis of host cells, and that they also release soluble cell material in polar waters, as indicated by seasonal changes in their abundances.

The picoplankton size range is highly diverse on all taxonomic levels, encompassing prokaryotic and eukaryotic members of different phylogenetic groups, including many phototrophic (photosynthetic) species. The bacterioplankton (eubacteria) of polar seas are present at about the same abundances as those of temperate waters, usually within an order of magnitude of 10^6 cells ml^{-1} with seasonal changes correlated to phytoplankton biomass. They primarily utilize dissolved organic matter exuded by phytoplankton, and the release of organic solutes when organisms die and lyse. Exudation of organic matter by living phytoplankton can reach more than 50% of total primary production in Arctic and Antarctic waters, and bacteria play an essential role in transferring the dissolved organic matter back into the particulate organic carbon pool, which is in turn grazed mainly by flagellates and ciliates.

The rapidly increasing use of molecular tools, including metagenomics studies (Grzymski *et al.* 2006), has provided first insights into the functional and genetic diversity of this size fraction. Suprisingly, the actual growth rates of bacteria in polar seas is frequently comparable with those in more temperature waters. This is possible due to specific cold-temperature adaptions of protein and cell-membrane structural properties, making those functional and stable at low temperatures. This causes many polar bacteria to be psychrophiles or psychrotolerant (see Chapter 2), with the most remarkable examples coming from the sea-ice environment (Deming 2002, Mock and Thomas 2005). The extent to which bacteria can actually utilize the dissolved organic matter at these temperatures is still a matter of debate, specifically in the Arctic, where concentrations of dissolved organic matter in surface water can be very high. Wiebe *et al.* (1992) suggested that bacterial mineralization is low in cold-water environments due to increased substrate requirements. Other investigations have questioned this view (Yager and Demming 1999) and the discussion is still ongoing (Kirchman *et al.* 2005).

The occurrence of archaea in polar waters is a relatively new finding starting with observations in the 1990s. Archaea were once thought to be restricted to hypersaline, extremely hot or cold, or anoxic habitats, but are now known to occur in ordinary sea water. DeLong *et al.* (1994) reported that up to 34% of the prokaryotic biomass in samples taken in Arthur Harbour (64°46′S 64°05′W) in the late austral winter consisted of archaea. A more recent study (Bano *et al.* 2004) comparing Arctic with Antarctic archaean communities observed different community patterns for the Arctic and Antarctic but consistently lower diversity for Archaea compared to bacteria in both seas. The presence of such a substantial fraction of microorganisms with currently still unknown biochemical activities points towards exciting opportunities to explore and understand their ecology and contributions. Currently, only a minuscule fraction of the eubacteria and archaea in polar seas have been cultured and therefore many new findings can be expected over the next decade, including a better understanding of successional patterns in the microbe fraction (Ducklow *et al.* 2007).

In general the picophytoplankton includes prokaryotic forms—cyanobacteria and prochlorophytes—and unicellular eukaryotes belonging to the chlorophytes and prasinophytes. The nanophytoplankton comprises principally flagellates, belonging to various groups, and some diatoms.

The free-living marine phototrophic prokaryotes (including both oxychlorobacteria and cyanobacteria) have a clear preference for more temperate oceans (Gradinger and Lenz 1995, Griffith 2006). The few recorded instances of, for example, cyanobacteria in Arctic and Antarctic waters were explained by advection processes. Both in the Arctic and Southern Ocean abundance of pico-cyanobacteria of the genus *Synechococcus* (typical size of 0.8 μm) are related to temperature and drop dramatically with latitude. In the North Atlantic for example, numbers of picoplanktonic cyanobacteria fall from $1.8 \times 10^5 \, \text{ml}^{-1}$ at 38°00′N to $2.0 \times 10^3 \, \text{ml}^{-1}$ at 58°32′N. They are nearly absent in the Arctic Ocean. South of Australia cyanobacterial numbers are correlated with temperature, but not with day length, varying from about $10 \, \text{cells ml}^{-1}$ at −1°C near the Antarctic continent to about $10^4 \, \text{cells ml}^{-1}$ at 12°C, further north. In the Arctic, ocean currents and regional input of cyanobacteria by freshwater run-off can lead to locally higher abundances ($>4000 \, \text{cells ml}^{-1}$) of phototrophic prokaryotes (Waleron *et al.* 2007). The overall low marine abundance of phototrophic cyanobacteria is in contrast to their frequent occurrence in polar freshwater environments, which might be caused by the combination of higher grazing pressure through marine microbial grazers and slow growth rates.

Consequently, eukaryotic picophytoplankton dominate in diversity, biomass and productivity in this size class, and tend to increase in abundance with decreasing temperature, replacing the cyanobacteria as the major primary producers. They typically occur in abundances of 10^2–$10^4 \, \text{cells ml}^{-1}$ in all oceans, including the polar seas. It is an extremely diverse size fraction: a detailed analysis of the picoeucaryotes (<1.6 μm) of the Weddell and Scotia Seas (Diez *et al.* 2001) based on rDNA analyses found members of, for example prasinophytes, prymnesiophytes, stramenopiles, and cryptophytes, in their genetic libraries. The central permanently ice-covered Arctic basin harbours a unique eurkaryotic assemblage (Lovejoy *et al.* 2006) with a surprising low diversity compared to the adjacent Arctic seas.

The nanophytoplankton is mainly dominated by autotrophic nanoflagellates and solitary smaller diatoms. Nanophytoplankton can dominate the biomass and activity of phytoplankton mainly during periods of ice cover and outside the phytoplankton maxima (which are mostly formed by microphytoplankton). For example, cryptophytes alter in dominance with microphytoplankton in the Antarctic Peninsula region (Ducklow *et al.* 2007). Heterotrophic flagellates (including dinoflagellates) and ciliates are major contributors to the nanoplankton size class. These small protists feed mainly on bacteria and small phytoplankton. Their efficient top-down grazing control on bacteria keeps bacterial concentrations nearly constant (aroung 10^6 cells ml^{-1}) while bacterial production can fluctuate over larger ranges. Another potential group of bacterivore (and herbivore) protists, the marine amoeba, has hardly been studied from polar seas. Moran *et al.* (2007) described partially novel *gymnamoebae* in Antarctic marine waters, some of them with psychrophilic growth characteristics.

The smaller size fractions, from femto- to nanoplankton, represent a diverse and very dynamic component of the marine food web. Short generation times (less than 3 days) combined with effective utilization of food sources allow a strong top-down control of picoplankton biomass by bacteri- and herbivores. The heterotrophic bacteria play a key role in the mineralization of organic matter to regenerate the inorganic forms of carbon, nitrogen, and phosphorus on which the phototrophs depend, specifically in the deeper waters of the Arctic and Southern Ocean, while sediment reminerization is important on the shallow Arctic shelves (Codispoti *et al.* 2005).

Thus there is a highly dynamic and close-knit community, based on the photosynthesis and exudation by phytoplankton. The production of cell material by bacterioplankton comes from the release of soluble products of photosynthesis, lysis brought about by viral infection, and sloppy feeding and excretion by the phagotrophs. Transfer of organic substrates and mineral nutrients by molecular diffusion is rapid and efficient over the short distances for organisms living at low Reynolds numbers. Larger organisms are poor competitors for nutrients specifically at low nutrient concentrations because of their low surface area/volume ratios and the ultraplankton can thus form a self-sustaining community through the microbial network interactions within which materials are recycled.

The low sinking rate of ultraplankon cells (typically <1 m day^{-1}) result in low loss rates of nutrients in sedimenting particles, allowing the community to maintain itself in surface waters given adequate light. Consequently, ultraplankton communities are the typical representatives in stratified, more oligotrophic waters of the Arctic and Antarctic following spring blooms. The population densities of the component species are set by the kinetics of their growth and trophic relationships and through competition rather than by input of materials from outside (Fig. 6.3). Changes of temperature have differential effects on these kinetics and in this seems to lie the explanation for a striking alteration in the composition of the picophytoplankton with latitude.

The extent to which the ultraplankton contributes to the sustenance of larger organisms is still disputed. It was originally thought that the microbial loop is important in returning energy and materials via heterotrophic bacterioplankton to the classical pelagic food chain involving microphytoplankton and metazoans. New evidence shows a more complicated picture: direct fixation of inorganic carbon through picophototrophs is an

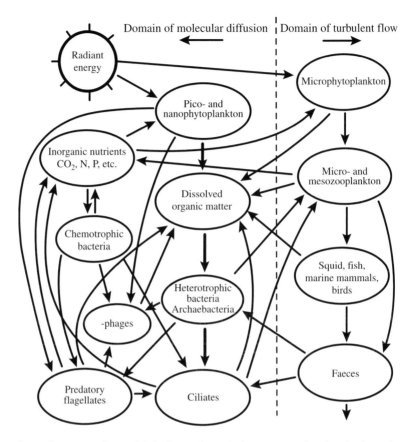

Fig. 6.3   Flows of energy and materials in the marine pelagic ecosystem showing the interrelations of the ultraplankton community with higher trophic levels.

important pathway in addition to the formation of particulate organic carbon by bacterial dissolved organic matter uptake. The low transfer efficiency along a multi-level food web (as typical for the microbial network) leaves only a small amount of primary production for higher trophic levels. On the other hand, some meso- and macropelagic grazers, specifically filter-feeding appendicularians, salps, and pteropods with partially short generation cycles, might be able to use a short cut by directly feeding on the smaller pico- and nanolankton size classes in a very efficient manner. These taxa might provide a means of tapping the productivity of the ultraplankton for the ultimate benefit of fish, seabirds, and mammals.

6.3.2 The microphytoplankton

The microphytoplankton are largely dominated by turbulent flow rather than by laminar flow with molecular diffusion. In oligotrophic waters it competes poorly for scarce nutrients with the picoplankton. Polar waters, however, usually contain inorganic nutrients (nitrate, phosphate, ammonium, silicate) in ample concentrations at least prior to the annual spring phytoplankton bloom and microphytoplankton blooms are able to develop if other conditions are suitable.

Because of its larger size microphytoplankton escapes the voracity of the rapidly multiplying phagotrophic flagellates. This size-based predator–prey relationship is frequently assumed in marine ecology, but not necessarily the case since heterotrophic dinoflagellates are capable of feeding on algal cells exceeding their own body size. Most planktonic metazoans, specifically the dominating herbivorous crustaceans, take weeks to years to grow and complete their life cycle, reducing their capability to react to fast increases in prey abundances through enhanced reproduction. Population densities of microphytoplankton are thus in general less tightly controlled than those of picophytoplankton and variations are much greater, up to five orders of magnitude. The community is essentially opportunistic and non-equilibrium in nature. Whereas the ultraplankton seems fairly uniformly distributed in polar seas, the microplankton is patchy in abundance, its growth being largely controlled by sea ice cover and water-column stability. These bloom scenarios were responsible for the development of the paradigm of short food webs in polar seas, which have now been modified significantly to include the smaller size fractions of the microbial network.

6.4 The physiological ecology of polar phytoplankton

The general relations of photosynthetic rate to light conditions have been outlined by Fogg and Thake (1987) and Smith and Sakshaug, in Smith (1990). The euphotic zone is the upper part of the water column that supports photosynthesis. The bottom of this zone is generally defined as the depth at which 1% of the surface irradiance is measured. However, a better representation of the bottom of the euphotic zone is the *compensation depth*. This is the depth at which the gross photosynthetic carbon assimilation by phytoplankton equals the respiratory carbon losses, or when the net photosynthesis is zero. Of course, phytoplankton cells are not at a static depth as they and/or the water may move. In fact they are being mixed either throughout the whole water column, or, where water stratification takes place, within surface-mixed water layers. Because of this phytoplankton cells will be mixed above and below the compensation depth, to depths as deep as the *mixed-layer depth*, which is the maximum depth of the layer between the ocean surface and a point where the density is the same as at the surface, usually ranging between 25 and 200 m. When considering net phytoplankton growth it is therefore more pertinent to relate the daily integrated photosynthetic gains to the integrated respiration losses over the water column (day and night) to the mixed-layer depth.

The *critical depth* is the water depth where the integrated daily photosynthetic carbon assimilation is balanced by the integrated daily respiratory carbon losses. As long as sufficient nutrients are present, net phytoplankton growth occurs when the mixed-layer depth is shallower than the critical depth. When the mixed layer extends below the critical depth, algal growth is limited by light, and there is no net phytoplankton growth (Smetacek and Passow 1990). The seasonal changes of mixed-layer depth and incident light play a key role in the seasonal dynamics of phytoplankton.

If the water column is stable so that a given cell remains at the same depth for a period commensurate with its generation time then adaptation can occur. Near the surface, cells become less susceptible to photoinhibition and photosynthesize at higher rates than they otherwise would. At depth, cells become shade-adapted and able to photosynthesize at higher rates than non-adapted cells. On the other hand, if the water column is mixed, a cell may be carried between the surface and the compensation depth and its photosynthetic capacity will adjust to some intermediate, low level of irradiance. This usually means that all cells become shade-adapted. An index of shade adaptation is the assimilation number, p_m^B, the amount of carbon fixed per hour per unit amount of chlorophyll *a* at light saturation. Assimilation number is low, with a value of about 1, in turbulent waters, in contrast to values of up to 10 or more in phytoplankton adapted to high irradiance.

In polar seas, substantial light changes can occur on time scales of seconds to minutes due to the sea ice cover and clouds. Phytoplankton in open leads between ice floes and polynyas (a word derived from the Russian for an area of open water within ice; see Chapter 7) is exposed to about an order of magnitutde higher irradiances than that under sea ice. Several important polar algal taxa, including diatoms, prymnesiophytes, and dinoflagellates, have evolved mechanisms to acclimate to such short-term changes with cycling energy between two xanthophyll pigments (diadinoxanthin and diatoxanthin) as one major mechanism. On longer time scales, algae respond to varying light conditions by altering the chlorophyll-specific absorption, the amount of chlorophyll per cell, and the size and number of photosynthetic units (Geider and MacIntyre, in Williams *et al.* 2002).

A chemical compound produced by polar phytoplankton has received increasing attention due to its potential direct impact on the Earth's climate: dimethylsulphoniopropionate (Simó and Vila-Costa 2006). Several Arctic and Antarctic microphytoplankton species, including diatoms, coccolithophorids (e.g. *Emiliana huxleyi*), and *Phaeocystis* spp., produce large pools of intracellular dimethylsulphoniopropionate, which are released at later stages of plankton blooms or during grazing. After its release into the water it is converted into dimethylsulphide (DMS), 1–10% of which ventilates into the atmosphere (the remaining 90% is recycled in the ocean). This DMS flux is substantial and may contribute about 50% of the global biogenic contribution of sulphur to the atmosphere. In the atmosphere DMS is oxidized to non-sea-salt sulphate and methane-sulphonic acid to form aerosol particles which act as cloud condensation nuclei (Fig. 6.4). The albedo of clouds strongly influences global climate and is itself determined by the concentration of these nuclei. An increased production of DMS by, for example, increased phytoplankton growth due to less sea ice, has been discussed as a possible negative-feedback loop countering Earth's warming. Indeed, one DMS model study for the Arctic proposes that DMS fluxes might increase by over 80% by 2080 compared to current conditions due to loss of sea ice and increased sea surface temperature (Gabric *et al.* 2005). The increased water vapour content and enhanced levels of cloud condensation nuclei could alter the radiation balance of the Arctic. The questions remains how changes in the composition of the phytoplankton communities caused by the environmental change will influence the described scenario.

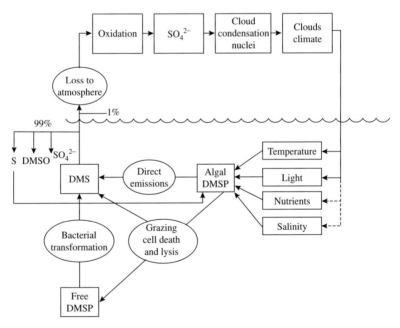

Fig. 6.4   Dimethylsulphide (DMS) released from phytoplankton and macroalgae (which contain dimethylsulphoniopropionate, DMSP) is oxidized to sulphur dioxide and the subsequent formation of aerosol particles and cloud condensation nuclei is part of a complex system of localized and global climate control. Note: most of the DMS in the water is converted to dimethyl sulphoxide (DMSO), SO_4^{2-} and sulphur, only 1% being exported to the atmosphere. Image after idea by Gunter Kirst, and Malin and Kirst (1997).

The maximum potential growth rate of phytoplankton species is determined by its genetic information (its acclimation potential) and by the actual water temperature. In addition to temperature, light and nutrients (traditionally nitrogen, phosphorous, and silicate) will modify the actual growth rate of the phytoplankton under conditions *in situ*. The individual nutrient, temperature, and light requirements of the individual species will determine the relative composition of the phytoplankton communities, following basic ecological competition theories (Sommer 1988). This creates a tight link between physical and chemical properties of the sea water and the composition of the phytoplankton communities.

Temperature is an important factor controlling metabolic activity of phytoplankton in the Arctic and Antarctic. Indeed, it appears to influence a process which is considered to be normally independent of temperature (i.e. light-limited photosynthesis; see Chapter 2). In the Scotia Sea and Bransfield Strait in an area centred on 61°30′S 57°0′W, assimilation numbers of phytoplankton at saturating light levels, $p_m{}^B$, and the slopes of the light-limited region of the photosynthesis against irradiance curves, α^B, have been found to be lower than in algae at lower latitudes. $p_m{}^B$, as expected, increased with a rise in temperature, having a Q_{10} of approximately 4.2 between −1.5 and +2°C, whereas, unexpectedly, α^B also increased, with a Q_{10} of approximately 2.6 between −1.5 and 5°C. Above 5°C there were no increases with rise in temperature in either rate. It seems that at extremely low temperatures some temperature-dependent reaction becomes rate-limiting for the photochemical reactions (Tilzer *et al.* 1986). One interesting observation of this study was a higher Q_{10} value for respiration than primary production, implying that Antarctic phytoplankton loses more fixed carbon to respiration at higher temperatures, causing slower growth.

Deep mixing may carry phytoplankton below the critical depth and there is also the darkness of the winter months, most noticeably in the high latitudes of the Arctic. Polar phytoplankton has evolved various strategies to deal with such unfavourable conditions. One mechanism is the reduction of basal respiration to a minimum so that cellular reserves suffice to carry them through. These resting phases are sometimes linked with the formation of morphologically distinct cell types (resting stages, spores), that sometimes, in the beginning of polar exploration, were described as separate species. Some planktonic algae are capable of heterotrophic nutrition, including phagotrophy of bacteria. The utilization of dissolved organic matter appears to be of minor significance. Experimental studies demonstrated that diatoms can survive long periods of darkness and cultures of algae from both Arctic and Antarctic kept in the dark at −2°C were still viable after 12 months. Survival was increased if darkness was imposed gradually, as would happen under natural conditions (Kirst and Wiencke 1995).

Psychrophilic algae have a low temperature optimum, usually below 15°C, for net photosynthesis. This has been explained by supposing the Q_{10} for gross photosynthesis to be less than that for respiration, which thus progressively overtakes the other as temperature is raised. In fact, the Q_{10} values for photosynthesis in entire samples of Antarctic plankton are about the same as those for respiration. However, it has to be remembered that in such samples the respiration of both phototrophs and heterotrophs is included in the overall determinations. The problem thus remains unresolved (Kirst and Wiencke 1995).

The products of photosynthesis can only be elaborated into cell material and result in growth if the essential elements and, in certain cases, specific organic moieties, are available. Some of these, such as carbon, hydrogen, oxygen, potassium, calcium, and sulphur, are in ample supply in sea water but other major nutrients, notably nitrogen, phosphorus, and silicon (necessary for diatoms and silicoflagellates), are often present in low concentrations and may seasonally limit growth (Arrigo 2005). Such limitation of phytoplankton growth by macronutrients (silicate, nitrogen) has been described during massive blooms of phytoplankton in both the Arctic and Southern Oceans, specifically in the marginal ice zones. The onset of limitations by silicate causes a shift in the species composition of phytoplankton communities with time. Large microphytoplankton diatoms initially dominate due to their high uptake rate of inorganic nutrients, followed by non-silicate-requiring taxa (e.g. *Phaeocystis* spp.) after silicate depletion. In areas with glacial run-off large blooms of small cryptophytes may form.

Both the Arctic (with the exception of the oligotrophic central Arctic deep basins) and Southern Ocean are characterized by high nutrient concentrations. In fact, the Southern Ocean receives a large supply of nutrient-rich water by upwelling at the Antarctic Divergence and has concentrations of these elements about twice as high as those of the most fertile areas elsewhere in the world's oceans, so they are rarely limiting. There are, however, indications that silicate may limit growth of diatoms immediately south of the Polar Front (Kirst and Wiencke 1995).

The situation in the Arctic is more complex, with no major input as in the Antarctic, but there are significant inputs of nutrients from the Siberian and Alaskan rivers, from the Atlantic via the Norwegian Current, from the Pacific via the Bering Strait, and from regeneration in the shelf regions. Nutrient concentrations are low in the surface waters of the central Arctic Ocean but increase at the halocline at depths where utilization by phytoplankton is minimal. Elsewhere, nutrient concentrations are high where upwelling of deeper water occurs at fronts, as in the Labrador Sea. High production in the western Bering Strait region is supported by a cross-shelf flow and upwelling of nutrient-rich water from the Bering Sea continental slope through the Bering Strait into the Chukchi Sea. In contrast to Antarctica, nutrient depletion is common in Arctic waters. Figure 6.5 shows a device used for sampling nutrient content in polar waters.

Fig. 6.5   Rossette bottle samplers are used to sample waters for species composition and nutrient content of the waters at different depths Each bottle is triggered by an electronic
signal when it is at the required depth (photograph by David N. Thomas).

The lack of large-scale nutrient depletion in the Southern Ocean had
been a puzzle for decades. Indeed, Antarctic waters are among the prime
examples for so-called HNLC (high nutrients, low chlorophyll) regions,
where low phytoplankton biomass is observed in regions with high nutrient concentrations and light. Several explanations have been brought
forward for this fascinating observation, including deep mixing and
consequent light limitation, efficient grazing, and limitation by micronutrients. Micronutrients include trace metals and organic growth factors
such as biotin, thiamine (vitamin B_1), and cyanocobalamin (vitamin B_{12}).
Vitamins are not required by all species and are unlikely to limit total
standing crop but may determine its species composition.

John Martin (Martin *et al.* 2002) initiated a discussion that focused attention on iron as the trace element most likely to affect phytoplankton
abundance. Iron is a constituent of several vital enzymes and is the fourth
most abundant element in the Earth's crust. River waters contain high
concentrations of iron in the form of complexes with organic matter and
coastal waters are correspondingly rich in this element. The chemistry of
iron in oceanic waters is complicated but ferric hydroxide, the main inorganic form, is sparingly soluble and, being readily adsorbed on particulate
matter, is removed by precipitation. Oceanic waters appear generally to
have low concentrations, their principal supply coming as fall-out of dust
derived from the land. Arctic seas get sufficient supplies from exposures
of rock and drylands in their vicinity, and through input of sediment by
the large river systems.

The Antarctic is much less favourably situated since dry exposed land lies far to the north and out of the path of the prevailing westerly winds, while most of the Antarctic continent is covered by ice. Sediment cores from the Atlantic sector of the Southern Ocean show periods of increased accumulation of biogenic detritus during glacial periods coinciding with increases in iron content of up to 5-fold. This stimulated research into whether the Southern Ocean may be deficient in micronutrients such as iron, starting an era of several large-scale, open-ocean iron fertilization experiments (Buesseler et al. 2004, Coale *et al.* 2004, Boyd *et al.* 2007). The basic design of these experiments was simlar, in that iron and an inert tracer (SF_6) were added to the surface waters over dozens of square kilometres and observations by ships with high temporal resolution and remote-sensing tools followed the reaction of the ecosystem over several weeks.

The results of these experiments were very consistent, showing a strongly enhanced growth of phytoplankton and an increased biomass in the area of iron enrichment after an initial lag phase. These blooms were primarily formed by diatoms which also lead to an increase in the DMS levels as described above. The increase in diatoms indicates a mis-match between grazers and algal growth during these short-term experiments, allowing the formation of elevated biomass levels in the iron-fertilized regions.

It appears that greater transport of dust from Patagonian deserts during the Last Glacial Maximum led to conspicuous stimulation of phytoplankton by increasing iron supply, leading to iron concentrations comparable to those in the enrichment experiments. It remains an open question whether long-term iron fertilization might actually allow increased primary production and burial of CO_2 in Antarctica, or whether it might lead to elevated levels of grazers, with consequently increased recycling of nutrients and iron by grazer-mediated processes. This uncertainty, including the potential changes in the composition of the regional plankton communities, could be addressed by long-term experiments to be conducted over the next few decades.

Interestingly it has been seen that drifting icebergs can introduce substantial terrigenous material, including high concentrations of inorganic nutrients and in particular trace metals (de Baar *et al.* 1995). Smith *et al.* (2007) describe free drifting icebergs in the Weddell Sea as being comparable to estuaries in that they supply coastal waters with inorganic nutrients. In their study the high amount of nutrients released from the icebergs was related to increased primary production, zooplankton numbers, and even seabirds up to 3.7 km around the icebergs. They scaled up their results to take into account all of the icebergs in the region and estimated that 30% of the surface ocean of the region is probably fertilized by material being released from the icebergs.

The seasonal cycle of microphytoplankton in polar waters is traditionally represented by a bell-shaped curve, rising from low levels in spring to a peak around midsummer and declining again to a low level by early autumn. This contrasts with the curve for temperate waters which has peaks in spring and autumn with a trough in the summer. In polar waters phytoplankton growth follows the seasonal cycle of radiation fairly closely, the slow development of herbivores at low temperatures delaying their impact until their prey is already in decline as a result of diminishing radiation and nutrient supplies and increasing turbulence. In temperate waters, the herbivores develop more quickly and their grazing, combined with exhaustion of nutrients, reduces the standing stock of phytoplankton by midsummer. Nutrient regeneration follows and temperature, water-column stability, and radiation allow a second peak in autumn (Fogg and Thake 1987).

Ultraplankton in Antarctic waters has a seasonal periodicity similar to that of the microplankton but the curve of standing crop is flatter and the peak occurs later in the summer, likely due to the higher grazing pressure through microzooplankton with its shorter generation times. Phytoplankton growth in polar seas is patchy, and vast areas of low productivity contrast with limited regions of intense productivity in marginal ice zones or on the ice shelves. Remote sensing retrieved by the Coastal Zone Color Scanner, Sea-viewing Wide Field-of-view Sensor, and Moderate Resolution Imaging Spectroradiometer sensors highlighted the extreme patchiness in Antarctica, with elevated phytoplankton concentrations in frontal zones and marginal ice zones.

A similar patchiness exists in the Arctic, where a seasonally stratified water column with a pycnocline at about 25 m, combined with initially lower nutrient concentrations, cause widespread macronutrient limitations after the intial spring biomass peak. Production then depends on regeneration within the surface layer and increase can only occur if disruption of the pycnocline introduces nutrient-richer water from below. This can be brought about by storms, shear forces between currents, banks, and islands, vertical motions induced by upwelling or eddies, or mixing induced by the tides in shallow waters. An example is a particularly rich area in the western Bering Strait where, in a 3-month summer season, most of an annual yield of $324\,\mathrm{g\,C\,m^{-2}}$ is produced, sustained by nutrient-rich Anadyr water and remineralization of nutrients on the shallow shelves (Codispoti *et al.* 2005). Growth during the winter is inhibited by lack of light, deep mixing, and ice cover, and nutrients accumulate. When the ice retreats and the mixed layer depth decreases conditions become favourable for phytoplankton growth. This growth is maintained by upwelling of nutrient rich water along the edge of the continental shelf, which is then carried across the shallow shelf by a current flowing northwards along the western coast of the Bering Strait. This produces a situation rather like that in a laboratory continuous culture, in which nutrients are replaced as rapidly as they are taken up. Similar processes maintain high productivity on the seasonally covered inflow shelves of the Barents and Kara Seas. Sakshaug (2004) estimated the total primary productivity in High Arctic waters (excluding sub-Arctic seas) at about 0.33 Gt C year[-1], which corresponds roughly to about 0.7% of the total for the world's ocean (Carr *et al.* 2006). This implies that the Arctic is less productive on average than other parts of the world's ocean due to the combination of permanent sea-ice cover and low nutrient concentration.

In the Southern Ocean, the reasons for patchiness are more obscure. Conditions of light and temperature are similar all around the continent and nutrient concentrations are high. The prevalent westerly winds generally mix the surface layers to an extent sufficient to prevent stratification. However, where ice melt produces local stratification, plankton production is intense. The importance of stability of the water column is well illustrated by studies in inshore waters around Signy Island. Phytoplankton begins to increase as fast ice disappears and reaches a peak a month or so later. There is then an abrupt decline although water temperature and light are still relatively high, sea ice is not yet re-forming, and nutrients, as exemplified by nitrate, are not exhausted (Fig. 6.6). The algal biomass is not grazed to any significant extent and the decline coincides with high

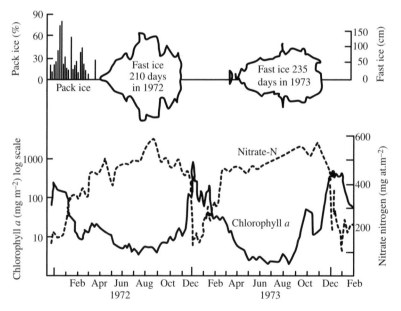

Fig. 6.6   Observations over 2 years of amounts of ice, phytoplankton chlorophyll, and nitrate concentration in the water column (16.5 m mean depth) in Borge Bay, Signy Island. Data courtesy of T.M. Whitaker.

winds, presumably extending mixing below the critical depth. The low impact of grazers on phytoplankton bloom development was also evident during iron-enrichment experiments (Schultes *et al.* 2006).

This observation raises the possibility that, in addition to micronutrient limitation, the low standing stock of phytoplankton generally found in the Southern Ocean is due to turbulent mixing. Growth in these circumstances can only take place where hydrographic features, such as eddies, maintain populations in the photic zone. The upward vertical velocity in an anticyclonic eddy is probably around $1 \, \text{m day}^{-1}$ so this would offset to some extent the settling of microplanktonic cells, the velocity of which is in the range of $1–10 \, \text{m day}^{-1}$. This idea is supported by a correlation of high phytoplankton densities with upwelling, as indicated by low surface temperature, or with features of the seabed, such as shelf breaks, submarine mountains or islands, which produce eddies (Fig. 6.7). It is unlikely that these effects are due to nutrient enrichment from deep water, although it could be that supply of iron or other micronutrients is sometimes important.

Where phytoplankton is densest, values of $40 \, \text{mg}$ chlorophyll $a \, \text{m}^{-3}$ or more with daily carbon fixation in excess of $2 \, \text{g C m}^{-2} \text{day}^{-1}$ are reported. These are comparable with values from the most fertile areas in upwelling region or daily production estimates for the Bering and Chukchi Sea shelves. On the other hand, in the open ocean values are usually around $0.5 \, \text{mg}$ chlorophyll $a \, \text{m}^{-3}$ and $0.2 \, \text{g C m}^{-2} \text{day}^{-1}$. The remoteness and harsh weather in Antarctica made most ship-based estimates episodic events and allowed for only tentative estimates of primary productivity. Carr *et al.* (2006) used a remote sensing approach, combining ocean colour variations with ecosystem models, for a global estimate of primary productivity. They concluded that the Southern Ocean might contribute 5.5% (2.6 Gt) of organic carbon per year to total productivity, also contributing less than average per unit ocean area to total ocean primary productivity.

6.4.1 The Arctic phytoplankton

In terms both of numbers of species and biomass diatoms predominate, with dinoflagellates, other flagellates, and green algae being present in much smaller proportions, although the already mentioned lack of taxonomic studies probably causes a hidden diversity in the ultraplankton size class. The species are not the same in different sea areas (Guillard and Kilham 1977). In the Barents Sea the important diatoms include *Chaetoceros diadema*, *Corethron criophilum*, *Skeletonema costatum*, and *Rhizosolenia styliformis*. Among the dinoflagellates are *Peridinium depressum* and *Ceratium longipes*, and of the green algae only *Halosphaera viridis* is distributed widely. In small bays and inlets and along the marginal ice zones there may be mass development of *Phaeocystis* spp. in spring. Some warmer-water forms (e.g. *R. styliformis* and *Ceratium tripos*) are brought

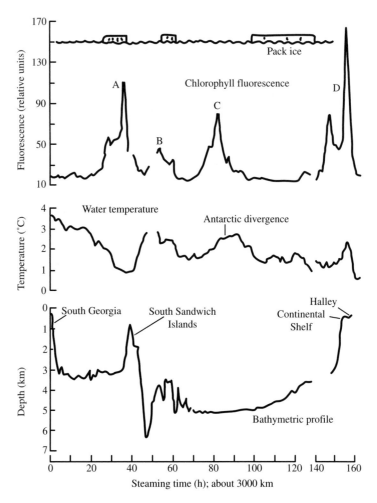

Fig. 6.7   Patchiness of phytoplankton in the Southern Ocean; observations of chlorophyll fluorescence, surface water temperature, and depth made during a cruise across the Weddell
Sea from South Georgia to Halley Station. The chlorophyll peak A is an example of the
'island effect'; B, of phytoplankton growth in the marginal ice zone; C, perhaps also in
the marginal ice zone and/or associated with the Antarctic Divergence at around 65°S;
and D, in inshore waters of the Brunt Ice Shelf. Data from Hayes et al. (1984).

in by the North Atlantic Current. The Bering Sea divides into two distinct
areas; cold-water plankton are found on the western side and warmer-
water forms on the eastern side. In the cold water the diatoms include
Thalassiosira nordenskiöldi, *Chaetoceros socialis*, and *Eucampia groenlandica*, whereas on the other side are *Thalassiosira japonica*, *Rhizosolenia
alata*, and *Ditylum brightwellii* with the dinoflagellate *Peridinium excen-
tricum*. This distribution is related to currents. The Laptev Sea is different
again, having the strong influence of the Siberian River Lena. The common

forms in the offshore, highly diluted, region include brackish–freshwater species of cyanobacteria, *Aphanizomenon flos-aquae*, and *Anabaena* spp., and the freshwater diatoms *Melosira italica* and *Asterionella gracillima*, with marine forms, *Thalassiosira baltica*, *Chaetoceros gracile*, and *Peridinium breve*, providing only about 5% of the population. There is no great degree of endemism among Arctic phytoplankton.

6.4.2 The Antarctic phytoplankton

As might be expected, the circumpolar circulation and sharp delineation by the Polar Front combine to given the Southern Ocean a relatively uniform plankton flora (Guillard and Kilham 1977). As in the Arctic the most prominent forms are diatoms (Fig. 6.2). Silicoflagellates are also abundant but dinoflagellates less so, although because their cells are larger they may sometimes exceed the diatoms in biomass. The predominance of diatoms is to be expected since they sink rapidly and hence are dependent on turbulence to maintain them in the photic zone. The larger flagellates, on the other hand, lose the biological advantage of being able to move to the optimum depth in the water column if there is vigorous mixing. There is no well-defined seasonal succession of species.

The number of species thought to be bipolar has diminished as taxonomic knowledge has advanced. The supposedly characteristic Antarctic species *Thalassiosira antarctica* appears to be bipolar since it has been found in the northern hemisphere but not in latitudes between 58°S and 58°N. It may yet be found in intermediate locations or, quite possibly, it may not be genetically isolated from other *Thalassiosira* spp. from which, in fact, it is distinguished only with difficulty. Some species appear to be cosmopolitan but the species of *Phaeocystis*, now distinguished as *Phaeocystis antarctica*, which is abundant in the Southern Ocean, has been shown by sequence data from 18 S small-subunit rDNA to be genetically distinct from *Phaeocystis pouchetii*, the northern cold-water form, and from *Phaeocystis globosa*, the warm-water species from which the other two seem to have evolved (Medlin *et al.* 1994). Endemism is high among the Antarctic microphytoplankton; 80–85% of dinoflagellates and around 37% of diatom species. However, at generic level most of the important taxa in the Antarctic phytoplankton are cosmopolitan but the two monospecific genera, *Charcotia* and *Micropodiscus*, seem to be endemic. In addition to *Thalassiosira*, *Corethron coriophyllum*, *Fragillariopsis* spp., *Pseudonitzschia* spp., and *Rhizosolenia* spp. are typical members of the diatom community in Antarctic waters.

6.5 The zooplankton

Nanozooplankton has been discussed in the section on ultraplankton (Section 6.3.1). Larger forms, which are nearly all metazoans, are best dealt with together although they range in size from microplankton (20–200 μm), through mesozooplankton (0.2–20 mm), to macrozooplankton (2–20 cm). Most of these, with the exception of appendicularians, salps. and pteropods (e.g. *Limacina* spp.) feed on prey usually a minimum of about two orders of magnitude smaller than themselves, which they either filter out or seize as individuals while actively swimming. They may be herbivorous, feeding on microphytoplankton, omnivorous, or carnivorous, feeding on other zooplankton. Some microzooplankton protists (e.g. *Myrionecta rubra, Laboea strobila*) harbour functional chloroplasts either permanently or seasonally in their cytoplasm, making them functionally mixotrophic. Most have life cycles with several developmental stages which are controlled by low temperatures and the brief abundance of food. Growth when food is available is fast but overall rates are slow and lifespans extended, because of long periods of zero or negative growth when basic metabolism is maintained at the expense of reserves.

The storage of energy reserves (mainly lipids) plays a crucial role in the life cycle of many dominant mesozooplankton taxa in polar seas, as it allows for overwintering and partially also reproduction (Lee *et al.* 2006). Lipids are characterized by very high energy content (39 compared with 17–18 kJ g^{-1} for proteins and carbohydrates). The high energy content also makes these plankton taxa a rich food source for higher trophic levels. Three major groups of lipids were observed in polar zooplankton: waxesters, triacylglycerols, and phospholipids. Waxesters are the primary long-term storage molecule of zooplankton that overwinter in a diapause stage: examples are *Calanus hyperboreus* and *Calanus glacialis* in the Arctic, where waxesters contribute more than 75% to total lipid mass during diapause. Triacylglycerols are less common in polar marine invertebrates (in contrast to terrestrial ecosystems). Phospholipids are typically components of biomembranes; however, one key polar crustacean species, the Antarctic krill *Euphausia superba*, uses phospholipids for energy storage (Lee *et al.* 2006).

Recent findings indicate that a fourth group of lipids, diacylglycerol esters are used by polar pteropods as major energy storage lipid. Current research also used the composition of these lipids, mainly the fatty acid composition, to trace food-web interactions. This was possible, as many herbivores incorporate the fatty acid signature of their prey unmodified. This allowed the relevance of diatoms compared with dinoflagellates to be distinguished in the nutrition of polar marine herbivores such as copepods and euphausiids.

In both hemispheres the chief groups represented are radiolarians, coelenterates, rotifers, chaetognaths, copepods, ostracods, euphausiids, amphipods, mysids, pteropods, and appendicularians. The species are, however, different (Zenkevitch 1963, Smith and Schnack-Schiel, in Smith 1990, Knox 1994).

6.5.1 The Arctic zooplankton

Zooplankton abundance broadly reflects phytoplankton abundance (Zenkevitch 1963). Thus, the central Arctic Ocean has sparse zooplankton with biomass around $1–3\,mg\,m^{-3}$, about a thousandth of that in the Bering Sea. The main input of water into the Arctic basin is from the North Atlantic so that one finds Atlantic species, such as *Calanus finmarchicus* and *Metridia lucens*, but the most frequent copepods are *C. hyperboreus*, *C. glacialis*, and *Metridia longa*. Long-term studies of these have been made using ice islands as research platforms in the centre portion of the basin. The copepods predominate near the surface in the summer. The life cycle of *C. hyperboreus* seems to take 3 years; 1 year is required for development from egg to copepodid stages II and III, the second for stages IV and V, and the third for adulthood and spawning. Copepodids stay below 300 m in winter, migrating to the surface 100 m in summer. Gravid females descend slowly from the surface down to 300 m in spring, while males, which are fewer in number, go below 400 m.

The marginal seas of the Arctic basin have a similar pelagic fauna to that of the central parts with the addition of neritic species dependent on the temperature and salinity conditions arising from the inflow of the rivers as well as additions through advection from the Pacific and the Atlantic. The Kara Sea shows particularly distinct zones with faunas indicating the origins of the waters—in the north from the Arctic basin, in the south from the Barents Sea, and an intermediate zone of brackish water with freshwater species occurs in waters from the rivers Ob and Yenisei—which is also evident in the phytoplankton composition.

The high primary productivity of the western Bering Strait is matched by secondary production (Sambrotto *et al.* 1984). The zooplankton species are characteristic of the North Pacific. The large copepods, *Neocalanus plumchrus* (up to 5 mm in length) and *Neocalanus cristatus* (up to 10 mm), make up 70–80% of the zooplankton biomass in summer. Arctic species are of no significance due to the northward flow in this region. The large copepods overwinter at depth and eggs are produced from autumn to spring and the young rise to the surface from early spring onwards. In the Bering Sea much of the annual growth takes place during the dense spring phytoplankton bloom.

The predominant herbivorous copepod genus in the Barents and Greenland Seas is *Calanus* but there has been confusion because this is an area of sympatry between two closely related species, *C. finmarchicus* and *C. glacialis*. The former is found mainly in temperate as well as sub-Arctic seas whereas the latter is endemic to the Arctic. There is morphological similarity between the two, although *C. glacialis* is the larger, but if collected alive their swimming patterns and behaviour are different enough to distinguish them. The situation is further complicated because *C. glacialis* has a variable life cycle, lasting 2 years with spawning in May and June in the Barents Sea but a 1-year cycle with spawning in December off western Greenland. In the southern part of the Barents Sea the development of *Calanus* spp. occurs in early summer but north of 72 or 73°N it occurs in the autumn. The biomass may attain as much as 300 mg fresh weight m^{-3}, of which 90% or more is *Calanus*. It is the key organism in the food web. Large numbers of herring, haddock, and the fry of various other fish feed on it in competition with jellyfish and ctenophores, which may reach as many as 123 individuals m^{-3}.

This identification problem with zooplankton taxa is specifically evident for all larval and juvenile stages which frequently miss the distinctive features required for species identification. For this reason genetic barcoding is a novel technique being applied for marine metazoan studies. This *barcode of life* approach (the use of DNA information) to identify species in samples was adapted from methods developed in determining species composition in the microbial world. These methods are currently being adapted to analyse plankton communities including all juvenile and larval stages of zooplankton encountered in polar seas.

6.5.2 Antarctic zooplankton

As with the phytoplankton, the zooplankton assemblages of the Antarctic have a generally circumpolar distribution. There is, however, latitudinal zonation as seen, for example, in the distribution of euphausiids (Fig. 6.8). As described for the phytoplankton communities, the Antarctic zooplankton have evolved in a certain framework of environmental characteristics; therefore, physical oceanographical features and the composition of communities are tightly linked.

A wide variety of species belonging to several taxonomic groups, including salps, pteropods, copepods, and even tomopterid polychaetes, are found but are often overlooked because of the dominance of one species which is generally regarded as the key organism in the food web and is likely one of the best-studied plankton organisms of all, with many investigations over the last 40 years. This is *Euphausia superba*, the Antarctic krill. The common name is derived from the Norwegian *kril*, originally used by whalers to denote small fish and thence transferred to cover crustaceans eaten by baleen whales. *E. superba* is a macroplanktonic euphausiid with a length of up to 60 mm and a fresh weight of about a gram. It is bigger than any corresponding Arctic organism (e.g. *Thysanoessa raschii*). It is stretching a point to describe it as 'planktonic', implying that it drifts at the mercy of tides and currents, since it can swim at a speed of around 13 cm s^{-1} and krill swarms have been seen to move against prevailing currents. This mobility and the often patchy distribution make for great difficulties in precise studies of krill in their natural environment.

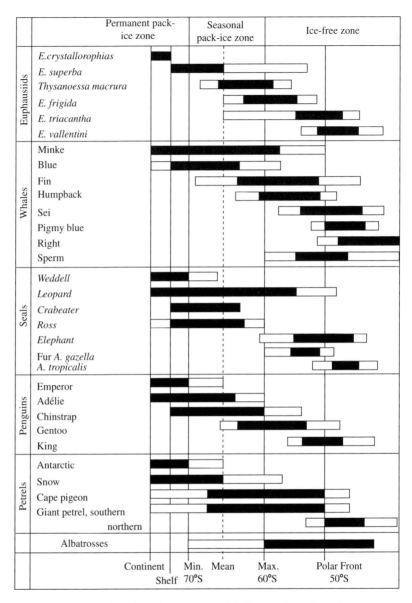

Fig. 6.8 Zones occupied by selected species of euphausiids (*Euphausia* spp.), marine mammals, and birds, from the Antarctic continent northwards. Each species has a circumpolar distribution and the average latitudinal range is given, with the shaded areas indicating the greatest densities. Min. and Max. refer to sea ice extent. Redrawn after Laws (1977).

Modern nets can yield reasonably quantitative results with micro- and mesozooplankton but not when, as with krill, the organism can detect the approach of a net and take rapid avoiding action. Large trawls without attachments directly in front of the net give the best results but even these are avoided effectively during the day. A sonic technique for sampling is proving more useful. This is a modification of echo-sounding, by which surfaces reflecting acoustic signals can be detected by the echoes received from them. The time elapsing between dispatch of the signal and receipt of the echo is a measure of the distance of the reflecting surface and its strength is a measure of the amount of material in it. Integrated data from an echo-sounder operating at an appropriate frequency give a mean volume backscattering strength from which the density of a krill swarm can be estimated, given a target strength determined experimentally (Fig. 6.9). However, krill in the surface layer, above the echo-sounder transducer, are not sampled and the technique functions poorly in pack ice, although transducers to map krill distribution have successfully deployed from autonomous vehicles operating under sea ice (see Chapter 7).

Euphausia superba was studied intensively by the British Discovery Investigations between 1925 and 1939 (Hardy 1967), and by BIOMASS, an international programme of Biological Investigations of Marine Antarctic Systems and Stocks, between 1977 and 1991 (El-Sayed 1994). The combination of experimental with field distribution studies lead to the development of several conceptual models describing the distribution and trophic role of *E. superba* (Nicol 2006). It occurs in surface waters south of the Polar Front, showing most activity at −1.5°C and much less at 4°C. In this it contrasts with its ecological counterpart north of the Polar Front, the lobster krill *Munida gregaria*, which is most active at 8–10°C and killed if transferred to 0°C.

E. superba is found down to 200 m, but mostly between 20 and 100 m. It typically undertakes daily vertical migrations, rising to the surface at night and sinking to concentrate in swarms during the day evidently in

Fig. 6.9   The Southern Ocean krill, *Euphausia superba*. Photograph courtesy of Alfred Wegener Institute.

response, although in no simple manner, to food levels and irradiance. The swarms are usually dense and often large, measuring up to 500 m or more horizontally. Occasionally, they are extremely large as, for example, a 'superswarm' found in March 1981 north of Elephant Island (approximately 61°S 55°W), which acoustic observations showed as extending over 450 km^2 with a biomass of 2.1×10^6 t. Large dense krill swarms contain up to 10^{12} individuals (30,000 animals m^{-2}) and each individual can eat up to 25% of its body weight per day (Smetacek and Nicol 2005). Hence a krill swarm can very effectively clear a water column of any food particles. The movement of individuals in a swarm, containing billions of animals, is coordinated and synchronized, the whole changing shape in an amoeboid fashion; this provides an entrancing sight for divers, especially at night because krill has a blue luminescence.

Krill swarms often differ in composition in terms of size of individuals, stage of maturity, sex ratio, and feeding state. Feeding tends to be more active when the krill is dispersed. The food is mainly phytoplankton but small zooplankton, even their own larval stages, may be taken. Prey is filtered out in a basket formed by the animal's fringed thoracic limbs. *E. superba* does not have great reserves of lipid to carry it through the winter as it uses phospholipids and not waxesters, but can withstand starvation for long periods by utilizing its body protein causing actual body shrinkage. The feeding ecology of *E. superba* is very flexible, exploiting sea-ice (see Chapter 7), pelagic and benthic resources.

Growth and longevity are difficult to establish from studies of natural populations and laboratory observations may not be reliable guides to what happens in the sea when animals are in swarms. Biochemical features indicate that *E. superba* has a lifespan of between 5 and 8 years. Spawning may take place three or more times during the life of a female, up to 8000 eggs being released at a time. This happens at depths around 100 m and the eggs immediately sink at a rate of 150–250 m day^{-1} and hatch at around 1000 m. Whereas adults are carried north east towards the Polar Front in Antarctic Surface Water and occur most frequently along the continental slope and deeper waters, eggs and young larvae are carried back south in the Circumpolar Deep Water and back to the shelf regions (Fig. 1.8), resulting in spatial seperation between juveniles and adults in summer. As the larvae develop they ascend slowly to the surface, so that different stages are found stratified in order, with the young adults eventually arriving back once more at the surface. In this way, *E. superba* is able to maintain itself over the latitudinal extent of the open Southern Ocean.

The distribution of krill, however, is not uniform. It is particularly abundant, for example, in the Scotia Sea, around South Georgia, north of the Ross Sea, and in the Weddell Sea. This pattern seems generally explicable in terms of the transport of larvae and young adults by currents and their collection in gyres and eddies. The total standing stock of *E. superba* has been variously estimated as between 14 and 1000 million t; this is clearly a large amount by global standards and probably exceeding the total biomass of the human race. Because krill is of ecological importance and a possible economic resource, it is necessary to know the rate at which this standing stock produces new biomass.

Other important zooplankters in the Southern Ocean are chaetognaths, such as *Sagitta gazellae*, copepods, such as *Calanoides acutus*, and euphausiids such as *Thysanoessa macrura* and the coastal *Euphausia crystallorophias*. Reproduction is usually timed to coincide with the seasonal pulse of phytoplankton production. The salp, *Salpa thompsoni*, a barrel-shaped organism which filters phytoplankton as it pumps the water through itself, shows remarkable adaptation to the situation. During the winter it exists in a solitary form but in the spring it buds off, asexually, large numbers of aggregate forms. Each solitary organism can produce about 800 aggregates when triggered, presumably by the phytoplankton bloom, so that dense swarms are built up during the summer. The mean biomass of zooplankton other than *E. superba* is estimated as 10–50 mg wet weight m^{-3}, going up to 300 mg wet weight m^{-3} at the Polar Front. This is probably about the same biomass as for *E. superba*.

A feature of Antarctic meso- and macro-zooplankton is that its biomass in the upper 1000 m scarcely fluctuates over the year, the bulk of the stock consisting of a few large species which overwinter in the deeper water. It may be that these feed on the ultraplanktonic community, which also tends to a steady standing stock over the year, as has been suggested for benthic filter feeders. These zooplankters are not a major food source for large predators such as prey on *E. superba* but they are eaten by carnivorous zooplankton, including the amphipod *Parathemisto gaudichaudii*. This has a life cycle of 1 year geared to producing juveniles at around the time when herbivores are producing theirs. It has marked diurnal vertical migration, collecting at the surface at night, but it does not swarm in the way that herbivores do.

6.6 Squid

Squid are abundant and ecologically important in both polar regions; they are a major food of the male sperm whales, which migrate into both Polar seas, and of albatrosses in the Southern Ocean as well as supporting important fisheries for some species around Alaska. Nevertheless, we know relatively little about them. Squid are fast-swimming, with well-developed nervous systems, and are difficult to catch in scientific research nets. However, because they live mostly in deep waters, only coming towards the surface waters at night, not very much is known about them. Squid are have muscular bodies with side fins that they use to swim, but they are also capable of very rapid *jet propulsion* when avoiding predators or trying to catch their prey. Both in species and size there is little correspondence between those caught in research nets and those whose remains are found in predators such a sperm whales (Rodhouse, in Kerry and Hempel 1990).

The food of squid evidently consists largely of euphausiids, other crustaceans, smaller cephalopods, and fish. They themselves are important in the diets of killer whales, seals, birds, such as the emperor penguin and albatross, and larger cephalopods. They are found in their stomachs of whales in vast quantities, or at least parts of the squid are found such as the hard horny beaks (jaws of the squid). Antarctic fish are estimated to provide about 15 million t of food for birds, seals, and whales, whereas squid are thought to provide 35 million t. For this to be possible there must be over 100 million t of squid in the region (Knox 2006).

There are about 20 Antarctic squid species that range in size from the small *Brachioteuthis* species (20 cm) to the large *Mesonychoteuthis hamiltoni* (6 m) which has large hooks, as well as suckers, on its arm and tentacles for capturing its prey which can include the giant Antarctic cod. It is thought that squid feed on a range of fish species and crustaceans including krill. They will also eat other squid.

6.7 Fish

Fish belonging to a variety of groups are found in Arctic waters but notothenioides, which are dominant in the Antarctic, are notably absent. In the High Arctic, the Arctic cod is a key species. It feeds on euphausiids, copepods, and pteropods, and, in coastal waters, benthic amphipods. In turn it is preyed on by a range of predators including the ringed seal, beluga, Arctic tern, glaucuos gull, black guillemot, and other seabirds. Where there is freshwater input around the Arctic basin many species of Salmonidae are abundant, particularly off the Alaskan and eastern Siberian coasts. Arctic char is one of these, abundant in streams but fattening in the sea in its fourth and fifth years before it spawns in fresh water. In subpolar regions, capelin (*Mallotus villosus*) winters in deep waters, rising to form huge surface shoals in summer. Eel pouts (Zoarcidae), polar cod (*Arctogadus glacialis*), and polar halibut (*Reinhardtius hippoglossoides*) occur in deeper waters. In the Low Arctic the fish faunas are derived from adjacent temperate seas; capelin, herring (*Clupea harengus*), and sandlance (*Ammodytes hexapterus*) in the north-east Atlantic; walleye pollock (*Theragra chalcogramma*), and herring (*Clupea pallasi*) in the north Pacific (Zenkevitch 1963, Ainley and DeMaster, in Smith 1990). These and other cold-water species provide the basis for commercial fisheries.

The Antarctic fish fauna is more homogeneous than that of the Arctic. Species diversity is so low that a sea area which amounts to 10% of the world's oceans contains only about 1% of the fish fauna and in shelf and slope waters 55% of the species fall into one group, the Notothenioidei. This has not always been so. Fossils from Seymour Island (64°17'S 56°46'W) show that in the late Eocene, warm seas over extensive shallow shelves supported a diverse and cosmopolitan assemblage of fishes, few of which are represented taxonomically in the region in the present day. One must assume that the establishment of the Circumantarctic Current some 25 million years ago and the subsequent onset of thermally stable cold-water conditions caused their extinction. No fossil record of notothenioides has yet been discovered and the origins and subsequent dominance of the group remains contentious. That 97% of species and 85% of genera are endemic points to their evolution in isolation. Possibly the notothenioides, originally a benthic nearshore group, survived or possibly they were most successful in invading across the Polar Front. This front seems to be a barrier preventing invasion by temperate water fish, but it may not be simply the temperature drop across this front which is the operative factor; after all, in the Arctic temperate species are able to establish themselves quite successfully in seasonally cold waters.

Survival in constantly cold water requires evolutionary adaptation rather than temporary acclimatization. For successful colonization it would be necessary to adapt to the short period when food is abundant and to develop the capacity to produce antifreeze substances. The exceptionally cold and oxygen-rich water of the Antarctic has produced the conditions in which the 15 species of *ice fish* (family *Channichthyidae*) can survive without haemoglobin, making them unique among the vertebrates. Haemoglobin and red blood cells are important for most animals for carrying oxygen around bodies. However, oxygen is very much more soluble in the cold Antarctic waters than elsewhere. Also, Antarctic fish generally have very low metabolisms, cutting down the need for high oxygen content to be carried in the blood. The two factors combined have resulted in many Antarctic fish having significantly reduced red blood cell counts and haemoglobin levels compared to fish from warmer waters. The Antarctic ice fish have no haemoglobin in their blood at all and appear a rather ghostly white. They also have a much greater volume of blood compared to similar-sized fish from elsewhere, thereby enabling them to them to transport oxygen without having to produce specialized blood pigments.

The freezing point of the blood of many fish is about 1°C higher than the freezing point of the surrounding seawater, i.e. without modification fish blood would freeze at around –0.8°C whereas seawater freezes at –1.8°C. Most Antarctic fish are able to produce antifreezes that lower the freezing point of the blood to values close to the freezing point of seawater. Various classes of macromolecular antifreezes have been found in Polar fish, and the content of antifreeze molecules varies greatly between species and depends upon the fish's habitat. Species living close to sea ice have the highest amounts. *Pagothenia borchgrevinki* produces eight different types of these antifreeze glycoproteins (AFGPs). The largest of the Antarctic fish, the giant Antarctic toothfish (*Dissostichus mawsoni*) produces large amounts of AFGPs and specimens can reach up to 140 kg and can live for 45 years. It is one of the best-studied fish for the production of AFGPs, but because its heart beats on once every 6 s, it is even used in medical studies to see how hearts behave during certain cardiac treatments. The Arctic cryopelagic gadiforms *Boreogadus saida* and *Arctogadus glacialis* have glycoproteins similar in structure to the Antarctic ones. In Antarctic fish species AFGPs are synthesized year-round, wheras in the Arctic species the synthesis of antifreeze molecules very often occurs only during winter (Cheng and DeVries 1991, DeVries 1997, Wöhrmann 1997).

Another feature of Antarctic fish is that few species frequent the upper 200 m of the water column. The notothenioides do not possess swim bladders and about 50% of Antarctic species are found in benthic habitats. They are only secondarily pelagic. Some species, including the Antarctic cod, *Notothenia rossii*, that was extremely overfished in the 1970s, shifts its habitat during development. Larvae and fingerlings are pelagic, fingerlings and juveniles are benthic in beds of macroalgae, and adults are semipelagic in offshore feeding and spawning grounds. These semi-pelagic fish do not have adaptations for neutral buoyancy.

Antarctic silverfish, an abundant ecologically important notothenioid with a circumantarctic distribution, is a truly pelagic species, shoaling, feeding, and spawning in open water down to 900 m. It achieves neutral buoyancy by reduction of skeletal material and accumulation of fat. Antarctic silverfish is the major fish species both in number and biomass in most shelf areas of the Antarctic Continent. It feeds on copepods, euphausiids, and chaetognaths and is itself preyed on by most of the larger carnivores. The other abundant pelagic fishes in the Southern Ocean are the Myctophidae, lantern fishes, so-called because of their luminescence. They are small fish, migrating vertically from depths between 200 and 1000 m in the day to the surface at night. They are opportunistic feeders on crustacea and larval fishes. It has been suggested that the biomass of these mesopelagic fish is said to exceed that of krill, benthic fish, birds, seals, and whales, and they are undoubtedly an important component of the food web (Knox 2006).

6.8 The polar marine pelagic systems

Pelagic ecosystems do not have a sharp physical definition. In the Arctic, the boundaries between polar and cold temperate waters are confused and although the Southern Ocean is demarcated by the definite line of the

Polar Front, birds, mid-water fish, and whales cross this frontier without hindrance. There are exchanges between the pelagic, sea ice, benthic, and terrestrial systems. Sinking phytoplankton, dead organisms, faecal pellets, and other detritus are usually the main source of food for the benthos (Schnack-Schiel and Isla 2005). This bentho-pelagic coupling (see Chapter 8) is specifically close on some Arctic shelves, allowing for the large abundance of benthic-feeding mammals like walrus or gray whales in, for example, the Chukchi Sea. Regeneration of mineral nutrients takes place on and in the sea bottom and they are returned to the water column above given suitable conditions, with the Bering Sea/Chukchi Sea complex as prime example (Codispoti *et al.* 2005). This inter-meshing of pelagic and benthic ecosystems is in general more extensive in the Arctic, where the areas of shallow shelf seas are greater and direct coupling of sinking phytoplankton and benthos is common.

The view of the polar pelagic food webs, as well as from other parts of the world, has considerably changed over the last decades away from the simplest food-chains approach—primary productivity (e.g. diatoms), to secondary producers (e.g. krill), to tertiary producers (e.g. whales)—to a more correct representation of the signifcance of the members of the ultraplankton size fraction (DeLong and Karl 2005). Simple food webs for Arctic and Southern Ocean pelagic ecosystems can be cast in similar form (Figs 6.10). The main groups of animals are mostly the same although the species are different. Both in the Arctic and Antarctic the primary productivity, which provides the input of energy into the food web, is low compared with other parts of the world's oceans. Primary productivity is greatest, and upper trophic level organisms are concentrated, in particular at shelf-break fronts, convergences, islands, marginal ice zones, and polynyas. In both polar regions the upper trophic levels (see Chapter 9) include few species but these are represented by large numbers of individuals that exploit concentrated food sources. The further concentration of these into restricted breeding sites gives an impression of high productivity that is misleading. There are about 4.5 mg of penguins per square metre of Southern Ocean, one-millionth of the concentration in breeding colonies. Estimates of ecological efficiency, the ratio of predator production to prey consumed, are difficult to make because of the paucity of data but tentative values are around 5%, less than the 10% which is usually assumed for temperate marine ecosystems. There is even greater difficulty in constructing mathematical models to describe energy flows as envisaged in Figs 6.3 and 6.10.

Many current ecosystem models (including those for polar seas) still simplify the complexity of pelagic food webs by reducing the number of interactions and trophic levels, restricting themselves frequently to nutrient–phytoplankton–zooplankton (NPZ) interactions within a hydrographical domain. Such targeted models have been successfully used

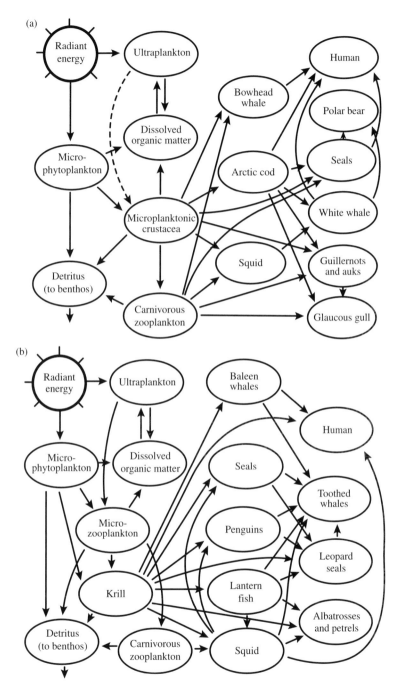

Fig. 6.10 Flows of energy and materials in the open seas of (a) the Arctic; and (b) the Antarctic. Ultraplankon connections are not shown (see Fig. 6.3).

in polar seas to estimate, for example, vertical particle fluxes (Jin *et al.* 2007), impact of iron limitation (Arrigo *et al.* 2003), or the production of DMS (Gabric *et al.* 2005). These models are particularly useful to provide basic estimates for the impact of climate change variations as outlined by the Intergovernmental Panel of Climate Change (IPCC). These reports provide scenarios for various CO_2 emission schemes and proposed substantial warming and loss of sea ice, specifically in the Arctic. How these changes will propagate through the marine food web is difficult to assess. Certainly the loss of summer sea ice, increased light and river run-off will alter the Arctic pelagic food web, likely strengthening the phytoplankton–zooplankton interaction and reducing direct phytoplankton sedimentation to the sea floor, specifically on the Arctic shelves (Bluhm and Gradinger 2007). However, certain limitations are inherent to simplified ecosystem models, as obvious from recent field observations.

Warming and increased glacial run-off along the Antarctic Peninsula induced substantial blooms of cryptophytes in coincidence with mass occurrences of salps, whereas diatoms and *Euphausia superba* were reduced (Moline *et al.* 2004). Similar shifts on decadal scales with linkages to the Pacific Decadal Oscillation index were seen in the Bering Sea with outbursts of coccolithophorids and jellyfish blooms during exceptionally warm years (Schumacher *et al.* 2003). Such sudden changes in ecosystem structure make predictions on the resilience of the ecosystem to changes on the species to community level difficult and they will remain partially not possible due to the lack of information from many Arctic and Antarctic regions also regarding the ecology of most species.

The International Polar Year 2007–2009 aims to fill many of these gaps. For example, the collecting and providing information regarding the distribution and ecology of species is the core task of two Census of Marine Life Projects (the Arctic Ocean Diversity Project, and the Census of Antarctic Marine Life), providing a snapshot about the current conditions. A combination of these mapping efforts with regional ecosystem studies (like the Circumpolar Flaw Lead System Study) will help to monitor, understand, and predict changes in the polar marine ecosystem. The complex intermeshing of organisms and processes which constitute the ecosystem is beginning to be apparent but there are more intricacies to be taken into account if we are to progress towards even an approximate understanding.

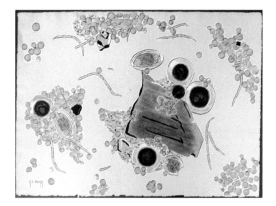

Dedication Watercolour painting by Tony Fogg of red snow algae, green algae, and cyanobacteria from Signy Island in the Antarctic.

Fig. 1.1 Contrasting climates north and south in *summer* situations inside the Polar Circles. (a) Longyearbyen, Svalbard (78°N) in July (photograph by David N. Thomas) and (b) Scott Base (78°S) in January (photograph by Jean-Louis Tison). Note that despite the comparable time of the year that both photographs were taken, the Antarctic landscape is dominated by sea ice cover off the coast whereas the northern seas at these latitudes are ice-free.

Fig. 2.5 The emperor penguin (*Aptenodytes forsteri*) is a supremely well-adapted ectotherm (photograph by David N. Thomas).

Fig. 3.7 A section at right angles to a sandstone surface from a Dry Valley in Victoria Land, Antarctica, showing layered endolithic growth. Courtesy of D. Wynn-Williams and the British Antarctic Survey.

Fig. 3.8 High Arctic vegetation: a mosaic of foliose and fruticose lichens with cushions of moss and flowering plants, Svalbard (photograph by David N. Thomas).

Fig. 3.10 Lush vegetation growth is possible through nutrient enrichment under cliffs used by nesting birds; this example in Krossfjord, Svalbard (photograph by Peter Convey).

Fig. 3.16 Megaherbs dominate native vegetation on Iles Kerguelen (photograph by M.R. Worland).

Fig. 4.1 The wind-swept surfaces of ice sheets and glaciers can appear to be barren wastelands devoid of life (photograph by David N. Thomas).

Fig. 4.5 Sampling a pond on the Markham Ice Shelf in the High Arctic in summer. See Vincent *et al.* (2004) (photograph by Warwick Vincent).

Fig. 5.8 Lake Frxyell, Dry Valleys. Note the perennial ice surface with its undulating surface and load of grit. The polar haven covers the sampling hole in the ice and helps prevent it from re-freezing during the field season. The Canada Glacier that feeds the lake with melt streams in summer is visible in the background.

Fig. 5.11 (a) Drilling the ice surface with a motorized Jiffy Drill prior to sampling on Crooked Lake; note the smooth glassy surface of the annual ice-cover. (b) Close up of the ice surface in early summer; note the transparency of the ice: the auger is clearly visible through the ice.

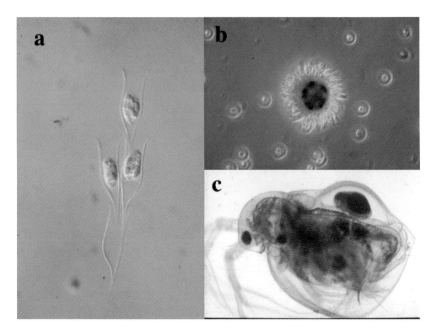

Fig. 5.2 (a) *Dinobyron* from a High Arctic Canadian lake; each lorica contains a single flagellate cell (courtesy of Warwick Vincent). (b) *Mesodinium rubrum*, a ciliate from Highway Lake in the Vestfold Hills; note the darkly stained region inside the cell: this is the endosymbiotic cryptophyte. (c) *Daphniopsis studeri*, a cladoceran from the freshwater and slightly brackish lakes of the Vestfold Hills; note the single juvenile in the dorsal brood pouch.

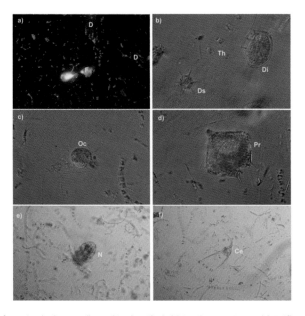

Fig. 6.2 Images of Arctic plankton collected in the Chukchi Sea in May 2004. (a) Epifluorescence image of SAPI (4,6 diamidino-2-phenylindole; a DNA stain)-stained bacteria (abundant small particles), diatoms (D), and phototrophic flagellates (F) (magnification 100x); (b) examples of chain-forming diatom (Th; *Thalassiosira* sp.), dinoflagellate (Di; *Dinophysis* sp.), and silicoflagellate (Ds; *Distephanus speculum*) (magnification 40x); (c) oligotrich ciliate (Oc) (magnification 40x); (d) hetertrophic dinoflagellate (Pr) (magnification 40x); (e) copepod nauplius (N) (magnification 10x); (f) phototrophic dinoflagellate (*Ceratium* sp.; Ce) (magnification 10x). Note the high abundance of chain-forming diatoms in e and f.

Fig. 7.5 A slab of artificial sea ice formed in an ice tank, held upside down to reveal the skeletal layer of crystals (photograph by Jean-Louis Tison).

Fig. 7.10 Microscope image of a brine channel system with pinnate diatoms contained within (photograph by Christopher Krembs).

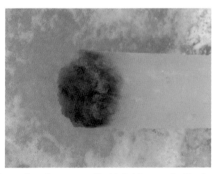

Fig. 7.11 Bottom of an ice core with high biomass in the lowermost 10cm (photograph by David N. Thomas).

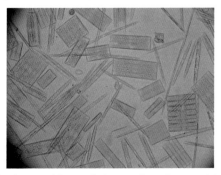

Fig. 7.14 Mixed assemblage of bottom-ice diatoms (photograph by Jacqueline Stefels).

Fig. 8.16 Seabed photographs of Antarctic and Arctic benthic communities. Assemblages dominated by suspension feeders, gorgonians (a) and hexactinellid sponges (b) from the shelf of the southern Weddell Sea, Antarctic at 250m depth. Assemblages dominated by suspension feeders from the waters off north-west Spitsbergen, Arctic, at 100m depth (c) and from the north-western Barents Sea, Arcitc, at 80m depth (d). Photographs by (a, b) Julian Gutt, Alfred Wegener Institut for Polar and Marine Research, Bremerhaven and (c, d) Dieter Piepenburg, Institute for Polar Ecology, University of Kiel.

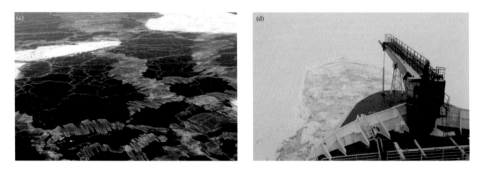

Fig. 7.2 Various stages of sea ice formation. (c) Nilas ice, and (d) closed ice requiring an icebreaker to ram the ice to break it (photographs by David N. Thomas).

Fig. 9.6 (a) King, and (b) Emperor penguins (photographs by David N. Thomas).

Fig. 9.8 A young Adélie penguin shedding juvenile plumage (photograph by David N. Thomas).

Fig. 9.10 (a) Antarctic fur seal and (b) Southern elephant seal on South Georgia (photographs by David N. Thomas).

Fig. 11.1 Whale catchers at Leith Harbour, South Georgia, at the start of the whaling season in 1957 (photograph by W.N. Bonner).

Fig. 11.3 Boreal forest in Alaska where melting permafrost is undermining the stability of many trees (photograph by David W.H. Walton).

Fig. 11.4 Community of Barentsberg on Svalbard (approximately 500 inhabitants) that was established mainly because of coal mining (photograph by David N. Thomas).

Fig. 1.10 Aerial view of Lake Fryxell in the McMurdo Dry Valleys, Antarctica (photograph by Dale Anderson).

Fig. 9.14 Polar bears are dependent on sea ice for hunting their prey (photograph by Finlo Cottier).

7 Frozen oceans in polar regions

7.1 Introduction

The advance and retreat of sea ice imposes a corresponding pattern on the entire polar marine ecosystem through its effects on water temperature, penetration of light, and stability of the water column. Rates of sea ice growth and decay are equivalent to movements of the sea ice edge at 1.6 and $0.9\,km\,h^{-1}$ respectively and with the exception perhaps of tides over mudflats, the passage from day to night, or the spread of forest fires, it is difficult to conceive of a faster-changing biological environment (Brierley and Thomas 2002). The seasonal formation, consolidation, and subsequent melt of vast parts of the Arctic and Southern Oceans have profound effects on climate and large-scale oceanography, as well as obvious implications on the organisms that live in the frozen realms. These include biological productivity, cycling of materials and energy, and, for many species, life cycles and behaviour.

The ice edge has attracted attention as one of the ecologically most interesting regions of the polar oceans. Terrestrial ecologists have long been aware of the special nature of transitions between two or more diverse communities, such as forest or grassland: so-called *tension belts* or *ecotones*. These communities usually contain many of the organisms from the neighbouring ones as well as those characteristic of, and sometimes restricted to, the ecotone. Ecotones commonly have greater species diversity and greater biomass density than do the flanking communities, evidently because the variety of niches is greater and organisms may be able to draw on the resources of the adjacent environments. The ice edge is an ecotone and, like the terrestrial examples, shows increased species diversity and greater productivity. The physical and chemical mechanisms underlying this so-called edge effect are, of course, different and, unlike ecotones on land, the ice edge has no fixed location.

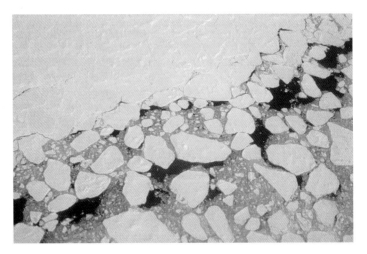

Fig. 7.1   Antarctic pack ice floes breaking up at the ice edge (photograph by David N. Thomas).

Sea ice has a unique ecological role: it interposes a solid interface between two fluid phases, the biologically productive sea water and air. It modifies environmental conditions in the sea below and also provides a platform on which air-breathing birds and mammals can live, breed, and base foraging forays into the water. Remarkably, it also forms a temporary habitat in which a diverse biology—recruited from the plankton—are able to thrive. Although its annual cycle of advance and retreat is a recurrent pattern, the distribution and local structure is irregular. Its variations from year to year affect all levels in the marine ecosystem.

Living organisms in sea ice were first described by the protozoologist C.C. Ehrenberg (1841, 1853) as part of his observations on Arctic pack ice and the underlying waters (Fig. 7.1). This was followed by the descriptions of sea ice diatoms from Antarctic sea ice by J.D. Hooker (1847). Even the great polar explorer Fridtjof Nansen (1897) expressed excitement when observing psychrophiles within sea ice, as his ship the *Fram* drifted through the Arctic Ocean: '…and these are unicellular pieces of slime that live by the million in pools on very nearly every ice-floe all over this endless sea of ice, which we like to call a place of death! Mother earth has a strange ability to produce life everywhere. Even this ice is fertile ground to her'.

7.2 The physical characteristics of sea ice

The formation of sea ice begins with *frazil ice*, consisting of ice crystals, minute platelets or needles at first but growing to as much as 2–15 cm across and 2 mm thick (Eicken, in Thomas and Dieckmann 2003). Continued freezing and clumping of crystals produces a slushy mixture called *grease ice*. Under calm conditions the frazil crystals freeze together to form a solid continuous cover called *Nilas ice*, between 1 and 10 cm thick. Usually, however, wind-mixing prevents this and frazil is driven downwind to accumulate in thicknesses of up to 1 m in the convergences between wave circulations. Sustained wave action moulds frazil into *pancake ice*, circular masses up to 3 m in diameter, of half-consolidated crystals, which come to have upturned rims through constant bumping against each other. Eventually the pancakes raft together to form a continuous sheet (Fig. 7.2).

As temperatures decrease, both consolidated pancake ice and Nilas ice thickens, not necessarily by the accumulation of more frazil ice crystals,

(a)

(b)

(*Continued*)

(c)

(d)

Fig. 7.2   Various stages of sea ice formation. (a) Grease ice slicks, (b) pancake ice, (c) Nilas ice, and (d) closed ice requiring an icebreaker to ram the ice to break it (photographs by David N. Thomas) (see colour plate).

but also by the quiescent growth of *columnar ice*. This ice is made up from vertically elongated crystals that can reach diameters of several centimetres and lengths of tens of centimetres, and these grow to add layer upon layer of ice on the underside of the side of the frazil surface ice layer. The growth of columnar ice is very much slower than the growth of frazil ice, and is greatest in less-turbulent waters. The proportion of frazil ice to columnar ice in any one ice field is obviously heavily dependent on the turbulence of the waters in which it was formed: the more turbulent the waters, the higher will be the proportion of frazil ice. These differences are so stark, that for example between 60 and 80% of much of the Arctic pack ice is made up from columnar ice. In contrast in the more turbulent Southern Ocean, frazil ice forms between 60 and 80% of the total ice.

Ice formed on open waters moves on surface water currents and as such its distribution is largely determined by the prevailing winds. Although a 'skin' of ice on the surface of the ocean effectively dampens wave activity, it is by no means a static zone. Constant motion caused by wind, currents, and wave action breaks up ice and alters its morphology. The enormous pressures that can be generated by wind blowing across an ice field produce ridges, hummocks, and rafting, in which floes override each other. *Pack ice*—the term denotes any area of floating ice that is not *fast* to land—may be *open*, with plenty of water visible, or *close*, with channels or *leads* between large areas of continuous ice (Fig. 7.3). Leads are ten to hundreds of metres wide and may be several kilometres long. However, wind and ocean currents do not just pull ice floes apart, they can also cause them to collide. This results in the converging edges of the ice masses breaking into boulders of ice as they collide, and the rubble collects in ridges of ice called *pressure ridges*. Tremendous volumes of ice, up to thousands of tons, can collide along transects that can extend for many tens to hundreds of kilometres. These jumbled up blocks of ice can in extreme cases extend down (keel) to 50 m under the water and tens of metres into the air (sail), although these dimensions are extreme. Such substantial sails and keels clearly are important in determining the effects of currents and wind on the subsequent transport of that ice field.

So, both convergent (pressure-ridge formation) and divergent (lead formation) processes result in pack ice having a non-uniform thickness (Fig. 7.3). In fact, because 30–80% of the volume of an ice field may be contained within pressure ridges, the dynamics of that ice field caused by wind and ocean currents can be more important to the amount ice produced than the simple freezing of water by thermodynamic processes.

In coastal regions a substantial proportion of the sea ice is actually attached to floating ice shelves and/or land. This is *land-fast ice* and is often characterized by vast areas of level ice of a relatively uniform thickness (Fig. 7.4). This is because since initial formation it has remained intact and because the ice is actually anchored to the land the processes causing deformation are not able to act on the ice. These sheets of sea ice tend to be important areas where penguin colonies are formed and seals haul-out, since they can exist over many years and so provide a stable habitat.

Dense clouds of *ice platelets* are a common feature under land-fast ice in many places around the Antarctic continent. Ice platelets are discs (Fig 7.4b) of ice formed when sea water flowing underneath floating ice shelves is supercooled. Therefore the platelets form and grow in depths of more than 200 m, and dense accumulations of platelet ice have been found at

(a)

(b)

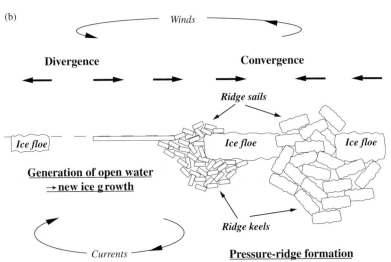

Fig. 7.3   (a) Pressure ridges can throw up huge slabs of ice (photograph by David N. Thomas). (b) Schematic of divergent and convergent processes acting within pack ice zones (image by Christian Haas).

depths of 250 m. The ice platelets that can be up to 15 cm in diameter and 3 mm thick, then rise up thorough the water towards the surface. These loose platelets can accumulate under overlying sea ice, and trap sea water between the plates. These accumulations can be up to 20 m thick, although generally the platelet layers are 1–5 m thick. With time a proportion of platelets do become frozen into the underside of overlying ice.

(a)

(b)

Fig. 7.4   Areas of Antarctic land-fast ice like the ice covering this inlet in this ice shelf (a) can have large amounts of platelet ice (b) underneath it (photographs by David N. Thomas).

Anchor ice is another form of plate-like ice that forms at water depths between 0 and 30 m. This ice grows attached to an object that is not frozen itself, and sheets of the ice fasten to submerged objects such as rocks, gravel, and even animals. It is a major physical disturbance in shallow benthic polar regions, and can entrap large benthic organisms including fish, as well as seaweeds, rocks, and sediments. Anchor ice is buoyant when detached from its attachment site and it rises from the seafloor or structure on which it formed, even carrying with it entrapped organisms to the surface. Masses of seaweeds and animals weighing up to 25 kg have been observed under (and even incorporated into) overlying sea ice, evidently carried there by the anchor ice.

As soon as an ice layer is formed on the surface of the water there is an immediate temperature gradient set up across the ice layer. Thickening of the ice, or rather growth of the ice, basically depends on the temperature of the air at the upper ice surface and the temperature of the water at the underside of the ice. Ice will only continue to grow if the seawater below the ice is at or below the freezing point of sea water (–1.9°C). Ice growth takes place at this ice–water interface and the growing ice crystals form a region called the *skeletal ice layer*. This is extremely fragile, and hard to sample intact from normal ice coring, and realistically can only be sampled undamaged by divers. Generally these growing edges are comprised of blades or plates of ice aligned vertically from the underside of the overlying ice floe. As more ice is added to the lowermost growing edges of the blades (Fig. 7.5) higher up they consolidate by the formation of cross walls and linkages to form a more solid structure.

Theoretically ice could continue growing, which is clearly not the case. The combination of the heat of the ocean, snow cover, and air temperatures all interact to limit Antarctic sea-ice growth by ice-crystal growth alone to around 1 m and approximately 2 m in the Arctic. Deformation processes play a role in altering the thickness, since in both the Antarctic and Arctic ice floes in excess of 10 m are encountered.

The salts in sea water cannot be incorporated into ice crystals, so that when ice crystals grow the salt contained in the water from which the crystal forms is expelled. Therefore in the growing skeletal ice layer the blades of ice are separated by grooves filled with collected brines expelled from the ice crystals. The brines are either released into the waters below, or trapped into inclusions in the ice as the ice consolidates. The input of cold

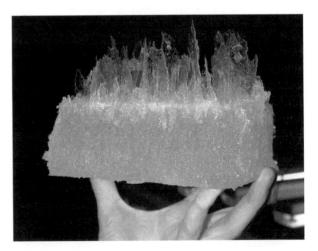

Fig. 7.5   A slab of artificial sea ice formed in an ice tank, held upside down to reveal the skeletal layer of crystals (photograph by Jean-Louis Tison) (see colour plate).

salty water is fundamental to the production of cold heavy water masses that sink to the bottom of the ocean and help to drive global ocean circulation patterns (Chapter 1).

A complex interaction between temperature, salinity, ice crystal growth pattern, and the distribution of the concentrated brines produced all interact in a complex way to give the resulting block of ice a specific microstructure. In a rather crude way, sea ice can be considered to be rather like that of a sponge or Swiss cheese, an intricate solid matrix permeated with a labyrinth of channels and pores that contain the highly concentrated brines. The brine channels vary in size from a few micrometres through to several millimetres in diameter, and they connect the brine inclusions in pores formed when the ice was forming. The structure of channels in either type of ice may be visualized by making casts using water-soluble resin which can be polymerized by ultraviolet (UV) irradiation at −12°C (Weissenberger *et al.* 1992 and Fig. 7.6).

The volume of ice occupied by the brine channels is directly proportional to the temperature of the ice, as is the brine concentration in the channels (Fig. 7.6; Eicken *et al.* 2000). At −6°C the brine salinity is 100, at −10°C it is 145, and at −21°C it is 216. The brines are not static and gravity drainage results in a gradual desalination of sea ice as it ages and brines are expelled into the underlying waters.

The temperature of the upper surface of an ice floe is determined by the air temperature and the extent of insulating snow cover. In contrast the temperature at the underside of an ice floe will be at or close to the freezing point of the underlying sea water. This results in gradients of temperature, brine salinity, and volume of brine channels and pores throughout an ice floe. During autumn and winter the ice is generally colder, brine salinities higher, and brine volume lower in surface ice, compared with underlying ice (Fig. 7.7). Naturally as ice begins to warm and melt in spring and early summer these gradients break down (Eicken 1992).

With daily variations in surface air temperature, the concentration and distribution of the trapped brine alter, often abruptly. Warm conditions promote the coalescence of originally unconnected channels so that eventually the brine can drain away. When large pockets of brine empty an ice stalactite may be produced. In addition to formation of new ice on its underside, the sheet is added to by snow accumulating on top and by freezing of any interstitial melt water or seawater. Snow ice is coarser-grained and bubbly compared with that formed from frazil (Haas *et al.* 2001).

Sea ice is generally less transparent to radiation than clear lake ice. Peak transmission is in the blue-green part of the visible spectrum, around 500 nm, but this is shifted if algae are present towards the yellow-orange, around 600 nm. Both the intensity and quality of irradiance varies from

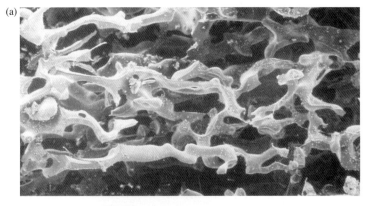

	Ice temperature	Brine salinity
	–6°C	100
	–10°C	145
	–21°C	216

5 mm

Fig. 7.6   (a) Cast of brine channels (see Weissenberger *et al.* 1992). (b) This series of images of the same piece of sea ice shows how the pore space and size reduces with decreasing temperature, with a corresponding increase in salinity of brines contained within the pores. Adapted from Eicken *et al.* (2000).

place to place in relation to patchiness in the ice, its surface configuration, and proximity to open water channels (Eicken, in Thomas and Dieckmann 2003). Whereas the temperature at the bottom of the ice is near the freezing point of sea water, –1.9°C, that near the surface is approximately that of the mean air temperature and may be as low as –20°C (Figs 7.6 and 7.7).

It is surprising that although Arctic and Antarctic sea ices are subject to the same physical laws in their formation and ablation, there are significant differences between them. These arise because the Arctic Ocean is nearly landlocked whereas Antarctic ice extends over an open sea on which cold katabatic winds, blowing off the continent, can rapidly thicken inshore ice but at the seaward ice margin warmer winds, warmer water, and wave

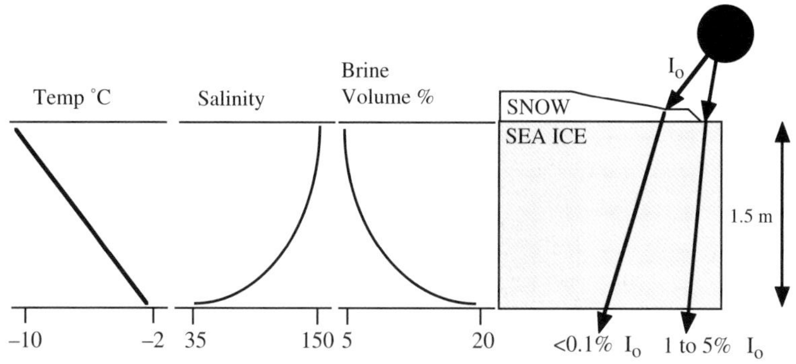

Fig. 7.7   Gradients of temperature, salinity, and brine volume are established across an ice floe. The underside is always at the freezing point of sea water, −1.9°C, and the top of the ice close to air temperature, although this is largely dependent on snow cover. The illustration shows how snow cover can significantly reduce the amount of incident irradiance (I_o). Adapted from Eicken (1992).

action work together to break the ice into small floes. The unconstrained nature of the Antarctic ice edge allows wave action to extend further into the pack than it does in the Arctic. Antarctic sea ice has few melt pools, evidently because the drier air and higher wind speeds increase heat losses to the atmosphere so that surface air temperatures must be well above zero for melting to begin. In contrast, Arctic ice melts at the surface in summer to the extent that 60% of the surface may be covered by pools.

Once a melt pond has established on the surface of an ice floe, because of its lower albedo than the surrounding ice, it absorbs more energy and therefore enhances further melting (Fig. 7.8). In this fashion melt ponds grow both in area and also deepen. Because these ponds are largely derived from snow and low-salinity surface ice, they really are akin to having lakes floating on the surface of the ocean. Eventually all the ice underlying the pond melts away and there is a direct connection between the ocean and the fresh water contained within the pond. This can either mix with the surrounding/ underlying sea water, or form a freshwater layer at the surface of the ocean. In the Arctic, sometimes these freshwater layers collect as lenses, or pools, of freshwater under large sea-ice floes. These can become refrozen into the ice floe in the following autumn, forming a distinctive layer in the ice.

Not all ice melting takes place by absorption of solar radiation, but from below by warming by the underlying sea water. As the ice warms it can take on a rather fragile appearance, melting first in brine channels containing high-salinity brines. At the extreme case the ice becomes so porous that only an extremely fragile skeleton of ice remains.

Melt ponds are not necessarily started due to melting by absorption of solar radiation alone. Many melt ponds are initiated by the flooding of ice-floe surfaces by the surrounding sea water. These floods either wash

(a)

(b)

Fig. 7.8 (a) Melt pond and pressure ridge; (b) melt pond on level ice (photographs by David N. Thomas).

the snow away or moisten the snow so that the surface albedo is reduced. Flood-induced surface ponds are also often associated with uneven ice surfaces on floes that cause flood water to collect into ponds.

A common feature of Antarctic sea ice in austral summer is the formation of rotten surface ice layers, between the snow–ice interface or in the top 50 cm of the ice floes. These are given a number of names—such as infiltration layers or gap layers—and can be continuous gaps or voids filled with rotten ice slush, frequently mixed with sea water that percolates into the floe interior through these gap layers. These gap layers are frequently associated with the flooding and refreezing of snow layers to form snow ice, and differential melting properties between the hard freshwater snow

ice and more saline sea ice below results in these layers forming (Haas
et al. 2001).

A described above there are several notable differences between Antarctic
and Arctic sea ice that have large implications for the biology it contains.
These are summarized in Table 7.1. Just as physical conditions differ,
sea-ice habitats in the Arctic and Antarctic are similar but not identical. Surface melt pools are common in the Arctic but infrequent in the
Antarctic. The snow ice or infiltration ice-layer assemblage (Fig. 7.9) has

Table 7.1 Examples of major differences between Arctic and Antarctic sea ice.

Feature	Arctic	Antarctic
Maximum extent	$15.7 \times 10^6\,km^2$ (February–March)	$18.8 \times 10^6\,km^2$ (September)
Minimum extent	$9.3 \times 10^6\,km^2$ (September)	$3.6 \times 10^6\,km^2$ (February)
Mean thickness	1996: 3 m 1976: 5 m	0.5–0.6 m
Annual ice cover (first-year ice)	$7 \times 10^6\,km^2$, <50%	$15.5 \times 10^6\,km^2$, >80%
Latitudinal range	0–44°N	55–75°S
Multiyear ice	$9 \times 10^6\,km^2$	$3.5 \times 10^6\,km^2$
Extent of land-fast ice	Not known	5% ($0.8 \times 10^6\,km^2$)
Annual average heat flux	$2\,W\,m^{-2}$	$5-30\,W\,m^{-2}$
Platelet ice	Common under fast ice	Sporadic, fresh water under floes
Polynyas	Coastal	Large, open ocean
Melt ponds	Significant feature	Insignificant feature
Flooding	Not extensive	Extensive
Sea-ice residence time	5–7 years	1–2 years
Drift velocity	$2\,km\,day^{-1}$	$>20\,km\,day^{-1}$
Texture	5–20% frazil	50–60% frazil
Salinity	Generally low	Higher
Pollution	Considerable riverine and aeolian	Insignificant aeolian
Sediment inclusion	Considerable	Insignificant
Top predators	Polar bear, polar fox	Leopard seal
Seals	Walrus and ringed, harp, bearded, and hooded seal	Weddell seal, crabeater Seal, Ross seal, fur seal
Flightless birds	None (except during moulting)	Emperor penguin, Adélie penguin,
Fish associated with sea ice	Arctic cod (*Boreogadus saida*)	Broadhead fish (*Pagothenia borchgrevincki*)
Central crustaceans associated with sea ice	Amphipods	Krill (*Euphausia superba*)
Foraminifers	The planktonic foraminifer *Neogloboquadrina pachyderma*	The planktonic foraminifer *Neogloboquadrina pachyderma*
Nematodes	Several species, common	One record

From Dieckmann and Hellmer, in Thomas and Dieckmann (2003).

Fig. 7.9   Sub-surface gap layer supporting dense biological growth (photograph by Christian Haas).

only been reported from the Antarctic. Communities in the ice are common in the Antarctic but less well developed in the Arctic. The increase in thickness of Antarctic ice by accumulation of frazil ice rising beneath it tends to incorporate plankton whereas the extension of stable Arctic pack ice downwards by growth of congelation ice does not. The biology reported from the sea ice of the two polar regions has generic resemblances but the species are not the same. However, Arctic and Antarctic can here be considered together by concentrating on general physiological aspects and ecological interrelations (Horner 1985, Palmisano and Garrison, in Friedmann 1993, Brierly and Thomas 2002, Thomas and Dieckmann 2002, Lizotte, in Thomas and Dieckmann 2003).

7.3 The biology of sea ice

Heterotrophic bacteria and unicellular algae represent the two major groups in sea ice assemblages that have been best studied to date. However, there is a wide diversity of prokaryotes and eukaryotes, and much of the attention of sea ice studies over the past 50 years has focused on identifying the diversity of organisms that are able to survive the transition from the open water to the semi-solid ice matrix (Table 7.2; Lizotte, in Thomas and Dieckmann 2003). The interest in these organisms is of cause based on a fundamental desire to understand how organisms can both survive and thrive in extremes of temperature, salinity, and low light. However, we still only have a limited understanding of the biochemical and physiological mechanisms by which these organisms survive, and even less about the molecular controls of these.

Table 7.2 Microbiology reported from sea ice.

Eukaryotes (single celled)	Diatoms
	Dinoflagellates
	Chrysophytes
	Prasinophytes
	Silicoflagellates
	Prymnesiophytes (including coccolithophorids)
	Cryptophytes
	Chlorophytes
	Euglenophytes
	Kinetoplastids
	Choanoflagellates
	Ciliates
	Heliozoans
	Foraminiferans
	Amoebae
	Chytrids
	incertae sedis (unknown affinity)
Eubacteria	Proteobacter group (alpha, beta and gamma subgroups)
	Flexibacter-Bacterioides-Cytophaga group
	Gram-positive bacteria (groups with high and low GC contents in their bacteria)
	Cyanobacteria
Archaea	Crenarchaeota
	Euryarchaeota

Based on detailed lists in Bowman *et al.* (1997), Horner (1985), Garrison (1991), Ikävalko and Gradinger (1997), Ikävalko and Thomsen (1997), Maranger *et al.* (1994), and Staley and Gosink (1999). From Lizotte, in Thomas and Dieckmann (2003).

The size range of ice-associated biology ranges from the micrometre (viruses, archaea, bacteria, and algae) to the centimetre (amphipods, krill, and even some fish) range. Ultimately it is the physical and chemical constraints of the ice that can determine whether or not an organism can live within the ice matrix, and the most important of these is obviously space (Fig. 7.10). Brine channel walls constitute large surface areas that can be colonized by algae and bacteria and used as sites for attachment, locomotion, and grazing. Krembs *et al.* (2000) quantified the ice surfaces of interconnected brine channels, and found that the total surface area of internal brine channels ranged from 0.6 to 4.0 $m^{-2}kg$ of ice^{-1}. They estimated that between 6 and 41% of the brine network surface area at $-2°C$ may be covered by microorganisms.

In contrast, the most dynamic part of sea ice in terms of changing microstructures is the interface between ice and the surrounding sea water, often referred to as the ice–water interface. In this region of the ice space is not limiting, and the highest biomass of sea-ice biology are recorded in these bottom assemblages (Fig. 7.11).

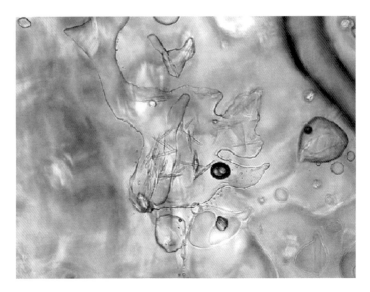

Fig. 7.10   Microscope image of a brine channel system with pennate diatoms contained within (photograph by Christopher Krembs) (see colour plate).

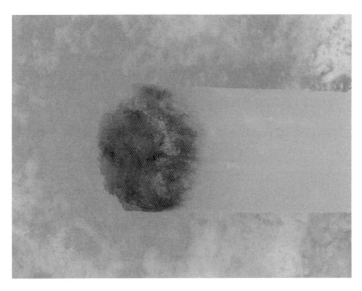

Fig. 7.11   Bottom of a sea ice core with high biomass in the lowermost 10 cm (photograph by David N. Thomas) (see colour plate).

Apart from the logistical difficulties of study *in situ* of sea ice communities there are profound problems in sampling. One arises from the large heterogeneity of the ice: the biology contained within an ice core can vary considerably from that in an ice core taken less than a metre away. Another problem is that much of the sea ice biota lives in the brine phase and when a sample is melted for examination the organisms are subjected to severe osmotic shock. In fact, if samples which are treated in this way are compared with pore-water samples collected without allowing the ice to melt, there can be a considerable difference between the two. Dinoflagellates, other autotrophic and heterotrophic flagellates, and ciliates are particularly susceptible to osmotic shock, with losses amounting to as much as 70%, whereas diatoms are little affected. The effect may be minimized by allowing samples to melt in larger volumes of sea water to buffer salinity and osmotic changes (Garrison and Buck 1986).

A schematic representation of the structure of sea ice in relation to the biota it supports is given in Fig. 7.12. There are three principal assemblages associated with sea ice, one within the ice itself, one on and in the underside

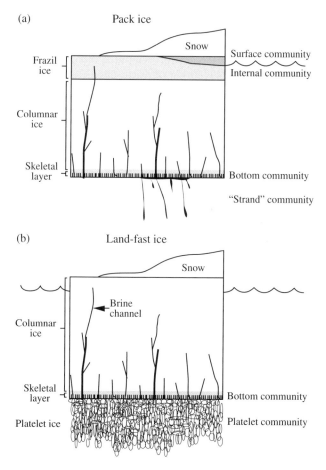

Fig. 7.12   Distribution of sea ice assemblages in (a) pack ice and (b) land-fast ice. Based on Arrigo and Thomas (2004).

of the ice, and thirdly that comprising the warm-blooded animals that live on its surface. The assemblages found in sea ice are often termed *epontic* or *sea-ice microbial communities* (SIMCOs) and comprise microalgae, bacteria, protozoa, and small metazoans. Broadly, the microalgal assemblages, which are the most conspicuous, are classified into surface, interior, and bottom but the scheme in Fig. 7.12 gives a better impression.

7.3.1 Colonization of sea ice

Most ice organisms colonize the ice during formation of new ice and any fauna and flora that cannot swim out of the way are scavenged by rising ice crystals as they form into grease-ice slicks. These loose accumulations of ice crystals are further innoculated by the pumping of water through these effective filters (Weissenberger and Grossmann 1998). However, for some organisms the vectors or mechanisms leading to the ice phase of the organisms' life remain unknown. For coastal regions with shallow water it is not difficult to imagine colonization of the sea ice from the benthos by larval stages, even in species with poor swimming capabilities. Another possible vector in shallow water is lifting of organisms from the benthos attached to anchor ice (Schnack-Schiel *et al.* 1995). Most pack ice formation occurs over water several thousand metres deep, however, and here mechanisms of colonization by organisms remain enigmatic. Ice platelets can be formed at great depths (Dieckmann *et al.* 1986) in large quantities potentially acting as a vector for lifting organisms to overlying waters. This phenomenon will be limited to localized patches, though, and is hardly a widespread process that could explain how organisms colonize sea ice overlying deep waters. The most obvious vector is that residual multiyear ice contains populations of organisms that act as innocula for newly formed ice. Whereas this is fine in the Arctic, in the Antarctic only a very small percentage of the sea ice lasts for more than one season, so this seems unlikely.

Nematodes are apparently not present in Antarctic sea ice (there is only one record), a stark contrast to Arctic sea ice, where free-living species, especially belonging to the superfamily *Monhysteroida*, are found in abundance (Riemann and Sime-Ngando 1997). No rotifers have been found in Antarctic sea ice either, even though these too are also common in Arctic sea ice samples. The reasons for this Arctic/Antarctic difference are unclear, and it is possible that they are simply sampling artefacts: in time, more-comprehensive sampling may produce more complete faunal records for Antarctic sea ice, for example for foraminifers, which are very abundant in Antarctic sea ice, had for many years remained unknown from Arctic sea ice but since the late 1990s have now been found there, albeit only in a few samples.

Antarctic turbellarians spawn in sea ice in austral summer. Eggs, juveniles, and adults will be released into the water column upon ice melt. Although sea ice turbellarian species can swim, none have been reported in the plankton and it is presumed that they sink to the sea floor. It has been suggested that sea ice turbellarians may have an adhesive disc by which they attach to crustaceans before being released from the ice. Swimming crustaceans, including amphipods that migrate from the sea floor to the ice peripheries, or the common ice copepods, may act as vectors to transfer flatworms to different ice floes (Janssen and Gradinger 1999).

Many ciliate species have been described from sea ice, with no equivalent planktonic form being described (Petz *et al.* 1995). It is speculated that for many of these species colonization takes place via resting spores, although there is no direct evidence for this. There is a similar conundrum for the Arctic nematodes, and even whale baleen plates that contain thriving populations of nematodes have been cited as possible vectors for bringing nematodes into close contact with sea-ice floes.

7.3.2 Viruses and bacteria in sea ice

Heterotrophic bacteria are the main group of prokaryotes in sea ice, and most sea-ice bacterial strains have been found to be cold-adapted and halotolerant (Brown and Bowman 2001, Deming 2002, Junge *et al.* 2002). Archaea have also been found in both Arctic and Antarctic sea ice (DeLong 1998, Junge *et al.* 2004). Cyanobacteria have been recorded in Arctic sea ice (Laurion *et al.* 1995, Gradinger and Ikävalko 1998), but are mostly associated with surface melt features such as melt ponds, and are more likely to be found in coastal regions influenced by freshwater run-off, and for this reason they are not found in sea ice in the Southern Ocean.

The mechanisms described above by which many organisms are incorporated into sea ice do not apply to bacteria cells. Instead it appears that bacteria are incorporated into new sea ice primarily by adhering to other organisms, especially algae (Fig. 7.13) and algal aggregates (Grossmann 1994, Grossmann and Dieckmann 1994). Some epiphytic bacteria have modifications enabling them to attach to diatom surfaces, and bacteria have even been shown to penetrate diatom hosts. Another strategy invoked for the incorporation of bacteria into new sea ice is ice nucleation: here the surfaces of bacteria themselves act as catalysts promoting the formation of ice crystals. A number of bacterial strains, some of which are abundant in sea ice, have strong ice-nucleating capabilities (Nichols *et al.* 1995). Some other sea-ice bacteria have been shown to produce gas vacuoles within their cells. Although their function is still unclear it has been hypothesized that it is a mechanism enabling bacteria to rise in the water column bringing them into contact with sea ice algae or even causing them to rise into the ice itself.

Studies have shown that larger bacteria may be incorporated preferentially into sea ice. Whether or not this is true sea ice certainly contains bacterial

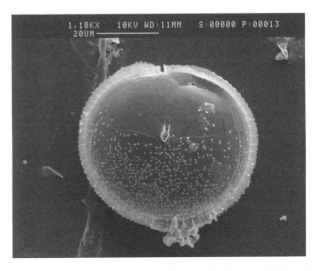

Fig. 7.13   Sea ice dinoflagellate covered with bacteria (photograph by Anna Pienkowski-Furze).

cells that are larger than those in the plankton. These large cell sizes are most likely to result from low bacterial mortality, with individual bacteria being largely protected from organisms grazing on bacteria and consequently enjoying a longer life, even though metazoan and protozoan grazers have a remarkable capacity for moving in small channels (see above).

There is a gradual transition in bacterial community composition from the dominance of psychrotolerant to psychrophilic bacteria as sea ice progresses from grease to pancake ice and through to a consolidated ice sheet. However, it seems that low temperature alone does not account for the selective enrichment of psychrophiles (Helmke and Weyland 1995, Pomeroy and Wiebe 2001). Rather, it has been proposed that a bacterial species' ability to utilize dissolved organic matter and nutrients at low temperatures may play a major role in determining the bacterial composition in the ice (Reay *et al.* 1999, Pomeroy and Wiebe 2001). Nichols *et al.* (1995) consider salinity to be a primary factor controlling bacterial growth and survival in sea-ice brines. They suggest that the frequency, magnitude, and rate of salinity variation may be the selective factor in the control of psychrophilic bacterial populations.

There are indications that some sea ice bacteria are active down to −20°C (Junge *et al.* 2004), and measurements in Arctic sea ice imply that sea ice bacterial assemblages can be more active than those in the water column (Junge *et al.* 2002). This may in part be due to highly concentrated biological assemblages being confined to small spatial habitats (and so closer to nutrient sources), and also the high concentrations of dissolved organic matter that have been reported for sea ice in both Arctic and Antarctic sea ice (Thomas *et al.* 2001), exceeding surface water concentration by factors of up to 500. The nature of this organic matter remains largely uncharacterized, although much of it may be produced in the form of extracellular polymeric substances (EPSs) and mucopolysaccharide gels produced by algae and/or bacteria (Krembs *et al.* 2002, Meiners *et al.* 2004, Mancuso-Nichols *et al.* 2004, 2005). These studies indicate that most EPS production is algal (diatom-)-derived, and Antarctic sea ice can contain large amounts of particulate EPSs, which can be densely colonized by bacteria. Therefore EPSs may increase sea ice bacterial diversity by providing microhabitats for distinct bacterial groups.

Studies of viruses in sea ice habitats are scarce. Viruses are concentrated in sea ice by the same factors as bacteria when compared with the underlying water column, and occur in greatest abundance in those parts of the sea ice where bacteria are most active. Gowing *et al.* (2002) and Gowing (2003) reported large (>110–424 nm capsid diameter) viruses in Antarctic sea ice. Although these viruses are of the size and form to infect a range of algae and protozoans, only a few protozoans and algae were observed to be infected, and no diatoms were infected. Interestingly in the latter study virus numbers were positively correlated with algal biomass in the ice, even though no algae were found to be infected. Psychrophilic viral phage–host systems from Arctic sea ice have been reported where the phages had more pronounced adaptation to cold temperatures than the bacteria (Borriss *et al.* 2003).

7.3.3 Algae in sea ice

Diatoms are the most-studied group of eukaryotes in sea ice, and their photosynthetic pigments cause the brown coloration of sea ice, in particular at the ice–water interface. However, diatoms are not the only eukaryotic protists to be found in sea ice, and others include prymnesiophytes, dinoflagellates, chrysophytes, cilliates, formanifera, and chlorophytes (Lizotte, in Thomas and Dieckmann 2003). There are sea ice habitats where diatoms do not dominate, in particular summer ice-surface assemblages (including melt pools) where *Phaeocystis* species proliferate. Most taxonomic studies on sea-ice samples are typically identified using microscopy following melting of ice-core sections, and groups lacking a robust cell wall, lorica, or frustule were poorly represented in early surveys because of the damage to delicate cells from the hypo-osmotic stress during the melting process. In particular, this will be true for the picoeukaryotes, which we know almost nothing about in sea ice, although they are undoubtedly present in high numbers.

Studies *in situ* and in the laboratory reveal that in sea ice diatoms are able to be physiologically active at low temperatures similar to those observed for bacteria in Arctic wintertime sea ice. Active photosynthesis was observed in autumn Antarctic sea-ice algal communities at −8°C and salinities of approximately 110 (Mock 2002), and laboratory studies with diatom species indicated that the centric sea ice diatom *Chaetoceros* sp. is able to grow at least at −17°C and a salinity of 196 (Plettner 2002).

The ability for diatoms to acclimate to hyperosmotic brine solutions is based on accumulation of free amino acids such as proline and other cryo-protectants such as dimethylsulphoniopropionate (DMSP; see Chapter 6; DiTullio *et al.* 1998, Lee *et al.* 2001). Some diatom species are able to produce more of these cryoprotectants and are therefore able to grow under lower temperatures and higher salinities, in contrast to pelagic diatoms that are less prolific producers of DSMP and proline. Intracellular concentrations of DMSP are also known to be controlled by light, nutrients, and pH (Chapter 6; Malin and Kirst 1997), and may even involved in protection against potentially toxic hydroxyl radicals (Sunda *et al.* 2002) and other reactive oxygen species within brines with high oxygen concentrations. Sea-ice studies have shown that very high concentrations of DMSP can be produced by ice algal assemblages (Trevena *et al.* 2003, Trevena and Jones 2006) reaching concentrations of over 2900 nM (seawater values are typically from 0 to 50 nM). All the environmental factors mentioned above will contribute to sea-ice algae producing high DMSP reserves, although the high brine salinities are most likely the predominant controlling factor.

There is increasing evidence that some sea-ice organisms are able to reduce the formation of ice crystals (Raymond and Knight 2003, Janech *et al.* 2006). Studies with sea ice diatoms reveal that some species exude ice-binding proteins (IBPs), which cause so-called ice-pitting, where the growth of ice crystals is reduced and the ice-crystal shape is altered (Raymond 2000). It has been postulated that the IBPs may prevent freezing injury to membranes by inhibiting the recrystallization of ice, which is a process where large grains of ice grow preferentially over small grains. By having protein-recrystallization inhibitors, such as these IBPs, physical disruption of cell membranes by large crystals may be reduced. It is also speculated that when released extracellularly they will prevent the recrystallization of the surrounding ice.

The algal growing on the underside of ice floes tend to be mainly colonial or chain-forming diatoms (Fig. 7.14). Although the irradiance which reaches them is less than 1% of that at the surface, the standing stock can be high, up to 1000 μg chlorophyll $a\,l^{-1}$ (typical water values are <5 μg chlorophyll $a\,l^{-1}$). These algae are strongly shade-adapted, and light saturation of photosynthesis occurs around 20 μmol photons $m^{-2}\,s^{-1}$, and photoinhibition sometimes sets in at only 25 μmol photons $m^{-2}\,s^{-1}$ (Arrigo, in Thomas and Dieckmann 2003).

A localized, but important, ice habitat under Antarctic fast ice is the platelet layer. Ice platelets are formed in supercooled water flowing from

Fig. 7.14   Mixed assemblage of sea ice diatoms from bottom of a core taken at the same site where the core in Fig. 7.11 was taken (photograph by Jacqueline Stefels) (see colour plate).

under ice shelves and float upwards to accumulate in layers up to 10 m thick, although in general they are about 2 m thick. Algae and bacteria grow well at the interface between this layer and the fast ice above and, although nutrients may become limiting, there is protection from grazers and standing stocks of up to 190 mg chlorophyll a m^{-2} have been recorded. The light reaching the platelet layer may be about 3% of that incident at the surface level but photosynthesis by the algae, mainly diatoms, is saturated at about 10 µmol photons m^{-2} s^{-1}. The highest concentrations of sea ice algae in platelet ice layers are there because of the large surface areas that the platelets provide on which algae can grow and the nutrients within the semi-solid clouds of platelet ice are replaced by periodic pumping of water through the platelet ice by tidal movement of water or water currents (Arrigo *et al.* 1995, Günther and Dieckmann 1999).

One of the major issues of studying these bottom sea ice assemblages is that the skeletal layer of ice is very fragile and often lost during standard coring activity. However, with the introduction of microelectrode systems for measuring photosynthesis and respiration in aquatic systems there have been dramatic advances in being able to set reasonable *in situ* measurements of primary productivity in such layers without disturbing the ice. These electrode arrays can be deployed either by divers, or from rigs that can be controlled from above. One advantage of such systems is that they can be set up to have good light sensors deployed under the ice so that primary production can be linked to actual light fields (McMinn *et al.* 2000, Trenerry *et al.* 2001, Glud *et al.* 2002). But even using divers this has to be done with caution since bubbles from traditional scuba gear can disrupt the fragile underside of the ice. The introduction of rebreathing apparatus avoids these problems.

As in the Antarctic, diatoms predominate in the Arctic ice–water interface communities but the species are largely different (Fig. 7.14). As well as planktonic forms, there are characteristic species, such as *Nitzschia frigida*, which dominates sub ice assemblages in the central Arctic Ocean and also occurs in the Antarctic, and *Melosira arctica*, which forms large mucilaginous masses (Hegseth 1992, Gutt 1995, Melnikov 1997). Diatoms *in situ* are shade-adapted with maximum rates of photosynthesis at about 50 μmol photons m^{-2} s^{-1} but in laboratory culture *N. frigida* adapts and maintains maximum rates up to 400 μmol photons m^{-2} s^{-1}. Other species are not so adaptable. By late May there is light limitation as a result of self-shading. Growth rates, varying between 0.15 and 0.8 divisions per day, are in the same range as found for Antarctic species.

In contrast, melt ponds on the surfaces of summer ice floes are habitats of high light intensity. Melt pools may be exposed or covered by a crust of ice. In the Arctic, where the pools are usually of fresh water, the flora, which includes species of snow algae, is mostly of land origin (Melnikov 1997). In the Antarctic the algae seem mainly derived from the marine plankton, primarily small diatoms and flagellates but often the colonial *Phaeocystis* spp. (Palmisano and Garrison, in Friedmann 1993). Some pools have almost no growth but in others the density may reach over 10^9 cells ml^{-1}. Standing crops of between 4 and 244 mg chlorophyll *a* m^{-2} have been recorded, although much of this growth is actually on the upper surface of the ice forming the bottom of the pool. Microalgae from the surface of Weddell Sea pack ice achieve saturated rates of photosynthesis at higher irradiances, 100–150 μmol photons m^{-2} s^{-1}, than do those within or at the bottom of the ice.

Sea ice algae get little or no light during the winter but some species are able to survive in darkness for up to 84 days. The possibility of a switch to heterotrophy, in particular uptake of exogenous amino acids and glucose in the pool of dissolved organic matter, as a means for winter survival by sea-ice algae has been postulated (Palmisano and Sullivan 1985, Rivkin and Putt 1987), although the concept is still open to much speculation. However, dark survival by algae (up to years) is well documented in both temperate and polar algae, especially in diatom species, and many species have been shown to produce physiological resting cells that morphologically are similar to the vegetative cells, but that are physiologically dormant. Cyst and spore formation is well documented for sea-ice dinoflagellates and chrysophytes (Stoecker *et al.* 1997, 1998, 2000), and physiological resting stages are likely to be used by several algal species, including diatoms, within the ice during adverse winter conditions.

Much of the early work on the ecophysiology of sea ice organisms has revolved around the effects of adaption to low light and light quality in sea ice. However, naturally algal growth is also limited by the supply of dissolved inorganic nutrients. Clearly the rate of re-supply of a nutrient is greatly influenced by whether or not there is possible exchange with the surrounding sea water or not. In biological assemblages at the periphery of ice floes, especially bottom-ice assemblages, nutrient depletion is only seldom reported as being a growth-limiting factor, because exchange with the underlying water is readily possible, and this exchange is promoted by the uneven surfaces of the skeletal ice layer which induces small-scale turbulence in the water as it passes under the ice and therefore promotes exchange of nutrients and gases. This is the reason why bottom-ice assemblages support the highest standing stocks of sea ice algae. However, even just 5–10 cm from the ice–water interface there can be significant nutrient depletion and limitation of algal growth, due to lack of water exchange (McMinn *et al.* 1999). Therefore interior ice assemblages that are effectively cut off from this exchange utilize nutrients rapidly and their growth quickly becomes inhibited once they form.

However, ice floes are often broken and rafted together. Such large-scale physical transformations will at times expose these internal regions of the ice to nutrients in sea water. The deformations may also result in stress fractures and fissures within the ice. This may act as conduits for nutrient-rich water to reach deep within ice floes, and supporting the growth of internal algal assemblages. In surface gap layers and layers close to the ice freeboard, large ice algal assemblages are often found, supported by re-supply of nutrients from the surrounding water (Kattner *et al.* 2004). These layers are often not continuous throughout a floe (Haas *et al.* 2001). In some regions in the ice, such layers can be cut off from nutrient re-supply, resulting in highly localized nutrient depletion, even within a few centimetres from that part of the floe which exchanges fully with the surrounding sea water.

It is not only the diffusion of obvious nutrients that is affected by ice porosity, but of course the exchange of oxygen and inorganic carbon (carbon dioxide and bicarbonate). These are the prerequisite for photosyntheis and respiration activity in the ice and will limit growth and activity if they are not present. Typically the concentrations of these gases in sea ice brines show depletion in inorganic carbon and elevated concentrations of oxygen, a signal resulting from high levels of photosynthetic activity (Gleitz *et al.* 1995). However, physical chemical processes also clearly influence the inorganic carbon dynamics at low temperature and high salinities in brines, including the precipitation of calcium carbonate (Papadimitriou *et al.* 2007).

Diatom frustules compose a large fraction of the material in Antarctic sediments, and the importance in the durability of the diatoms for palaeoenvironmental indicators is discussed by Leventer in Arrigo and Lizotte (1998). Stable carbon isotopic values of diatoms have been proposed as another palaeoenvironmental indicator of sea ice conditions. Several studies have illustrated an enrichment of ^{13}C in sea ice diatoms, and have linked this to carbon dioxide limitation in sea ice assemblages. If photosynthetic organisms enriched in ^{13}C are being produced in sea ice, it is reasonable to predict that material sinking out from sea ice may result in ^{13}C-enriched sediments (Gibson et al. 1999, Kennedy *et al.* 2002). However, it should be noted that often the most prolific growth of algae in sea ice is at the peripheries of the ice floes where of course inorganic carbon will not be limited. This material will therefore not show this stable isotopic signal. Another biomarker thought to be specific to sea ice is the recently described C_{25} monounsaturated hydrocarbon (IP_{25}) which has been found in sea ice diatoms and in sediments in regions covered with sea ice (Belt *et al.* 2007). It is possible if this marker proves to be a robust sea ice indicator that the position of ice edges, at least throughout the Holocene, will be able to be determined.

7.3.4 Grazers in sea ice

Microorganisms in sea ice form complex microbial networks, in which some organims such as the bacteria and diatoms are food for grazing protozoans like the heterotrophic dinoflagellates, ciliates, foraminifers, euglenoids, nanoflagellates, and amoebae. These are the same organisms that together with the algae, bacteria, and archaea are collectively known as the microbial network (Kirchman 2000). These organisms are a major focus of much research in open waters (Chapter 6), but within the ice there is the interesting phenomenum that whereas in the open water interactions between these organisms are probably spread out over a relatively wide spatial scale, in the ice system the biology is greatly confined spatially, and hence the interactions and resulting biogeochemical signals will be greatly amplified. However, the interactions and deciphering what is eating what is a complex task, especially since we do not know many of the organisms that are actually there, and very little about their life cycles and food preferences (Lizotte, in Thomas and Dieckmann 2003).

The grazing protozoans mentioned above are found in numbers many orders of magnitude greater than in the surrounding waters, and feed on bacteria, algae, and other protozoans in the rich biological assemblages. It is thought that the grazing by protozoans is highly significant in controlling the development and even species composition of sea ice algae and bacteria. However, to date this has been a largely neglected area of research. For instance, it is known that heterotrophic nanoflagellates are present in sea ice in high numbers, but unfortunately most studies have neglected to look for them, let alone enumerate them (Ikävalko and Gradinger 1997).

Ultimately the food source of these grazing protists is proportional to the size of the grazer. Heterotrophic nanoflagellates range in size from 2 to

20 µm and generally graze on bacteria, althougth they will ingest small algae. There are a few nanoflagellate species that can photosynthesize as well as ingest bacteria and these are examples of mixotrophic species in that they can switch modes of organic matter accumulation and energy production. In sea ice chanoflagellates (Fig. 7.15) and small dinoflagellates have been shown to reach high biomasses, and these will feed largely on bacteria, and possibly take up dissolved organic matter (Garrison and Close 1993).

Between 20 to 200 µm there is a large diversity of protozoans, although the most studied in sea ice are the ciliates and dinoflagellates. Sea ice ciliates have many forms, but in general they are spherical, oval, or conical filter feeders (Petz *et al.* 1995, Song and Wilbert 2000) with an array of cilia that are used for movement and wafting prey into the body cavity where they are digested. The flagellates, dinoflagellates, and ciliates are all characterized by being highly motile, and are quite able to move through brine channels, down to diameters of their body sizes. Therefore even in brine channels below 5 µm in diameter, if there are nanoflagellates present bacteria within the channels will potentially be grazed upon (Scott *et al.* 2001).

In general, therefore, sea ice ciliates and dinoflagellates are restrained by the size of their food items. Dinoflagellate can generally ingest food items about the same size as themselves, whereas ciliates can only take in particles about 45% of their oral diameter, or about one-tenth of their body volume. The consequences of this are highlighted by a study conducted

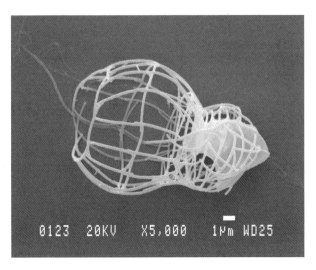

0123 20KV X5,000 1µm WD25

Fig. 7.15   Chanoflagellates are seldom recorded, but are probably very important grazers in sea ice (photograph by Elanor Bell).

within Arctic sea ice by Michel *et al.* (2002). The dominant dinoflagellates were between 20 and 40 µm in diameter, and since the bulk of the sea ice diatoms had lengths around 30 µm they were able to eat them. However, the main ciliate species ranged in size from 30 to 60 µm although most were 30–40 µm. Therefore the sea ice algae at this site could not be ingested by these ciliates, and presumably they were feeding on bacteria associated with the algal assemblages. Some Antarctic sea ice ciliates have been shown to preferentially ingest bacteria, despite the fact that they could ingest much larger organisms (Scott *et al.* 2001).

Another protozoan that is commonly found in Antarctic sea ice is the foraminifer, *Neogloboboquadrina pachyderma*, which is the only species of foraminifer to be reported from sea ice. One reason for this is a high tolerance to increased salinities, and it is the only known species of planktonic foraminifer that can tolerate salinities up to 80. It is striking that although it is commonly reported in Antarctic sea ice in very high numbers, especially in bottom-ice assemblages, it has been reported from Arctic sea ice in just a handful of studies, even though it is found in Arctic waters. One possibly reason for this anomaly is that *N. pachyderma* is not frequently found in waters of low salinity, such as those on the shallow Siberian shelf region where a high proportion of Arctic sea ice is first formed. Numbers of *N. pachyderma* can reach over 1000 individuals l of melted ice^{-1}. Although the foraminfers feed on the sea ice algae, there is not always a good correlation between the distribution of algal biomass and foraminifer numbers. Stronger relationships seem to exist between the type of ice and foraminifer distribution, with much higher numbers found in frazil ice than in congelation ice (Spindler and Dieckmann 1986, Dieckmann *et al.* 1991, Thomas *et al.* 1998).

Sea ice can be inhabited by both pelagic metazoans such as calanoid copepods and rotifers as well as organisms more typical of benthic habitats including turbellarians, nematodes, and harpacticoid copepods. The spectrum is very different between the Arctic and Antarctic (Fig. 7.16): Whereas copepods and acoel turbellarians dominate the sea ice metazoan fauna in the Antarctic, in contrast, although these taxa are also abundant in the Arctic, the dominant are the rotifers and nematodes. Some grazers opportunistically graze on the food reserves stored in the ice, but there are several species of amphipod and copepod that have life histories closely associated with the seasonal dynamics of sea ice (see discussion about krill and ice below and Schnack-Schiel, in Thomas and Dieckmann 2003).

Within the copepods three species clearly dominate the Antarctic sea ice fauna: the harpacticoid *Drescheriella glacialis* and the calanoids *Paralabidocera antarctica* and *Stephos longipes*. Other copepods such as *Ctenocalanus citer*, *Oncaea curvata*, *Oithona similis*, and various other harpacticoid species have occasionally been reported in Antarctic sea ice

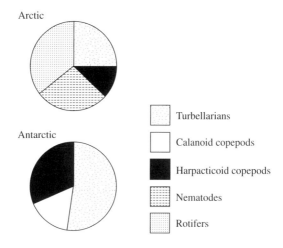

Fig. 7.16   Comparative distribution of major sea ice grazers in Antarctic/Arctic. Image adapted from Schnack-Schiel, in Thomas and Dieckmann (2003).

cores. In the Arctic the cyclopoids *Cyclopina gracilis* and *Cyclopina schneideri* as well as the harpacticoids *Harpacticus* spp., *Halectinosoma* sp., and *Tisbe furcata* are the most common species to be reported from sea ice (Schnack-Schiel, in Thomas and Dieckmann 2003, Gradinger *et al.* 2005, Werner 2006).

In some sea ice habitats high numbers of grazers are found in association with high algal standing stocks (Fig. 7.16; Garrison and Buck 1991, Stoecker *et al.* 1993, Archer *et al.* 1996, Schnack-Schiel in Thomas and Dieckmann 2003, Werner 2006). These include surface melt ponds, gap layers, and bottom-ice assemblages where the mobility of grazers is unconstrained (Schnack-Schiel *et al.* 2004). The greatest restriction, for protozoan and metazoan grazers attempting to exploit the high standing stocks of algae and bacteria that accumulate within the sea ice matrix is space. Since the brine channel system is space limited, grazing pressure is not as high inside this system as the algal stocks could otherwise support. Krembs *et al.* (2000) used glass capillaries ranging in size from micrometres to millimetres as model systems to mimic the brine-channel habitat. They were able to monitor the movement and colonization of brine-channel proxies by turbellarians, rotifers, nematodes, harpacticoid copepods, flagellates, amoebae, diatoms, and bacteria. Only rotifers and turbellarians were able to traverse channels significantly smaller than their body diameter. Turbellarians apparently changed their body dimensions in response to salinity changes. Rotifers traversed channels just 57% of their body diameter. Larger amphipods avoided narrow passages and indeed most of the other organisms tested simply congregated in the narrowest of tubes that they could physically fit into as determined by their body size. Krembs *et al.* (2000) concluded that pore spaces within sea ice of less than or equal to 200 μm in diameter are refugia in which bacteria, pennate diatoms, flagellates, and small protozoans benefit from very much reduced grazing pressure.

There are a large number of grazers that do not enter the ice, but rather graze on the underside of ice floes. Some graze on the ice algae and bacteria, whereas some feed on other grazers that accumulate at this ice–water interface. These *cryopelagic* assemblages are composed of both adult and larval stages of amphipods, euphausiids, and copepods, and also some fish. Siphonophores, appendicularians, chaetognaths, pteropods, foraminifers, mysids, and the pelagic polychaete *Tomopteris carpenteri* have all been observed in the sea ice–water boundary layer in the Arctic. Benthic invertebrate larvae such as cirripede nauplii can also inhabit the underside of ice floes in high numbers in shallow, sea ice-covered waters in the Arctic (Gradinger *et al.* 2005, Werner 2006). Organisms are being added to the lists routinely and Bluhm *et al.* (2007) have recently described Cnidarians (0.2–1.1mm) living within Arctic sea ice. Recent unpublished observations from expeditions in 2004/2005 and 2006 have shown that nudibranchs, polychaetes, and coelenterates may be very abundant at the ice–water interface below late winter/early spring sea ice of the Weddell Sea (R. Kiko, personal communication).

In the Antarctic the silver fish, *Pleurogramma antarcticum*, is the species most commonly recorded at the ice–water interface, especially the larval stage which feeds on nauplii and copepods. Another small pelagic fish, *Pagothenia borchgrevinki*, can also be found in shoals in waters immediately under the ice. This species has a varied diet including *Pleurogramma* larvae, amphipods, copepods, and euphausiids. Young Antarctic giantfish toothfish (*Dissostichus mawsonii*) have also been found to be feeding at the ice–water interface. All these fish are protected against freezing by antifreeze glycopeptides in their body fluids (see Chapter 6).

In the Arctic the grazers on the underside of ice floes are dominated by the amphipods *Apherusa glacialis*, *Onisimus* sp., and *Gammarus wilkitzkii*, which seem to be dependent on permanent ice cover (Lønne and Gulliksen 1991, Werner 2006). The amphipods together with the polar cod, *Boreogadus saida*, which frequent the underside of the ice in large numbers (Gradinger and Bluhm 2004), are important food sources for many seabirds and marine mammals. Glacial cod (*Arctogadus glacialis*) is also found in large numbers with polar cod living within the cavities in spring and summer ice. Polar cod is an opportunistic feeder with a wide range of food items, feeding on copepods and amphipods. However, it avoids *G. wilkitzkii* because it is large and spiny and presumably difficult to eat (Gulliksen and Lønne 1989).

Seasonal Arctic ice has to be recolonized every year, resulting in lower faunal densities than in perennial ice. It seems that the movement of ice from the Polar Basin through the Fram Strait leads to an annual loss of 7×10^5 t of biomass from the perennial ice zone. Seasonal ice off the northeast coast of Greenland carries heavy but patchy growths of the diatom *Melosira arctica* with no significant associated fauna (Gutt 1995). After ice melt these diatom mats sink to the sea floor where they may provide a substantial input the benthic system. Although perennial Arctic sea ice is in a state of continual vertical flux the organisms inhabiting it maintain themselves as a distinct ecosystem (Melnikov 1997).

7.4 The ice edge

The ice edge (or rather the marginal ice zone, often abbreviated to MIZ) is not simply a line where sea ice and its associated organisms disappear to be replaced by open water and pelagic biota. It is a dynamic zone (between 10 and 100 km broad), constantly shifting in position and interacting with the atmosphere to produce characteristic weather. The physical processes taking place modify biological processes to a profound extent. Because the situation in the Antarctic is somewhat more straightforward it will be discussed first.

A large amount of melt water is produced in a relatively short period in the spring and summer. Of the 20×10^6 km^2 of ice which cover the Southern Ocean by the end of winter about 16×10^6 km^2 melt during the summer, implying that a layer of low-salinity water of about the same thickness as the ice is produced over the same area. Although the stabilized water column has often been observed at melting ice edges, it is not always the case. When there are strong prevailing winds and pronounced wave activity in the MIZ the released fresh water will be effectively mixed with surface waters (Murphy *et al.* 1998). However, because of the short fetch of wind over the water between floes, wind mixing is minimized, and lenses of nearly fresh water, with salinities as low as zero, occur near melting ice edges. The marginal zone in which melting takes place may be several hundred kilometres wide but, of course, it is not fixed and travels polewards each summer. The resulting stabilized area may extend 100 km into the open sea but is not necessarily uniformly stratified, and there may be a mosaic of mixed and stratified patches. Local hydrographic features, such as fronts, may override the ice-edge effect, since the density gradients at the fronts between saline and less-saline waters set up both vertical circulation and horizontal currents paralleling the fronts.

Organisms released from melting ice undergo osmotic shock as the brine in which they were living is diluted and there is also an abrupt increase in irradiance to cope with. Some forms may be killed and those which survive may suffer temporary alterations in cell-membrane permeability, causing them to release soluble cell components, making more organic material available to heterotrophs. During the winter there was deep vertical mixing, associated with cooling and ice formation, to perhaps 125 m below the ice. This, together with low light penetration, keeps phytoplankton production at a low level. The stabilization accompanying melting confines mixing to the surface 10 m or so and conditions become much more favourable for photosynthesis.

The inoculum for plankton growth may come from the ice, with opportunistic species growing to dominate the community until mixing brings in species able to compete with them. For example, in the marginal ice community off southern Victoria Land the predominant phytoplankton species is *Fragilariopsis curta*, a member of the local sea ice community. It is active in photosynthesis under these conditions. Not all of the material released from the ice is incorporated into ice-edge blooms (Riebesell *et al.* 1991). There have been studies to show that much of the biology actually aggregates into large accumulations and sinks quickly to the depths as marine snow. The feeding of zooplankton on the ice organisms can also effectively package a high percentage of the ice biota into rapidly sinking faecal pellets. Mean settling velocities for faecal pellets of between 60 and 200 m day^{-1} are common although values up to 1500 m day^{-1} have been reported (Leventer, in Arrigo and Lizotte 1998). There are numerous reports of faecal pellets from copepods and protozoans containing unbroken ice-diatom frustules, often of monospecific origin, reaching the sediments. Krill faecal pellets contain mostly broken/digested diatom frustules and are easily broken. Therefore their efficiency as a major flux mediator to great depths is questionable, despite them having potentially high settling velocities. However, a sediment trap under a krill swarm recorded a flux of 660 mg carbon m^{-2} day^{-1}, which is the greatest flux recorded for faecal matter of herbivorous plankton (Cadée 1992).

Phaeocystis is often abundant at the ice edge and responsible for chlorophyll concentrations as high as 14 mg m^{-3}. In this area the bulk, around 74%, of the photosynthetic biomass is in nanophytoplankton like *Phaeocystis*, less than 20 μm in size. In the Southern Ocean nutrient concentrations are nearly always ample to support dense growths.

Often, the density of the phytoplankton is high enough to show up in satellite images as a belt along the ice edge. Rates of primary production are correspondingly high with a mean value of as much as 1.76 g carbon m^{-2} day^{-1} in spring, but declining to about a quarter of this by autumn and back to less than a tenth in winter. This compares with 0.36 and 0.87 g carbon m^{-2} day^{-1}, which are the mean spring values for water under close pack and in open ocean, respectively (Mathot *el al.* 1992). In turn swarms of krill feeding in 'frenzies' have been recorded at ice edges, and in turn the MIZ is often an area of very intense feeding for seabirds, whales, and seals. Birds and mammals also utilize the ice edges in their migrations, exploiting the rich food sources as they migrate.

Roughly co-extensive with the zone of phytoplankton concentration is one with high bacterial densities. This growth is presumably sustained by organic matter released first by melting of the ice and then by algae and zooplankton. In the Weddell Sea, bacterial biomass is about 16% of that of the phytoplankton and 7% of the total microbial biomass. Protozoa, such as choanoflagellates, phagotrophic dinoflagellates, ciliates, and amoebae, contribute rather more to total biomass, 23%, than do bacteria. There are close interrelations between the population dynamics of these organisms with the phagotrophs ingesting perhaps half of the daily primary production with concomitant recycling of mineral nutrients.

There has been debate about how much energy the ultraplankton contributes to higher trophic levels (see Chapter 6). Whether the consensus that it is not much applies to the MIZ is uncertain. There has been a suggestion that comparatively large zooplankton forms, such as krill, can benefit from food concentration by protozoa such as tintinnids and choanoflagellates. Be this as it may, higher trophic levels have greater biomass and activity in the MIZ, paralleling those in the lower levels (Fig. 7.17). There is greater productivity at the microplankton level, sedimentation of organic detritus fuels more benthic production, and decaying ice floes provide refuges for the larger crustacea which attract fish, seabirds, and seals.

As well as by changes in biomass, the marginal zone is marked by differences in species, particularly of seabirds. The pack-ice community is dominated in terms of biomass by emperor, chinstrap (*Pygoscelis antarctica*), and Adélie penguins, and in numbers by snow (*Pagodroma nivea*) and Antarctic (*Thalassoica antarctica*) petrels. The highest density of this assemblage lies around 7–10 km north of the ice edge, which it follows as it moves south to take advantage of the surge in production. Further north the abundant seabirds are southern fulmars (*Fulmarus glacialoides*) and cape pigeons (*Daption capense*). There is a difference in the prey of these two groups of birds. Within the pack ice, crustacea of the genera *Pasiphaea* and *Eurythenes* are taken. North of the ice edge, the prey is mainly krill and small lantern fish (*Electrona antarctica* and *Gymnoscopelus braueri*), which rise to the surface at night to feed on krill.

The processes at ice edges in the Arctic are not as clear-cut. In the Southern Ocean the MIZ lies over deep water, whereas in the Arctic there is year-round ice cover in the deep-water regions and the ice margins lie over broad, shallow, continental shelves. The melt season is shorter than in the Antarctic and ice margins more subject to disturbance by currents. In the Chukchi Sea, for example, the summer ice-edge system has upper- and lower-layer fronts, the upper arising from melt water and the lower, in a depth of only 50 m, marking the boundary between cold Chukchi Sea water and northern-flowing warmer water intruding from the Bering Sea. The extent of the MIZ depends on mesoscale dynamics and it is usually

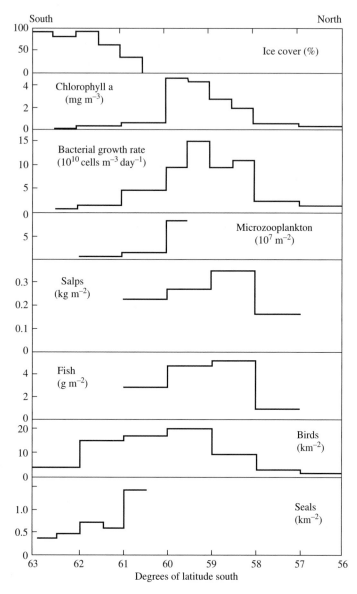

Fig. 7.17   Distribution of organisms along a south–north transect across the marginal ice zone in the Weddell Sea, November 1983. Redrawn from Vincent (1988).

about 50 km wide as compared with 250 km in the Antarctic, where large-scale dynamics have more effect.

The relative paucity of microbial growth in the interior of Arctic ice means that on melting there is relatively less of an inoculum into the sea compared to the Southern Ocean. Whereas phytoplankton growth at the Antarctic ice edge is sustained by high nutrient concentrations, the lower concentrations in Arctic sea water are quickly depleted and wind-driven upwelling of nutrient-rich water is necessary to support significant algal growth. Phytoplankton growth in the productive shallow-water shelf areas often masks increased production at the ice edge.

7.5 Polynyas

Wind is a very effective force for breaking open fields of pack ice. As soon as the ice breaks apart, the water underneath is exposed, to form a *lead*. Depending on the strength of the wind these leads can cover areas of just a few square metres, before being closed over again by moving ice floes, or extend for many hundreds of kilometres, effectively producing a sea surrounded by a coastline of ice.

Sometimes large bodies of water established within the pack ice persist longer than leads. These *polynyas* (a word derived from the Russian for an area of open water within ice) can persist throughout the whole of the winter, and may occur in the same region over a number of years. Sometimes, recurring polynyas open up at the same time each spring, and are so predictable that they are important features for the seasonal hunting activities of Inuit people. Polynyas vary greatly in size from a few square kilometres to huge areas with large implications for the oceanography and wildlife living in the area.

Polynyas are of great oceanographic importance as pathways for heat losses to the atmosphere (Muench, in Smith 1990), and they provide open water to birds and sea mammals in winter and an ice edge with enhanced productivity (Ainley *et al.*, in Thomas and Dieckmann 2003). Polynyas arise by two main processes: from the upwelling of warmer water (*sensible-heat polynyas*), or by mechanical divergence of the pack ice (*latent-heat polynyas*; Fig. 7.18), as with those in the Bering Sea or off the west coast of Greenland. Sensible-heat polynyas are so called because the heat that is transferred from the ocean to the atmosphere causes a decrease in water temperature and so a reduction in the sensible heat content of the water. They occur in regions of upwelling water due to tidal activities and in regions of steep sea-floor topography such as around sea mounts. One of the most well known Antarctic polynyas was the *Weddell Sea Polynya* that opened up in the Weddell Sea each year from 1974 to 1976, and covered an area of 200 000 km^2. It is thought to have been associated with the underlying sea mount, Maud Rise (Holland 2001).

Latent-heat polynyas are maintained by continued removal of ice by wind or currents, a windward shore preventing its replacement from elsewhere. They normally form on the lee side of coasts, land-fast ice, glacier tongues, and

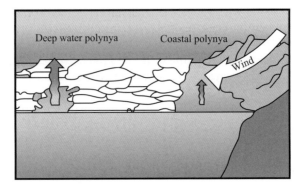

Fig. 7.18   Polynyas can either be coastal (latent-heat) or deep water (sensible-heat polynyas), or sometimes a mixture of the two.

grounded icebergs. The latent heat released by ice formation is dissipated into the atmosphere and the brine-produced sinks, thereby modifying the circulation of the adjacent sea area. These polynyas are a prolific source of ice; for example, it is estimated that the Bering Sea ice cover is regenerated between two and six times from this source in one season. Many polynyas can form by a combination of latent- and sensible-heat processes. To avoid confusion about the differences between latent- and sensible-heat polynyas at times there has been some attempt to refer to polynyas as deep-water (open-ocean) polynyas or shelf-water (coastal) polynyas.

Recurring polynyas are those that remain open throughout the winter, or open at the same time each spring, at the same location and time every year. One of the best-studied recurring polynyas is the *North East Water* (NEW) polynya, with an area of 45 000 km², which opens in May/June over the continental shelf of north-east Greenland. It is one of the largest and most consistently recurring polynyas in the Arctic and extensive plankton communities develop in the region during summer (Booth and Smith 1997). The consistent recurrence and enhanced primary production within the NEW polynya is reflected in large numbers of migratory birds and mammals. The largest Arctic polynya is the *North Water* (NOW) polynya, which covers an area of about 80 000 km² in northern Baffin Bay. This is also associated with increased primary production (Mostajir *et al.* 2001) and corresponding higher activity of higher trophics (Ainley *et al.*, in Thomas and Dieckmann 2003). Such biological activity has led some researchers to refer to these polynayas as being equivalent to the oases of pack-ice zones.

7.6 The wider significance of sea ice biology

Global ocean net primary productivity estimates are numerous and varied (Geider *et al.* 2001). Most of the variability is largely to do with the methods used to measure ocean primary production, and the different metabolic processes that these methods quantify. However, the most recent estimates based on satellite images of phytoplankton biomass in surface waters tend to range between 40 and 60 Pg C year^{-1} (P = peta, 1 Pg is equivalent to 10^{15} g). Estimates of terrestrial primary production vary between 50 and 60 Pg C year^{-1}, which combined with the oceanic primary production gives a total primary production on the planet of approximately 10^{17} g C year^{-1}. The polar regions are estimated to contribute 6–7 Pg C year^{-1} to this overall budget.

7.6.1 Primary production in sea ice

High standing stocks of sea ice algae that are so rich that they turn ice floes the colour of coffee would initially suggest that sea ice-based primary production must be a high contribution to the overall productivity of the polar oceans. However, comparing the annual water column primary production to that estimated for sea ice suggests that ice microalgae may in fact only contribute a small fraction to total production in the Southern and Arctic Oceans.

Despite all the work on sea ice there are few studies that attempt to estimate annual primary production in sea ice. This is based on there being relatively few measurements, and the uncertainty of how to scale up these measurements to take into account the great spatially variability in the distribution of organisms in the ice, and how production rates vary with season. However Arrigo (in Thomas and Dieckmann 2003) has calculated that the annual production of sea ice algae in the Arctic and Antarctic are similar in magnitude, ranging from 5 to 15 g C m^{-2} year^{-1} and from 0.3 to 34 g C m^{-2} year^{-1}, respectively. He goes on to conclude that even in the most productive sea ice habitats, annual production is below 50 g C m^{-2} year^{-1}, an amount similar to that estimated for the oligotrophic central gyres of the open oceans.

However, to understand the importance of sea ice primary production to the polar regions it is obviously important to integrate such estimates of primary production over the whole of the Arctic and Southern Ocean systems. Because of the paucity of data used to make such estimates, such large-scale integrations are crude and as yet provide little or no information about regional patterns. However, despite this Arrigo *et al.* (2003) and Legendre *et al.* (1992) estimate basin-wide production of sea ice algae in both the Arctic and Antarctic regions to be rather similar, varying between 30 and 70 Tg C year^{-1} (T = tera; 1 Tg is equivalent to 10^{12} g).

More detailed work has been done on the annual primary production estimates in waters of the Southern Ocean (south of 50°S), calculated from satellite ocean colour, sea ice cover, and sea-surface temperature data. For the whole of this area the estimates are between 2900 and 4414 Tg

C year^{-1}. Annual primary production is greatest in the permanently open ocean zone, the region of the Southern Ocean not impacted by sea ice, which contributes approximately 88% of the annual primary production. The MIZ and the continental shelf contribute 10 and 2%, respectively. The annual rate of primary production within Antarctic sea ice is therefore less than 5% of the overall production in the Southern Ocean.

At the maximum extent of sea ice in October, because primary production is highly light-limited, the spatially integrated rate of production is only 4 Tg C month^{-1}. This increases to 13 Tg C month^{-1} in November since the primary production increases by a factor of four and sea ice extent is still high. Annually approximately 60% of Antarctic sea ice primary production is thought to be produced during November and December (Legendre *et al.* 1992, Arrigo *et al.* 1997, 1998, Moore and Abbot 2000, Arrigo, in Thomas and Dieckmann 2003).

Sea ice primary production is a much larger fraction (10–28%) of total production in the ice-covered waters of the Southern Ocean, which ranges from 141 to 383 Tg C year^{-1}, especially since this includes the highly productive MIZs. Therefore, in those waters that are covered with ice for part of the year, algae growing in sea ice can be an important component of the marine food web, not only in the ice but in pelagic and benthic systems. The release of the high concentrations of biological matter (Fig. 7.19)

Fig. 7.19   Faecal pellets full of ice diatoms released from melting sea ice and collected at a depth of 150 m (photograph by David N. Thomas)

contained in the ice upon ice melt is an important event in the seasonal sea ice cycle (Leventer, in Thomas and Dieckmann 2003), and the fate of this material has consequences for biogeochemical cycling, bentho–pelagic coupling (see the MIZ discussion above and Chapter 8), and ultimately the sequestration of organic carbon into sediments (Armand and Leventer, in Thomas and Dieckmann 2003).

7.6.2 Sea ice biology as a diet

Naturally, the biochemical composition of the ice organisms has profound implications for their quality as a food source for protozoan and metazoan grazers. This is particularly true of polyunsaturated fatty acids (PUFAs) produced by both bacteria and microalgae in the ice. Many marine organisms cannot produce PUFAs, and require them to be supplied in their diet (Nichols 2003). Enhanced PUFA production within sea ice algae and bacteria has been measured, induced by low irradiance, low temperature, and high salinities within sea ice, and therefore sea ice assemblages will be a richer source of essential PUFAs for grazing organisms. Since the conditions that stimulate PUFA production in sea ice microbes are most extreme in winter, it is probable that PUFA production in the ice organisms will be maximal at that time. Organisms grazing within winter sea ice will therefore have a diet greatly enriched in PUFA which may be a significant factor in maintaining viable stocks, especially of larval stages of zooplankton species, and maintaining the fitness of these to exploit more favourable feeding conditions upon ice melt in spring.

Sea ice algae have been shown to produce high amounts of UV-protecting mycosporine-like amino acids (MAAs), and it is thought that by incorporating sea ice algae into their diet some organisms may significantly increase their resistance to damage due to UV radiation. Levels of UV radiation typically found within the upper 15 m of the water column can have significant effects on zooplankton, and have even been found to kill krill which, due to their nucleotide base composition, are particularly susceptible to UV radiation damage (Jarman *et al.* 1999).

Marine zooplankton, including krill, can obtain MAAs from their algal diets (Newman *et al.* 2000, Whitehead *et al.* 2001), and these grazers can obtain at least 10 times more MAAs per unit of chlorophyll ingested by consuming sea ice algae than they can from consumption of Antarctic phytoplankton. An ice algal MAA source may be particularly important to krill during the austral spring when ozone levels are still low, sea ice algae are growing actively, and the major phytoplankton blooms of the Southern Ocean have yet to develop. Sea ice algae also may be the source of the MAAs found in the tissues of shallow benthic organisms in areas like McMurdo Sound, where no other likely MAA source has been found (McClintock and Karentz 1997). Because MAAs can be transferred from grazers to their predators, MAAs produced by ice algae have the potential to benefit organisms at a variety of trophic levels.

7.6.3 Sea ice biology and krill

The wider significance of the primary production in sea ice for a whole of the associated ecosystem is well illustrated by the effects of sea ice on the distribution of the Southern Ocean krill (see Chapter 6). In winter *Euphausia superba* is primarily restricted to the region covered by ice, and there is a close correspondence of the life cycle of krill with the oscillation of the sea ice cover (Brierley and Thomas 2002, Atkinson *et al.* 2004, Smetacek and Nicol 2005). Krill larvae, juveniles, and adults have all been observed directly beneath sea ice, although the degree of association with the sea ice differs between the developmental stages.

Adults of *E. superba* do not necessarily depend directly on sea ice algae for winter survival (Quetin *et al.* 1996), although they can feed on ice organisms by very effectively scraping off ice algae growing on ice surfaces. Adult krill are also capable of surviving long periods of starvation (211 days) through utilization of body reserves and shrinking, but it is not certain how frequently krill really encounter food shortages sufficient to induce shrinkage in nature (Nicol 2006). Krill are also reported to switching to carnivorous feeding of lipid-rich copepods (Huntley *et al.* 1994), or migrate downwards to the sea bed (Quetin *et al.* 1996). According to Quetin and Ross (1991) lowered metabolic rate is by far the most important strategy of the adult krill for successful surviving during winter.

In spring and summer aggregations of adult krill feeding on the undersides of sea ice floes are reported frequently (reviewed by Brierley and Thomas 2002, Hofmann and Murphy 2004). Acoustic survey data gathered with the autonomous underwater vehicle Autosub-2 (Fig. 7.20) showed a significantly higher krill density under sea ice than in open water in the northern Weddell Sea in summer (Brierley *et al.* 2002). The krill were concentrated within a narrow band under sea ice, and the abundance increased southward from the point where ice concentration was more than 40%, and extended from 1 to 13 km south of the ice edge, after which the abundance decreased rapidly.

The distribution of krill larvae and juveniles in winter and spring is, in contrast to adults, closely coupled to sea ice, where they occur in large aggregations feeding on ice algae. Larvae and juvenile krill cannot accumulate sufficient lipid stores, and so cannot undergoe periods of starvation. By feeding on ice algae, larvae can ingest between 2 and 44% of their body carbon per day, which probably covers the metabolic requirements for growth and development (Hofmann and Lascara 2000). It has been argued that larval and juvenile krill will be in a better physiological

Fig. 7.20   Autosub-2, an autonomous underwater vehicle (floating behind the whale) used to measure ice thickness and krill distribution along transects under sea ice (see Brierley *et al.* 2002; photograph by Mark Brandon).

condition, have higher growth rates, and higher survival in winters with a greater extent in pack ice than during winters of low ice cover (Quetin *et al.* 1996).

Larvae and juveniles, and adults are more often concentrated under pressure ridges and deformed ice which provide better refuges from predators. It is now largely recognized that long duration of heavy sea ice cover during winter, and late opening of the seasonal pack ice in spring favours earlier onset of the krill spawning and high krill recruitment (Siegel 2000), and that changes in krill abundance have can predicted on the basis of cyclical variations in sea ice extent (Brierley *et al.* 1999). However, there are regional differences in this relationship (Smetacek and Nicol 2005).

This interplay between ice extent and krill abundance has been extended further. The salp *Salpa thompsoni* is thought to reach high densities in years following reduced ice extent. Salps live for less than 1 year, and feed by filter-feeding phytoplankton. They do not feed on ice organisms. In the absence of krill (i.e. following a poor ice year), the salps are able to exploit the spring phytoplankton bloom and undergo explosive population growth. In good sea ice years, the krill have the upper hand over the salps because the sea ice provides good feeding grounds over the winter, resulting in good gonad development and possibly allowing multiple spawning to take place. In these years the krill exploit the phytoplankton bloom, resulting in poor food stocks for the salp populations (Siegel and Loeb 1995, Loeb *et al.* 1997).

8 Marine benthos in polar regions

8.1 Introduction

Polar terrestrial habitats experience extreme variation of conditions, with desiccation or mechanical damage by wind, unstable substrates, or ice movement, as the usual limiting factors. Shallow inland waters have similar disadvantages. Deeper inland waters afford more stable conditions with steady temperatures and ample water, but tend to be poor in inorganic nutrients. Benthic vegetation is frequently the most successful form of life in them. The sea similarly provides stable conditions and temperatures which cannot fall below −1.9°C (the freezing point of full salinity sea water). This is reflected in the circumstance that a majority of polar invertebrates is *stenothermal* (i.e. unable to survive outside a narrow temperature range), and standing in contrast to the terrestrial organisms such as *Nanorchestes*, which have remarkably wide thermal tolerances. The sea has the additional advantages that there are no solid barriers to transport of nutrients and movement of organisms. On the other hand, mechanical damage by drifting ice can be severe and polar shores usually seem barren, with all exposed life down to several metres being battered and scraped off (Fig. 8.1). Icebergs may plough up the seabed down to 300 m or more. Nevertheless, benthic life is abundant in polar seas.

These considerations apply equally to Arctic and Antarctic and, of course, benthic habitats in the two regions have similar trends in irradiance and ice cover. Nevertheless, there are some striking differences, as shown below.

- Although extents of pack ice are of the same order, that in the Arctic is largely multi-year ice, covering the benthos year round, whereas in the Antarctic most of the ice melts each year. This distinction is important for abyssal benthos. Part of the Southern Ocean is covered by extensive permanent ice shelves, such as do not exist in the Arctic.

Fig. 8.1   In rough seas or waves floating sea ice and fragments of glacial ice can very effectively scour rocks in the littoral and sublittoral zones down to as much as 10 m (photograph by David N. Thomas).

- The Southern Ocean has open connections with the Atlantic, Pacific, and Indian Oceans, but the Arctic Ocean has only limited connection with the Atlantic and even more limited connection with the Pacific.
- The coasts of Eurasia, western North America, and Greenland lead continuously from well below the Arctic Circle into the high Arctic whereas there is a gap of 1100 km between the most northerly point of the Antarctic continent and the nearest large land mass.
- The continental shelf of Antarctica is narrow and 400–600 m deep whereas that in the Arctic Ocean is broad and <100–500 m. Shallow epicontinental seas—Barents, Kara, Laptev, and Chukchi—make up nearly 36% of the Arctic Ocean area, but contain only 2% of its water.
- The Antarctic has abundant, large, tabular icebergs whereas the Arctic has fewer, small, irregular bergs, mainly in the Greenland Sea, fewer in the Bering Sea, and very few in the Arctic Basin.
- The Southern Ocean has generally high levels of nitrate, phosphate, and silicate in the euphotic zone whereas Arctic waters have lower levels which are regularly depleted each summer.
- The sediments around Antarctica are a mosaic of muds, fine, and coarse sands, and large and small boulders, all of glacial origin. In contrast the Arctic has a large input of river-borne material with muds and clays predominating.

Marine benthos is not easy to study: sampling the bottom with dredges or grabs is a hit-or-miss procedure which at the best is only semi-quantitative. However, these methods have been reasonably effective in gathering the larger species, and biologists with early expeditions made some

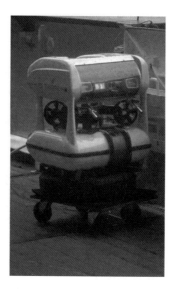

Fig. 8.2   Remotely operated vehicle (ROV) used to film benthic assemblages of organisms (photograph by David N. Thomas).

surprisingly complete collections. Observations by underwater video and still cameras provide a valuable adjunct but scuba diving, which was introduced into polar waters by marine biologists in the early 1960s, is an unrivalled means for observation and experiment *in situ*. Clearly there are considerable operational difficulties for diving in ice-covered seas, and most observations in recent years are made using remotely operated vehicles (Fig. 8.2).

8.2 The littoral zone

The littoral, or intertidal, zone extends between extreme high-water spring tides and extreme low-water spring tides. These two limits are more theoretical than actual water levels, which vary not only with the relative motions of Moon, Sun, and Earth, but with wind direction and atmospheric pressure. It is best to look on them as marking the centres of supralittoral and infralittoral fringes, respectively. Usually tides are semi-diurnal (i.e. with two more or less equal cycles in 24h 50min, as on nearly all Arctic coasts. The Antarctic region has diurnal tides, with only one cycle in this period, or mixed tides intermediate between the two. The tides around South Georgia, for example, are described as irregular. Over much of the Arctic the tidal amplitude is less than 2m, as in the Antarctic, but around the southern part of Greenland it is in the range 2–4m. The ecologically most significant feature is that the duration of immersion in one annual cycle oscillates between 100% at the bottom and zero at the top of the littoral zone, the slowest change in duration being centred on mid-tide.

To varying degrees, then, intertidal organisms alternate between the relatively stable temperature and salinity of the sea and the highly variable conditions of temperature, desiccation, and salinity encountered above sea level. It is, however, an oversimplification to regard the distribution of organisms in the littoral zone as a response to gradients in physical factors and to assume that the absence of a species from a particular level can be put down to its lack of tolerance for the conditions prevailing there. Chance plays a large part and only on a basis of study over many years of the population and community dynamics, life histories, and behaviours, of the species involved can we come to a full understanding of the zonation of organisms which is such an obvious feature of most shores.

Unfortunately, few studies in polar regions have extended over complete annual cycles, let alone several of them. It is clear, however, that frequent storms and the overwhelming effects of ice render tidal exposure of somewhat less importance than it is elsewhere. When the sea is frozen, sea ice becomes juxtaposed to land ice but moves with the tide and is fractured into a series of parallel tide cracks. These get filled with snow and the resultant slush provides a nutrient-rich and well-illuminated medium for growth of microalgae.

8.2.1 Arctic littoral zones

Nearly all Arctic shores are ice-bound from late autumn until summer thaw. Once air temperatures have fallen below freezing, spray freezes on the shore. The littoral zone consequently supports no active life in winter and when the thaw comes it is scoured by floating ice. The only large organisms to be found in this zone at this time are those, such as red or brown seaweeds, finding shelter in crevices or under overhangs, or those which can move up from the sublittoral. Rapidly growing microalgae, such as filamentous green forms, are able to colonize rock surfaces in the intertidal zone in the summer.

The marine Arctic has been defined as those areas in which unmodified water of polar origin is found at the surface and down to a depth of at least 200 m. The marine sub-Arctic has waters of mixed polar and non-polar origins. It is most extensive in the Atlantic sector, including the Scotian and Hudson Bay shelves, Newfoundland, the whole coast of west Greenland, the water around Iceland, and the Barents and Kara Seas off north-west Russia. These areas are still cold and occasionally have drifting ice. Spells of unusual cold can cause mortality in the littoral zone, the extent of the damage depending on how quickly and how much the temperature falls and subsequently rises, on the state of the tides, and on the prevailing weather. Species that can dry out may survive very low temperatures and mobile animals can take evasive action, retreating to crevices, pools, or the sea, or under seaweed.

Tidal pools may provide refuge but thaw water draining into them can lower salinity to the detriment of many species. Pools high up in the littoral characteristically contain green algae, which are more tolerant of salinity changes, whereas brown shore weeds occupy those in the mid-littoral zone. A succession of mild winters may allow the littoral zone to become occupied by brown seaweeds, barnacles, or mussels, all of which require several seasons to become established, then a return of cold and ice can obliterate them. The boundaries between Arctic and sub-Arctic shores may be determined, allowing for this type of season-to-season variation, by the sudden disappearance of common organisms of this type, for example, the mussel (*Mytilus edulis*), the barnacle (*Balanus balanoides*), and the periwinkle (*Littorina saxatilis* var. *groenlandica*).

Sub-Arctic littoral zones have many species in common with temperate waters in the same ocean further south. Thus Upernavik at 72°42′N on the west Greenland coast, which receives a mixture of polar water and North Atlantic water giving June temperatures of 0.2–1°C, has brown seaweeds (*Fucus* spp.) and a comparatively rich fauna, including *Littorina*. Mussels extend from the mid-littoral into the infralittoral fringe and in the upper part of this range have large specimens of the sea anenome *Actinia* attached to them (Dayton, in Smith 1990).

Fragmentary saltmarshes are found in sheltered spots on the coasts of the Canadian and American Arctic, Greenland, Iceland, northernmost Scandinavia, and Arctic Russia. Saltmarsh is a community, dominated by halophytic flowering plants, which establishes itself on stable intertidal silt and mud. It reaches its maximum development in temperate regions. In the Arctic, where intertidal deposits are frequently reworked by ice action, well-developed saltmarsh is lacking and only a mosaic of usually depauperate saltmarsh species is found. A grass, *Puccinellia phryganodes*, is always the primary colonist. It is not known to set seed but propagates vigorously by means of surface stolons. Accompanying it there is often *Stellaria humifusa*, *Cochlearia groenlandica*, and various sedges such as *Carex glareosa*. On the landward side saltmarsh may merge into brackish, freshwater or terrestrial vegetation. On the coast of Kotel'nyy Ostrov (75°59′N 138°00′E), where there are only 150 days in the year with mean temperatures over −10°C, *P. phryganodes* is accompanied by half-a-dozen other flowering plants in areas which are only inundated during winter storms. In the sub-Arctic, as for example between Anchorage and the Queen Charlotte Islands, 60–54°N, there is overlap with the northern ranges of temperate saltmarsh plants such as *Salicornia* and *Suaeda* (Chapman 1977).

The productivity of Arctic saltmarshes is low. Values based on above-ground dry matter harvested at the peak of the growing season for the dominant species are 0.02–0.24 kg m^{-2} year^{-1} for *Carex ramenskii* and 0.05–0.14 kg m^{-2} year^{-1} for *P. phryganodes*, both on the Alaskan coast at 69–71°N. Values for temperate saltmarsh plants are around 1 kg m^{-2} year^{-1}. The animal life of Arctic saltmarshes has scarcely been studied. Geese are the main herbivores (Bazely and Jefferies, in Woodin and Marquiss 1997).

8.2.2 The Antarctic littoral zones

As in the Arctic, the littoral zone of exposed Antarctic coasts supports relatively little life. On the shore of Terre Adélie, at 66°40′S 140°0′E, which is ice-covered for 10 months in the year, patches which become temporarily free of ice are colonized by algae from the sublittoral, such as benthic diatoms and the green alga *Monostroma pariotii*, but no animals have been reported as present. The lichen *Verrucaria* occurs in the supralittoral fringe and may extend lower down (Arnaud 1974).

Further north, on Signy Island, the littoral zone becomes ice-free for about half the year. In summer, the red alga *Porphyra* occurs on vertical surfaces around the level of extreme low-water springs. The colonizers which get highest are the green algae, *Ulothrix*, which has a particularly broad salinity tolerance, and *Urospora*, on firm substrates at about mid-tide level. Below this is a band devoid of larger algae, down to low-water springs. The red seaweed, *Leptosarca*, encrusting coralline red algae, and *Monostroma* are found in crevices in the more sheltered areas. Species found in these situations contain osmoregulatory solutes and antifreezes (Wiencke 1996). A limpet, *Nacella concinna* (*Patinigera polaris*), is the dominant invertebrate in the shallow waters and in winter is normally confined to the sublittoral but in summer moves up into the littoral zone. There it grazes on diatoms and green algae and is itself preyed on by the dominican gull, *Larus dominicanus*, and sheathbill, *Chionis alba*. This species withstands freezing temperatures for a limited time and its return to the sublittoral is correlated with the fall of air temperature in the autumn, the operative factor perhaps being the ice film that forms on exposed rocks. Other organisms, such as coralline algae and species of spirorbid polychaete worms, grow on the limpet shells and are thus carried willy-nilly into the littoral (Heywood and Whitaker, in Laws 1984). No obvious invertebrate other than this limpet is found in this zone, although crevices and the undersides of boulders may harbour a variety of animals. Where mudflats occur, both epifauna and infauna can exist intertidally.

Tide cracks, which occur around grounded icebergs as well as along the shore, have been studied on Signy Island. Sea water percolates freely through the slush of snow, which remains frozen, the winter temperature always being below zero. When snow is accumulating and sea water moving freely through it, there is abundant growth of the diatom, *Navicula glaciei*.

Appearing first at the end of May, this shows vigorous growth in early September when light conditions begin to improve, and a peak is reached in early November, when the chlorophyll *a* density reaches 7.5 mg l^{-1}, with a standing crop of 5.5 mg cm^{-2}. After this there is decline until the break-up of fast ice in December (Whitaker 1977).

The littoral zones of sub-Antarctic islands, all of which are outside the impact of pack ice, have more varied floras and faunas than those just described. Heard Island (53°05′S 73°30′E), a little south of the Polar Front, has air temperatures between −10.6 and 14°C and sea temperatures from −1.8 to 3.4°C. Mushy ice may persist for a few hours and ice boulders are frequent on some shores but the effects of these are small. The supralittoral fringe, wetted by splash or spray but exposed mainly to aerial conditions, is characterized by black lichen but has no littorinid molluscs, as would the equivalent zone in the northern hemisphere, although terrestrial arthropods, such as mites and beetles, are present. The top of the littoral zone proper is marked by a band of encrusting coralline red algae but barnacles, which usually define this zone, are absent.

The infralittoral fringe is occupied by the large brown kelp, *Durvillea antarctica*. Its fronds shelter species (e.g. a small chiton *Hemiarthrum setulosum*, the amphipod *Hyale* sp., and various small red algae), some of which are more characteristic of the sublittoral. It seems that the limpet *Nacella kerguelenensis*, which is also abundant in this zone, is prevented from invading the littoral by predation by the dominican gull. Between the coralline algae and the *Durvillea* is a zone of mixed algal species, mostly red with animals (e.g. the littorinid *Laevilittorina heardensis* and the small bivalve mollusc *Kidderia bicolor*) being found in crevices. An unexplained feature is that this zone is separated from the lichen zone above it by a few centimetres of bare rock. Despite persistently cool, damp, weather with little sunshine, some filmy algae nevertheless die through desiccation in late summer. The zones of the supralittoral, littoral, and infralittoral are not fixed absolutely in relation to tide level but, as seen elsewhere in the world, shift upwards with increasing exposure to wave action. In winter, the exposed shore is glazed by freezing sea spray. Presumably this kills off much of the summer's growth of algae (Knox 1994).

The rocky shores of Marion, Macquarie, and Kerguelen Islands are generally similar in flora, fauna, and zonation to those of Heard Island. There appears to be nothing resembling saltmarsh on any of the sub-Antarctic islands.

8.3 The shallow sublittoral zone

A sublittoral benthos can only develop fully in polar regions out of reach of scouring sea ice, around 10 m below low tide level. However, even at and below these depths an extreme example of severe habitat transformation is caused by icebergs, which when grounded can cause considerable damage to benthic communities in coastal areas in both the Arctic and Antarctic (Gutt 2000, 2001, Gerdes *et al.* 2003). Transects in shallow Arctic seas have shown as many as a 1000 of these events per kilometre with an average width of 7.5 m. During autumnal storms, gouging may overturn a muddy bottom to a depth of 30 cm. The huge icebergs of the Southern Ocean can scour from the intertidal down to depths of at least 500 m. Sessile organisms are eradicated and pioneer species begin to grow in high abundances on the devastated seafloor. In some areas major iceberg scour events have been estimated to take place over periods of every 50–200 years and because of the very slow growth of many species, particularly in the Southern Ocean, areas disturbed in this manner are likely to be characterized by a continuous natural fluctuation between destruction and recovery. Communities can be held at early successional stages, or even completely destroyed by scouring.

Ice damage not only comes from above. *Anchor ice* forms on the bottom when the temperature of this is below freezing point and, being buoyant, will eventually break away, carrying organisms with it. Ice is also the cause of extreme variations in salinity. As the brine from the freezing surface layer sinks to the bottom, salinities may rise locally to 80 or 100 or even as high as 183. When the brine moves seaward it causes thermohaline circulation. On the other hand, during the summer melt and the resumption of river flow, the salinity of the shallow sublittoral may decrease.

When the sea is ice-free the penetration of light is affected mainly by turbidity, contributed by both inorganic matter in suspension and plankton. The precise evaluation of the radiation available at different depths in water is complex but for most biological purposes it suffices that irradiance falls off exponentially with depth, assuming that the water column is of uniform transparency (Kirk 1994, Falkowski and Raven 2007). Put simply, if irradiance is reduced by half in penetrating 1 m, it will be reduced to a quarter by 2 m, to an eighth by 3 m, and so on: 0.1% of total photosynthetically available radiation entering at the surface may be expected to penetrate to 100 m in the clearest sea waters. The 0.1% level roughly defines the *photic zone*, in which photosynthesis is possible. In clear oceanic waters blue light penetrates most but in inshore waters, because of selective absorption and scattering by humic substances and particulate matter, the orange wavelengths have the greatest penetration. Thus, there are gradients in light quality and these can be of biological importance, as, for example, in affecting morphogenesis and reproduction. The quantum ratio of blue to red irradiation (quanta at 450 nm as a percentage of the total at 450 and 660 nm) is about 48 at the water surface but shifts to 98 and 2, respectively at the bottom of the photic zones in clear and turbid waters. Penetration of radiation is reduced by sea ice; for example, 2 m of congelation ice reduces photosynthetically available radiation by about 90%, with peak transmission in the blue-green, around 500 nm. Snow cover reduces penetration still further, for example, a 70 cm thickness reduces radiation to 3% of its incident value (see Chapter 7).

The question of what is the maximum depth at which benthic algae can live is general to marine biology but it has been raised particularly with respect to polar waters. Below 40 m, growth is sparse but there are reports of macroalgae recovered from depths in excess of 100 m where irradiance is at best extremely low. For example, the green alga, *Monostroma kariotii*, has been recovered from 348 m off Possession Island (72°S 171°E). Such reports must be regarded with caution since ice may detach algae and carry them into deeper water where, because of the low temperature, they may survive in a viable state for some time. Proof must come from direct observation from a submersible. This has been done in other parts of the world and has shown that coralline red algae can live attached to the substratum at depths of 130 or even 268 m. Calculations by Raven (1984) suggest that photosynthetic growth is just possible at a photon flux density of 1 μmol photons m^{-2} s^{-1}, about the maximum which may be expected during the day at 100 m in clear water. Nevertheless, deep-water red algae seem to survive at 0.05 μmol photons m^{-2} s^{-1}. Indeed, their pigmentation is such as to give maximum absorption of the blue light available at depth and it is presumed that low temperature ensures that the basal rate of respiration is minimal. Further research may show that these algae possess mechanisms making for highly efficient utilization of very low irradiance. The possibility that deep-water algae in Arctic regions supplement photosynthesis by heterotrophic assimilation of dissolved organic matter has been suggested but there is no direct evidence of this (Kirst and Wiencke 1995).

Within the photic zone much of the organic matter on which the benthos depends may be supplied by benthic plants. However, at greater depths the community is dependent on allochthonous organic matter which may come from plankton in the water column above, or be advected from elsewhere, from ice algae, or from debris from the land.

8.3.1 The shallow sublittoral zone in the Arctic

The Beaufort Sea, with its coast at roughly 70°N, has a shelf which is generally muddy with sandy areas nearshore and patches of gravel at the shelf break at a depth of about 70 m. These deposits are ice-borne glacial debris. The Mackenzie River and other large rivers cause seasonal fluctuations in salinity and ice covers the shelf from September through to June or July. The nearshore waters, subject to disturbance by ice and with a freshwater influence, have ephemeral populations of chironomid larvae and oligochaete worms. In deeper water, down to 20 m, there are patches of different substrata with associated communities of many species of polychaetes, bivalves, and isopods in the sediments (*infauna*), and mysids, amphipods, isopods, copepods, and euphausiids on the surface (*epifauna*). The offshore zone down to the shelf break has polychaetes making up 32–87% of the total macrobenthos, with bivalves, ophiuroids, holothurians, and many crustacea. In the relative absence of suitable rocky substrata macroalgae are not abundant (Zenkevitch 1963, Dayton, in Smith 1990).

However, an isolated patch of cobbles, covering some $20\,km^2$, supports a stand of kelp, *Laminaria solidungula*. For eight months in the year, like other macroalgal species, it exists in virtual darkness under ice (Fig. 8.3), which is rendered almost opaque by wind-blown debris from the shore and sediment brought up by anchor ice. Surprisingly, during this period it achieves rapid growth, depleting its reserves of organic carbon in doing so. It is then ready to take full advantage of the brief summer to carry out its photosynthesis. The isolation and almost monospecific nature of this patch of vegetation gave the opportunity to use the ratio of the carbon isotopes, ^{12}C and ^{13}C, in the animals in the community, to determine the fate of the photosynthetic products. During photosynthesis there is discrimination against ^{13}C, the heavier isotope, to an extent that varies according to the type of plant and the conditions to which it is exposed. The values found in animals reflect those in the plant material they have eaten, even if it is second-hand. The Beaufort Sea observations showed more discrimination (measured as deviation, $\delta^{13}C$, from a standard and expressed in parts per thousand) in phytoplankton (i.e. around –26‰), than in the kelp, around –16‰, as compared with –7‰ for molecular carbon dioxide in air.

As Fig. 8.4 shows, the herbivorous gastropod, *Margarites vorticifera*, feeds mainly on the kelp but the filter-feeding bryozoans largely on phytoplankton

Fig. 8.3 *Laminaria* and *Fucus* spp. growing under late winter sea ice in the Arctic Russian White Sea (photograph by David N. Thomas).

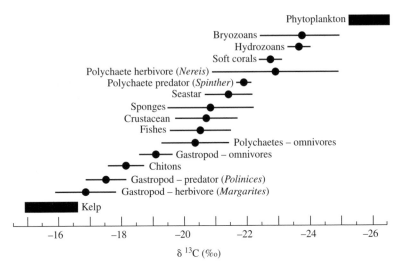

Fig. 8.4   Differences in contents of carbon isotopes, ^{12}C and ^{13}C ($\delta^{13}C$), of flora and fauna from Boulder Patch, Beaufort Sea. Redrawn from Dunton and Schell (1987).

(Dunton and Schell 1987). The benthic primary consumers have to rely on their own reserves and detritus to carry them through the winter. In regions where kelp grows, much of this detritus consists of decaying tissue worn away from the tips of its blades. Excretion of mucilage, a feature of many brown algae, is rather slight in *L. solidungula*. Not only herbivores but some filter-feeders, such as ascidians, and even carnivores, such as the gastropod *Polinices pallidus*, come to contain high proportions of kelp carbon. The opossum shrimp (*Mysis littoralis*) incorporates much kelp carbon and since it is a major food for many vertebrates, including marine mammals, the macroalgae contribute appreciably to the higher trophic levels. Ice algae also contribute to benthic secondary production, 1–10% of their biomass eventually sedimenting to the bottom. However, the total primary production, including that by phytoplankton, of the Beaufort Sea is low, between 10 and 25 g $Cm^{-2}year^{-1}$, and rather less than the input of organic carbon in the form of peat, eroded from the shores or brought down by rivers, which amounts to around 30 g $Cm^{-2}year^{-1}$. Little of this peat is utilized by the macrofauna but the fact that it does not accumulate suggests that it is decomposed by bacteria. In this case, there may be input of peat carbon into higher trophic levels via meiofauna feeding on bacteria (Dayton, in Smith 1990).

Contrast is provided by the adjacent Bering and Chukchi Seas, which together have a continental shelf of $1.5 \times 10^6 km^2$, stretching between 58° and 75°N and around 50 m deep. They are among the largest and most productive shelf habitats in the world. The deposits are poorly sorted mud, sand, pebbles, and cobbles. Most of the area is ice-covered during the winter but open during the summer. The benthic fauna is varied and its biomass is high. Bivalve molluscs are the most important infauna but large areas are dominated by amphipods, particularly *Ampelisca* and *Byblis* spp. There are at least 211 species of epifaunal invertebrates, most of which are molluscs, arthropods, and echinoderms. Among these, in the south-east Bering Sea, are four commercially important crabs. Species of starfish are abundant, making up some 70% of the epifaunal biomass.

Pink shrimp (*Pandalus borealis*) and crabs are important predators on smaller epifaunal species and in turn are themselves food for bigger crabs, fish, and marine mammals. Asteroids, including starfish, are generalized predators but affect bivalve populations in particular. Fin-fishes, such as flatfishes, cods, and sculpins, because of their numbers and active searching abilities, have major impact on the benthic community (Zenkevitch 1963, Dayton, in Smith 1990).

The richness of animal life, with a mean standing stock of perhaps $300\,g\,m^{-2}$, is supported mainly by the primary productivity of phytoplankton in the pelagic. The hydrographic conditions in the Bering Strait are particularly favourable for algal growth during the three-month summer. Zooplankton grazing consumes little of what is produced and much of it sediments out and can therefore be used by the benthos (see Chapters 6 and 7).

Another highly productive area is the Barents Sea. Upwards of 170 species of green, brown, and red seaweeds and 1700 invertebrates have been recorded from around its coasts and the standing stock of benthos varies from 10 to 15 up to $1000\,g\,m^{-2}$ or more on the south-eastern slope of the Spitsbergen Bank (approximately 75°N 20°E). As for several other Arctic and sub-Arctic shelf regions, brittle stars are dominant in the macrobenthic fauna. The trawling industry is active, the main catches being demersal fishes, cod, and haddock. Again this productivity is based on phytoplankton, the growth of which is particularly prolific along the fronts lying between Svalbard and the northern Scandinavian coast (Dayton, in Smith 1990).

Svalbard, 77–80°N and 10–30°E, at the north-west edge of the Barents Sea, has a rich algal flora which is associated with west Greenland and Arctic America rather than Siberia. Sublittoral algae are, however, little in evidence in some of the fjords, which because of shallow sills at their seaward ends have heavy siltation and reduced water exchange. The fauna is also sparse, with a bivalve, *Portlandia arctica*, characteristic of the coldest part of the Arctic basin, and a polychaete, *Lumbrineris* spp., as its most prominent components. Elsewhere, the rocky seabed supports luxuriant growths of *Laminaria solidungula*, associated with the red alga *Phyllophora*, down to a depth of at least 27 m.

The brown seaweed, *Scytosiphon lomentaria*, which occurs here, is of interest because its morphology is under photoperiodic control. It can develop either in the form of erect filaments under short days or as a prostrate crust under long days. A strain isolated at Helgoland, 54°N, has a critical day length between 12 and 13 h at 15°C. If a one-minute light break with a low irradiance of blue light is given in the middle of the dark period of 16 h in the short-day regime, the formation of erect filaments is completely inhibited. Isolates from different latitudes show a clear relation between latitude and the temperature range in which the erect form is produced under short-day conditions. Whereas strains from 32 to 48°N formed erect thalli at all temperatures from 5 to 20°C, one from Iceland, 66°N, was fully blocked in this respect above 15°C, and one from Tromsö, 69°N, from 10°C upwards. The Svalbard strain should be examined from this point of view.

The shelf areas of the Arctic Ocean have relatively nutrient-depleted waters and extensive permanent ice cover, which lead to low primary productivity and a paucity of benthic animal life. Sublittoral areas lack growth of macroalgae because of shortage of suitable substrates. In deeper waters, between 1000 and 2500 m, biomass is around 0.04 g m^{-2}, comparable with that in the central Pacific and much less than that found at similar depths in the Antarctic.

8.3.2 The shallow sublittoral zone in the Antarctic

The marine benthos extends as far south as liquid water and suitable substrata are available; that is, to at least 77°30′S in the Ross Sea, where there is ice cover 2 m thick for 10 months in the year (Dell 1972, White, in Laws 1984, Arntz *et al.* 1994). Three dominant red seaweeds show often extensive development and well-defined zonation; *Iridaea cordata* in water of around 3.5 m depth, providing that there is some protection from ice abrasion, *Phyllophora antarctica* at around 12 m, and *Leptophyllum coulmanicum* below 18 m. These zones shift downwards where thinner ice or less snow accumulation allow better light penetration. Brown seaweeds seem not to go quite as far south as the reds. In terms of biomass, benthic microalgae are more important than the macrophytes.

Diatoms are abundant in the top few millimetres of sediments and accumulations of sponge spicules are particularly favourable since they afford an easily penetrable substratum with protection from grazing. Although these habitats receive only around 1% of the light incident on the sea surface, recorded biomasses in terms of chlorophyll *a* range between 47 and 960 mg m^{-2}, higher values being found in the summer than in the winter. The diatom, *Trachyneis aspersa*, at depths of 20–30 m, with an irradiance of less than 0.6 μmol photons m^{-2} s^{-1}, is shade-adapted to the extent of becoming light-saturated at only 11 μmol photons m^{-2} s^{-1}. Surprisingly, it is not photoinhibited at 300 μmol photons m^{-2} s^{-1} whereas other algae from the same site become inhibited above 25 μmol photons m^{-2} s^{-1}. Primary productivity at the peak of development of the benthic microalgae is around $700\,mg\,Cm^{-2}day^{-1}$, about the same as that for the phytoplankton. Information on the annual benthic production is lacking but it seems to make an important contribution to the total in the area (Knox 1994).

The epifaunal benthos of McMurdo Sound shows three distinct vertical zones (Fig. 8.5). The top 0–15 m is a bare zone with a substratum of rock, pebbles, and volcanic debris, devoid of sessile animals because of ice scour

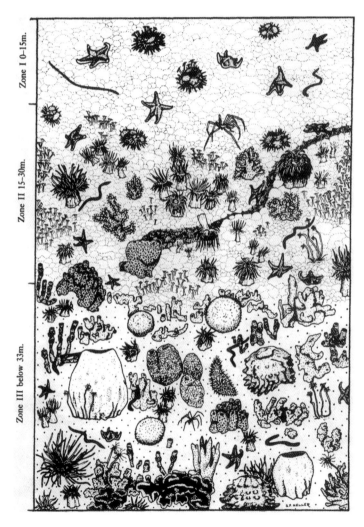

Fig. 8.5   Vertical zonation of fauna in the shallow-water benthic community of McMurdo Sound. A few mobile animals, but no sessile forms, are found in Zone I; the sessile animals in Zone II are mostly coelenterates and those in Zone III are predominantly sponges. From Dayton et al. (1970).

and disruption by anchor ice. It is briefly invaded during the summer by mobile animals including a detritus-feeding echinoid, a starfish, a necrophagous nemertine, isopods, occasional pycnogonids, and fish. The zone below has a cobbled bottom with coarse sediment in between. It is inhabited by abundant soft corals, anemones, hydroids, and ascidians. The most conspicuous among the many different forms are sessile coelenterates, the alcyonarian *Alcyonium paessleri*, and anemones such as *Artemidactis victrix* and *Hormathia lacunifera*. The mobile animals found on the bare zone are also found here, notably the fishes, *Pagothenia bernacchii* and *Trematomus pennellii*. Since the freezing point of the body fluids of teleost fish is normally above that of seawater—unlike invertebrates which are iso-osmotic or slightly hyperosmotic relative to seawater—this habitat is hazardous because the ambient water is near its freezing point and contact with ice is difficult to avoid. These benthic ice-foraging fish could not exist without antifreeze glycopeptides in their various fluid compartments. They remain mainly inactive on the bottom, hidden in crevices to avoid seals, or perched on sponges, thereby getting a better view of the water column.

Sponges of great variety of form cover up to 55% of the ground. The abundance of these gives a unique character to Antarctic benthos and provides cover, ecological niches, and food for a great variety of other animals. Glassy (siliceous) sponges, in other parts of the world confined to deep water, are particularly abundant, perhaps because oceanic water extends right to the edge of the Antarctic continent (Dell 1972). Among the most prominent are *Rosella nuda* and *Scolymastra joubini*, both known as white volcano sponges, which are up to 2 m tall and 1.5 m in diameter (Fig. 8.6).

Fig. 8.6   D.G. Lillie with siliceous sponges (the one he is holding was probably *Rosella villosa*) from the Ross Sea; *Terra Nova* expedition 1911–13. From Huxley (1913) *Scott's Last Expedition*, Smith, Elder & Co., London. Supplied by Scott Polar Research Institute.

Apart from these there are anemones, the alcyonarian already encountered in the zone above, hydroids, polychaetes, bryozoans, ascidians, and many molluscs. The total biomass of this sponge community is around $3\,kg\,m^{-2}$. Spicules from the sponges form a dense mat, varying in thickness from a few centimetres to more than $2\,m$, which provides a habitat for an abundant infauna. Some 12 500 individuals of an unspecified number of species, belonging to the crustacea, polychaeta, and other groups of worms, were counted in $1\,dm^3$ of this material.

The dynamics of these communities are complex. Sponges, with the exception of *Myacale acerata*, which increases its mass by as much as 67%, grow at rates which are imperceptible in the course of one year. *Myacale* has an advantage in competition for space and, indeed, sometimes overgrows other species, but this is offset by heavy selective predation. Sponges live by filtering out particulate organic matter and, probably, the concentration of phytoplankton in the southern Ross Sea is high enough to provide for them adequately.

A variety of predators, including starfish and the nudibranch *Austrodoris mcmurdensis*, feed on sponges, which nevertheless maintain large standing crops. The fish are generalist feeders, taking, among other things, polychaetes, fish eggs and small fish, and amphipods. In contrast to the situation in the Bering Sea, none of the marine mammals which frequent the Ross Sea feed on benthic infauna. The infauna in McMurdo Sound, comprising both deposit and suspension feeders, perhaps owes its luxuriance to the absence of disruption from this quarter. Another factor is certainly the high productivity of the benthic microflora. Differences between the infaunal biomasses on the east and west sides of the Sound are correlated with microalgal production. This in turn is related to currents, the east coast receiving water from the open sea whereas the west coast gets deoxygenated water flowing out from under the Ross Ice Shelf (Knox 1994, Arntz *et al.* 1994).

One general hypothesis about the relative importance of different environmental drivers in structuring these communities has been proposed recently (Thrush *et al.* 2006). The oceanographic circulation patterns and coastal topography/bathymetry are the main factors that regulate the shifts in physical factors such as light regime. Moreover, changes in the communities' diversity and biomass are predicted to be influenced by three main factors:

1. ice disturbance (e.g. via anchor ice and advection of supercooled water or icebergs);
2. photosynthetically available radiation (affected by ice and snow cover and water clarity);
3. the locations of polynyas and advection of planktonic production and larvae.

Interactions between these factors are expected to result in non-linear changes along the latitudinal gradient.

It is difficult to make precise comparison s between benthos in different parts of the high Antarctic. Sampling methods that have been employed vary greatly and the taxonomic basis still leaves much to be desired; identifications are mostly been made with preserved specimens by taxonomists who have not seen the living organisms, and there is a substantial backlog of undescribed collections. As far as one can tell, sublittoral communities all round the continent broadly resemble that just described. On the coast of Terre Adélie, almost on the Antarctic Circle at 66°33'S, the assemblages of organisms are generally similar to those in McMurdo Sound but brown seaweeds, such as *Phyllogigas* (*Himantothallus*) *grandifolius* and *Desmarestia menziesii*, are more in evidence and the diversity of fauna is greater (Arnaud 1974). Around the islands a tendency of the flora and fauna to become richer in species as one goes northwards is maintained. After the poverty in species of the terrestrial and freshwater habitats of Antarctica the richness of the marine benthos is astonishing. Roughly 100 species of macroalgae and over 4000 species of the more conspicuous kinds of benthic invertebrates have been recorded from the Southern Ocean (Fig. 8.7; Dell 1972, Arntz *et al.* 1994).

The species richness of Antarctic benthos varies widely between taxonomic groups. The most conspicuous groups of Antarctic benthos are the

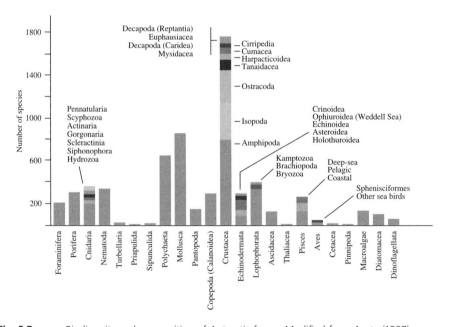

Fig. 8.7   Biodiversity and composition of Antarctic fauna. Modified from Arntz (1997).

polychaetes, gastropods, amphipods, bryozoans, isopods, and true sponges. In many studies, the number of species described from the Southern Ocean represents 8–12% of the world fauna (Clarke and Johnston 2003). One of the most important features of the Antarctic benthic fauna is the lack of durophagous (skeleton-breaking) predators, which is characteristic of shallow waters elsewhere: crabs, lobsters, and sharks are essentially absent, and there is only a very limited diversity of teleosts and skates (Aronson and Blake 2001).

The modern Antarctic benthic marine fauna inhabits a relatively atypical continental shelf environment (Clarke 2003a, 2003b). In contrast to all other continental shelves (including those of the Arctic) there is essentially no riverine input. Mudflats are rare and estuaries almost non-existent. Almost all terrestrial input comes via glacial processes or, in a few places, the wind. Most of the coastline is ice, with only 14% being rock (Clarke and Johnston 2003). This rocky coastline is subject to intense scour from floating ice and is consequently largely devoid of the traditional intertidal fauna. Clarke and Crame (1989, 1992) have proposed that periodic extensions and retreat of the Antarctic ice cap on Milankovitch frequencies may have been an important factor in driving speciation of the continental shelf fauna. As the ice sheet extended out over the shelf, distributions would have been fragmented with allopatric populations confined to refugia or even driven down the continental slope. Following retreat of the ice sheet, previously isolated populations would have mingled once more. But the effect of icebergs is widely accepted as one of the major environmental factors that influence the seabed landscape in the Antarctic (Fig. 8.8; Gutt 2001). Other factors to consider are the impacts of wave action, impact of sea ice in shallow water, and volcanic activities.

In general these communities are highly structured, with a high functional diversity and a considerable degree of patchiness in species composition at small or intermediate spatial scales (Gutt and Starmans 1998). Gutt (2007) has recently proposed a new classification of shelf-inhabiting Antarctic macro-zoobenthic communities. In general this classification consider two core communities, as follows.

1. The first is dominated by sessile suspension feeders supported by food entrained in high near-bottom currents. Variants of this community include assemblages without sponges, those that prefer sponge spicule mats as substratum and predator-driven systems.
2. The second core community is dominated by the infauna and mobile epifauna and controlled by vertical phytodetritus flux and soft sediments.

Between both core communities there is a broad range of a mixed assemblages that can be explained by a gradient in environmental conditions and trophic amensalism (interaction between two species in which one impedes or restricts the success of the other without being affected positively or negatively by the presence of the other).

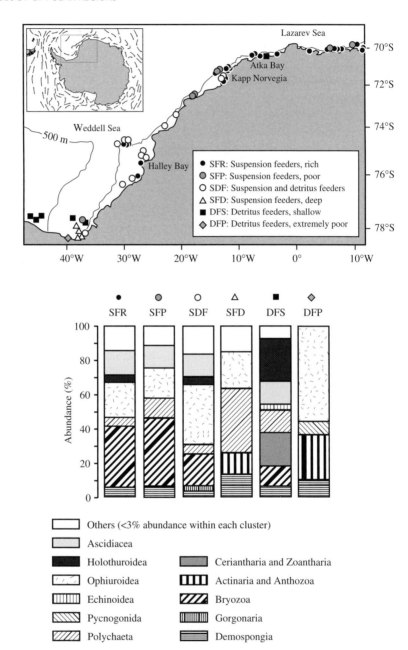

Fig. 8.8   Spatial distribution of the relative abundance proportions on a coarse taxonomic level of the main benthic communities identified in the eastern Weddell Sea. The main difference is the dominance of suspension feeders in the north sector of the open waters against the dominance of detritus feeders in the south sector. Modified from Gutt and Starmans (1998).

The sublittoral zone of Signey Island is a good example of benthos under moderately severe conditions which do not usually include formation of anchor ice (see Dell 1972, Arnaud 1974, Heywood and Whitaker, in Laws 1984, White, in Laws 1984, Dayton, in Smith 1990, Arntz *et al*. 1994, Knox 1994). Here the substrates vary from sand/silt, through gravel, to pebbles and cobbles. The ice abrasion zone extending down to 2 m bears an impoverished flora appearing only in the fast ice-free period and consisting of diatoms and the brown alga *Adenocystis utricularis*. Below this, from 2 to 8 m, growing on boulders are brown seaweeds, *Desmarestia* spp., and an underflora of red weeds. *Desmarestia* may be up to 1 m long and form dense stands with a cover of about 75% on boulders, falling to 22% on gravel. The mean biomass is in the range 230–830 g wet weight m^{-2}.

The next zone down, 8–11 m, is dominated by another brown seaweed, *Himanthothallus grandifolius*, which attains a cover of about 39% on boulders and 33.5% on gravel with biomasses of 150 and 40 g fresh weight m^{-2}, respectively. These standing crops are less than those of comparable growths off Anvers Island (64°S 64°W), which range from 1.64 to 6.34 kg m^{-2}, much the same as in temperate waters. Off King George Island (62°14′S 58°41′W), *Desmarestia* spp. are the most successful seaweeds under stable conditions but are replaced by *Himanthothallus* where substrates are exposed to turbulence or impact of icebergs. Below 11 m, around Signy, is a diminished flora of *Himanthothallus*, *Desmarestia*, and red algae.

Beds of seaweed provide shelter for various fish, *Notothenia coriiceps*, *Notothenia gibberifrons*, and *Trematomus newnesi*, together with molluscs and isopods. There are also annelids and nemerteans. Exposed rock supports a variety of echinoderms, pycnogonids, tunicates, sipunculids, sponges, hydroids, and bryozoa. The biomass of this fauna ranges from 4200 g wet weight m^{-2} on a rock overhang with mainly filter-feeding animals, to between 218 and 1723 g wet weight m^{-2} among the macroalgae. Estimates of the biomass of infaunal communities in mobile substrates, in which bivalves are dominant, range between 307 and 789 g wet weight m^{-2}. The large bivalve, *Laternula elliptica*, could not be dealt with by the sampling method used, but if it were included these values might increase up to as much as 2600 g wet weight m^{-2}.

Many of these benthic animals are suspension feeders and accordingly feed most actively during the summer peak of microphytoplankton. They have been presumed to cease feeding during winter but a study of bryozoans off Signy Island has shown that periods of near-zero activity during winter are actually short or non-existent. At this time they evidently feed on nanoplankton (Fig. 8.9; Clarke and Leakey 1996). The importance of the winter nanoplankton component of coastal plankton communities has also been shown at higher latitudes in a rare 8-year long-term study, the Rothera Oceanographic and Biological Time-Series (RaTs), in Marguerite Bay on

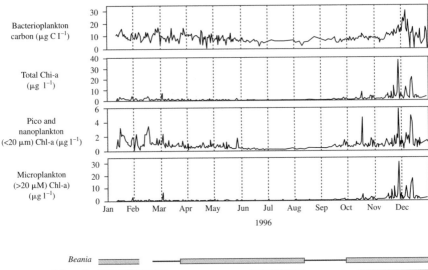

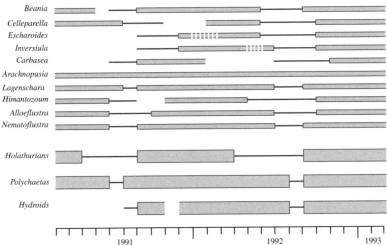

Fig. 8.9 (Top) Microplankton composition and biomass variability in nearshore waters of Maxwell Bay, Antarctica, during 1992/1993. Modified from Ahn *et al.* (1997). (Bottom) Feeding activity (blocks) and inactivity for a range of benthic suspension feeders during 1991–1993 near Signy Island. Broken sections indicate periods of likely feeding, with occasional cessation due to high current velocities. Modified from Barnes and Clarke (1995). Chl-a, chlorophyll a.

the Antarctic Peninsula at 67°37′S (Clarke *et al.* 2008). Rather surprisingly the length of the water phytoplankton bloom is actually longer at the southern site compared to Signy Island despite much shorter periods of favourable mixed-layer depths in Marguerite Bay. Benthic suspension feeders on Antarctic shelves feed on small-sized particles (e.g. ciliates, dinoflagellates, and other small phytoplankton) in contrast to species from other latitudes that mainly ingest zooplankton (Fig. 8.10; Orejas *et al.* 2003).

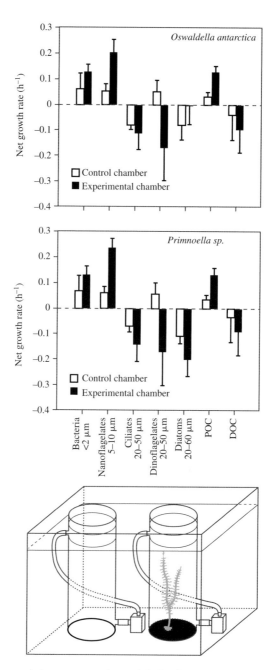

Fig. 8.10 The bottom panel shows an experimental device for conducting feeding experiments on two species of benthic cnidarian in an aquarium container with experimental and control chambers. (Top) Net growth rates (h−1) in the control (white bars) and experimental (black bars) chambers containing Oswaldella antarctica (hydroid) and Primnoella sp.(gorgonian) fedding on each plankton group and particulate (POC) and dissolved (DOC) organic carbon. The size range of each plankton group is indicated. Bars indicate standard error. Modified from Orejas et al. (2003).

Around the sub-Antarctic islands, in the absence of ice abrasion, really large kelps make their appearance. *Macrocystis pyrifera*, which may be up to 40 m in length and has fronds with gas bladders which float at the sea surface, is the most spectacular. Together with other large seaweeds, such as *Durvillea antarctica* and *Himanthothallus*, it is found around the coasts of South Georgia and Kerguelen. *Macrocystis*, especially, has an important effect on the littoral and sublittoral zones in general since it acts as a natural breakwater and reduces wave action. Additionally, the holdfasts and fronds of these kelps provide habitats for a rich variety of smaller algae and animals. Large numbers of the small bivalves of the genus *Gaimardia* occur attached to *Macrocystis*. Kelp beds are a favourite habitat for fish, where they prey primarily on crustacea, fish larvae, polychaetes, and molluscs. One species, *Notothenia neglecta*, grazes actively on the kelp itself. All this wealth of life provides food for large numbers of seabirds.

Some general characteristics of Antarctic benthic invertebrates should be mentioned. Slow, seasonal growth and delayed maturation are features which they share with Arctic species. One of the characteristics of many benthic organism groups, in particular molluscs and crustaceans, in polar regions is the fact that they reach much greater sizes than their counterparts in warmer waters. Again this seems anomalous considering the apparent scarcity of food. Of course small species are still present, and not all polar invertebrates are large. Antarctic sea spiders up to 40 cm across are a hundred times the size of the common European sea spider. The isopod *Glyptonotus antarcticus*, found throughout Antarctica, the Antarctic peninsula, and sub-Antarctic Islands from the intertidal to 790 m depth, grows up to 20 cm in length and weighs 70 g. For comparison, isopods in other parts of the world may reach a maximum size of just several centimetres. Other giants include sponges that are 2–4 m tall and ribbon worms over 3 m long. It is thought that this gigantism is brought about by a combination of factors (Peck *et al.* 2006). Low water temperatures certainly slow metabolic rates to the extent that growth rates are slow enough to enable organisms to live longer. Respiration rates are barely measurable in many of the benthic organisms by standard laboratory techniques. This results in polar organisms having life spans that are considerable longer than allied species from warmer waters. In fact some of the longest-living invertebrates are thought to be found in polar waters.

In the Antarctic sponges are often dominant organisms in benthic communities (Fig. 8.6). Some species, such as *Rosella nuda* (white volcano sponge), can grow up to 2 m high and weigh of up to 500 kg. It has been estimated that the Antarctic lollypop sponge (*Stylocordyla borealis*) can live up to 150 years. The 30–40-cm high *Rossellid* sponges are thought to be least 300 years old and the largest 2 m high are estimated to be 10 000 years old, which if true would make them the oldest living organism on the planet. Naturally these ages and/or growth rates cannot be measured directly, and these ages are mostly estimated from indirect measurements of the organism's metabolism combined with sophisticated mathematical models of growth rates. Therefore there is a very large degree of uncertainty in these estimates, but whatever the error in the methods it is clear that they are very old indeed.

However, many polar species are smaller than their counterparts in temperate regions, especially those that deposit calcium carbonate in exoskeletons or other structures, presumably because the solubility of calcite increases with a decrease in temperature so that its deposition becomes limiting. In temperate and tropical seas, benthic invertebrates typically have pelagic larval stages but it has been supposed that in polar regions few species do this, adopting instead some form of brood protection (Dell 1972). Large yolky eggs enclosed in capsules which are brooded are common in both Arctic and Antarctic molluscs. Eighty per cent of Antarctic sponges are either viviparous or show brood protection as against about 55% of those in tropical or temperate waters. Invertebrate groups that characteristically have pelagic larvae and which are dominant in the littoral and sublittoral elsewhere (e.g. prawns, lobsters, crabs, and barnacles), are poorly represented in polar regions. On the other hand, other crustacean groups, the amphipoda and isopoda, which can be regarded as preadapted to high latitudes in this respect, are abundant.

The relative absence of pelagic larval stages has various explanations. It avoids the hazard of surface waters where salinity may fall abruptly following ice melt and of turbulent seas with strong currents which might carry larvae away from suitable habitats. By the same token, of course, it limits dispersal and colonization. With the period of high phytoplankton production being so short it is also presumably of advantage to liberate juveniles in a stage of development at which they are best able to profit from the brief abundance of food. The recognition that ultraplankton production continues during the winter takes some of the force from this argument. Consideration of the total energy requirements of reproduction leads to the conclusion that production of larvae from large yolky eggs is more efficient under conditions of poor food supply and low temperature. Nevertheless, it appears that past surveys may not have been complete and that larvae of echinoderms, annelids, and nemerteans may sometimes occur in the plankton. There is evidence that these obtain particulate food from the water column—again it may be ultraplankton—especially near sea ice, during the winter. The 'rule' that protected development is more common towards the poles thus seems open to question (Knox 2006, Shreeve and Peck 1995).

There is clearly no uniform pattern in the degree of coupling of Antarctic invertebrate reproduction to the extreme seasonality of primary production in this environment, although uncoupling seems to prevail, and the percentage of largely uncoupled species seems to be higher than in temperate and Artic latitudes (Gili *et al.* 2001). For example, some species of cnidarians and sponges where coupling with the period of high primary production is indirect in this group, via a lecithotrophic larval stage in winter. However, frequent development of buds in Antarctic sponges under certain conditions has also recently observed in the Weddell Sea (Teixidó *et al.* 2006).

In bivalves, continuous reproduction seems to be a common pattern and many species have demersal lecithotrophic larvae or planktotrophic phases with demersal behaviour (Chiantore *et al.* 2000). In other groups, such as molluscs, metamorphosis takes place in egg capsules or some crustaceans attach their spawned eggs to the pleopods or brood their young in a marsupium (de Broyer *et al.* 2001). Uncoupling from the primary production cycle may be advantageous on evolutionary time scales because it should be easier to overcome glaciation periods when the ice shelves and the pack ice are extensive and open water blooms are scarce. On seasonal scales this decoupling may be important when short bloom periods that last perhaps a maximum of 8–12 weeks a year are followed by a much longer period with no major food input.

Lack of direct use of food sources does not preclude indirect use via seston and resuspended material from the seabed. The advantage is that these food sources are available to larvae, postlarvae, and juveniles all year round and can be advected from ice-free waters over long distances. Drift stages of some species have a tendency to cling close to the bottom (Arntz and Gili 2001).

In contrast to the Southern Ocean, planktotrophic larvae are more common in the Arctic (Pearse 1994). For example, in the north and east Greenland sector a strong influence of Atlantic water masses seems to transport meroplanktonic larvae into the region. Furthermore, the 'pelagic larvae' also include lecithotrophic larvae of the demersal type *Mya truncata*. *Laternula elliptica*, also with a pelagic stage, hatch from egg capsules as advanced juveniles, while Arctic ophiuroids have a tendency to reproduce via pelagic ophiopluteus larvae (Piepenburg 2000).

8.4 The benthos of deep waters

True deep-sea (abyssal) benthos has received little attention, and has often been viewed as a region of low biodiversity. However, three coordinated expeditions in the deep Weddell Sea (748–6348 m) between 2002 and 2005 have shown this not to be true (Brandt *et al.* 2007). Among the 13 000 specimens examined the team found: 200 polychaete species (81 previously unknown to science), 160 species of gastropods and bivalves, 76 species of sponge (17 previously undescribed), 674 isopods (585 new to science), 57 nematode species, and 158 species of foraminifera.

The deep-shelf mixed assemblage, found in the Ross Sea down to 523 m on fine sediments with erratic boulders, includes substantial populations of polychaetes, bryozoans, gorgonian corals, ophiuroid starfish, and crinoids. Other types of sediment have their own characteristic faunas (Dayton, in Smith 1990). On the south-western Weddell Sea shelf biomass wet weights up to 1.6 kg m^{-2} were found at around 250 m but at 2000 m biomass had dwindled to less than 1 g m^{-2} (Arntz et al. 1994).

Studies on the metazoan size fraction (32–1000 mm) from the South Sandwich Trench (Brandt et al. 2004) revealed unexpectedly high standing stocks, well above the predicted estimates from worldwide relationships of meiobenthos abundance and water depth. In particular, the greatest trench depth at 6300 m gave surprising results with regard to food supply and availability. Many species of meiofauna tend to be widespread and eurybathic in the Atlantic sector of the Southern Ocean. Nematodes predominated over the other taxonomic groups, as is the rule for all deep-sea communities.

Data from the Southern Ocean deep sea have shown that in very general terms, the macrofauna does not differ too much in composition on higher taxon level from that of other deep-sea regions of the world's oceans. For example, Southern Ocean deep-sea isopods show a high degree of endemism, probably due to the negligible sampling effort in the Southern Ocean deep sea in the past (Brandt et al. 2004). The most important taxa were polychaetes, peracarida (Crustacea), and molluscs (bivalves and gastropods). Within the peracarida it was the amphipods that comprised the major fraction (32% of all individuals), surpassed only by isopods (38%). This is in sharp contrast to other deep-sea samples where amphipods are much less important in terms of abundance.

Within the scavenging guild some 62 species of amphipods were collected and 98% of the individuals belong to *Lysianassoidea*, 31 species being collected from depths greater than 1000 m. There was a greater species richness in the eastern Weddell Sea shelf compared with other Antarctic areas. The Antarctic slope also seems to be richer in amphipods species compared with other regions of the world, while in the abyss, scavenger species richness appears to be lower in the Antarctic. A number of amphipod species extend their distribution from the shelf to the slope and only one was found to extend down into the abyssal zone. Therefore there is an apparent richness gradient from the shelf to the deep. Some interesting associations between cidaroid echinoids and amphipods have been described in the deep Antarctic benthos (de Broyer et al. 2004). At the same time, data on reproductive stages of some species including polychaetes suggest that species limited to abyssal depths are reproducing there.

The megafauna was less diverse than the other two size classes, and the most important taxa were the porifera, molluscs, echinoderms, and brachiopods. To date, 20 hexactinellid species have been reported from the deep Weddell Sea. This apparent high 'endemism' of Antarctic hexactinellid sponges is most likely to be an artefact resulting from the under-sampling of the Southern Ocean deep-sea fauna (Brandt *et al.* 2004).

The macrobenthic biodiversity in the deep-sea basins of the Norwegian and Greenland Seas in the Arctic Ocean shows the fauna is young and community development immature (it is no older than the Pleistocene). The impoverishment in species compared to the adjacent North Atlantic deep-sea basins is striking, with a high degree of endemism at species level but very low level of endemism at genus and family level (Clarke 2003a). In this context, local biodiversity are influenced by regional processes that are partially driven by ecology, together with unique historical events affecting large areas of the deep sea in both poles. Although the Arctic abyssal fauna contains elements suggesting ancient connection to the faunas of the Pacific and Atlantic, Quaternary glaciation, probably in concert with the massive Storegga slide, must have been the cause of considerable extinction within the deep sea as well as shelf faunas (Gage 2004). Because of their isolation and faunal impoverishment, it is unlikely that the Arctic and Nordic Seas basins have contributed in any way to diversification in the remaining global deep ocean.

8.5 Benthos under ice shelves

Ice shelves, which scarcely exist in the Arctic, are enormous in the Antarctic, covering a total of more than 1 400 000 km² or 7% of the total ice-covered area. The Ross Ice Shelf (530 000 km²) and the Filchner–Ronne Ice Shelf in the Weddell Sea (400 000 km²) are the largest. They vary in thickness from about 200 m at the seaward, floating, edge to about 600 m where they join the inland ice sheet. That life can exist, at least for a limited time, under these shelves is shown by the presence of marine fish in epishelf lakes. The question is whether resident communities are present. They do certainly exist near the seaward edge of shelves. Near White Island and Black Island (approximately 78°S 166°E), which emerge from the Ross Ice Shelf in McMurdo Sound, there are dense populations of filter-feeders living in darkness 30–50 km from the seaward edge (Knox 1994).

Sampling through a hole drilled in the Ross Ice Shelf more than 400 km from the open sea confirmed the presence of living organisms but evidence that these were permanent residents below the shelf, rather than strays, is equivocal. Some crustacea were found in the water column but the sediments seem devoid of infauna or sessile epifauna. This barrenness was confirmed by underwater video footage. Microbial biomass was highest at the sea floor, where it was around 1 mg bacterial $C m^{-2}$, much lower than found in sediments of the continental shelf and similar to that reported from impoverished abyssal regions. One cannot take results from one sampling point as representative but, considering possible sources of organic carbon for support of undershelf life, one is not encouraged to think that there can be much more life than that already found. Photosynthesis is obviously impossible. Chemosynthesis may occur. There is volcanic activity in the vicinity of the Ross Ice Shelf so that seepage of water carrying hydrogen sulphide might sustain populations based on bacteria and archaea. However, hydrothermal vents such as provide for this type of community in midocean are unlikely to be present under this ice shelf since they would produce obvious thinning of the ice. Another possibility is that seepages of methane, which has been detected in Antarctic marine sediments, might support methane-oxidizing bacteria and archaea (see below).

In late 2006 Antarctic researchers were able to gain a rare glimpse of potential life under an ice shelf. The RV Polarstern was the first research vessel to be able to extensively survey the benthos at the former position of the Larsen A and B ice shelves that had both collapsed since 1974. The collapse of the shelves had opened up an area of around $10\,000\,km^2$ that had previously been covered by ice shelf for at least 5000 years and possibly 12 000 years. Obviously it is not clear whether the lifeforms seen with underwater video or collected with benthic grabs were present before the shelves' collapse or migrants subsequent to collapse. However, despite these uncertainties the results of this study are the clearest picture we have to date of under-shelf marine benthos.

The Larsen zone seafloor sediments were extremely varied, ranging from bedrock to pure mud, and animals living on the sediment were highly varied as well. However, the species diversity was far less abundant in the Larsen A and B areas (perhaps only 1% animal abundance) compared with sea beds at similar depths in the eastern part of the Weddell Sea. In the relatively shallow waters of the Larsen zone, scientists found abundant deep-sea crinoids, echinoderms, and holuthurians. These species are more commonly found around 2000 m or so, so able to adapt to life where resources far more scarce: conditions similar to those under an ice shelf.

Very slow-growing glass sponges were also found with greatest densities in the Larsen A area, where lifeforms have had 7 more years to re-colonize than at Larsen B. The high number of juvenile forms of glass sponges observed probably indicates a shifting species composition and abundance in the past 12 years. Fifteen new amphipod species were found among the 400 specimens collected and four new species of cnidarians were found. Dense patches of fast-growing gelatinous sea squirts were found, which are thought to have grown in the Larsen B area since the ice shelf collapsed in 2002.

The 2006 expedition confirmed previous findings of American scientists in a 2005 expedition (Domack *et al.* 2005) that there is *cold seep* activity in the Larsen zone of the seabed, the first to be recorded in Antarctic waters. This supports the idea that the restricted environment below ice shelves may be suitable for chemotrophic systems. Cold seeps are regions where methane and hydrogen sulphide-rich waters are vented that in turn support a rich microbial activity and also bivalve molluscs. Both expeditions found dead bivalves and microbial mat systems common for such vent systems in other parts of the world.

In 2001 joint USA and German oceanoagraphic campaigns also first identified extensive hydrothermal activity in the Arctic basin, namely the Gakkel Ridge which is a gigantic volcanic mountain chain stretching beneath the Arctic Ocean (Edmonds *et al.* 2003, Jokat *et al.* 2003). The Gakkel ridge extends from north of Greenland to Siberia. It is the northernmost portion of the mid-ocean ridge system, the global 75 000-km-long volcanic chain. The Gakkel Ridge is of particular interest for scientists, because it spreads extremely slowly at about 1 cm year^{-1}, the slowest rate of any mid-ocean ridge and 20 times slower than the better-explored East Pacific Ridge. With deep valleys 5500 m beneath the sea surface and 5000-m-high summits, the Gakkel ridge is higher than the Alps. As with other hydrothermal systems in the world's deep oceans it is presumed that the microbial and vent fauna associated with this ridge will be the source of considerable scientific interest in the coming decades. In particular much effort will be made to compare and contrast the vent organisms found in the long-time-isolated Arctic vent system with those in other oceans which have been less isolated.

8.6 Seasonality and dynamics of benthic communities

The Antarctic spring is considered one of the planet's principal episodes of oceanic primary production (Hense *et al.* 2003), reaching values in excess of 1 mg chlorophyll *a* l^{-1} in just a few weeks. More than 10^7 km^2 of sea ice containing a huge trapped biomass melt, whereas the sunlight period considerably increases, driving notable changes within an ecosystem just emerging from a long, dark winter. This explosion of life is immediately followed by a growth spurt in the life cycle of the krill and other zooplankton grazers. Most of the large predators abandon the High Antarctic at the start of the long austral winter, when the continental shelf and large areas of the open ocean pass through a seasonal coverage of ice more than a metre thick. This general model set the conditions for one of the long-lasting questions in Antarctic marine science; that is, the pronounced seasonality in marine benthos, referred to by some benthic researchers as the *Antarctic paradox* (Clarke 1988). For many years there were a series of misconceptions, chief among them the notion that the High Antarctic benthic fauna undergoes a period of low activity in winter as a consequence of reduced food availability, given that the seawater temperature remains practically constant all year round.

While the marked environmental seasonality naturally does influence and condition life in the water column, the first inklings that the Antarctic paradox might not be entirely accurate arose after the discovery of the rich marine fauna dwelling on the continental shelves in the High Antarctic. Over the past 20 years, the region has been shown to host one of the most diverse, high-biomass benthic communities in the world's oceans. Suspension feeders constitute the bulk of these communities, which depend on the particles settling down from the upper layers of the water column or laterally advected to them by the currents.

Slow metabolic rates associated with a low energy demand, longevity, and gigantism (Peck *et al.* 2006) and other traits connected with reproduction patterns at first glance appear to be in consistent with the tenets of the Antarctic paradox, with the dormant state prevailing in winter. However, recent discoveries have forced researchers to reconsider the paradox. For instance, many species exhibit reproduction rates similar to those in other regions of the worlds oceans, while others quickly occupy areas scraped clean by icebergs, showing higher growth rates than expected (Teixidó *et al.* 2004). A series of experimental observations have given solid evidence to support the suspicion that benthic organisms during the Antarctic winter may not be as inactive as hitherto thought.

There is strong evidence that the Antarctic shelf benthos experiences a seasonal cycle of particulate organic matter (POM) flux resembling the patterns of primary production and phytoplankton biomass in the water column (Smith *et al.* 2006). Downward particle flux are especially strong on the Antarctic shelf if the melting of sea ice and the intense, but often short-lived, summer phytoplankton blooms cause rapid export of POM to the shelf floor. Several factors controlling phytoplankton blooms cause coupling or decoupling export flux (Fig. 8.11). These include wind-driven dispersal of sea ice prior to its melting, advective processes, development of nekton and zooplankton grazer assemblages, and wind-driven surface mixing that increases sedimentation rates or reduces primary production (Leventer, in Thomas and Dieckmann 2003).

Organic matter deposited during the summer may provide a food source, or *food bank*, for benthic organisms during winter months when primary production is low. The existence of food banks extending over hundreds of kilometres offers a potential food source for numerous bottom-dwelling organisms (Mincks *et al.* 2005). This pattern known as *green carpets* tends to form at the beginning of the spring, when

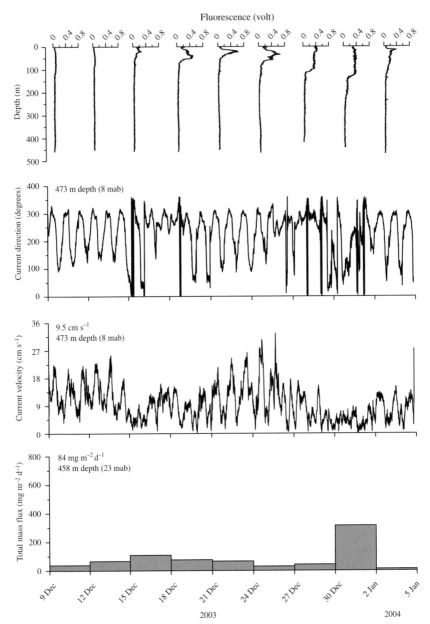

Fig. 8.11 Schematic representation of a 14-day cycle of particle sinking, current velocity, current direction, and fluorescence (top) recorded by a mooring equipped with a sediment trap and a current meter located near the sea floor and in the Weddell Sea at 485 m depth. It is possible to observe that the diurnal tides (spring tides) influence particle fluxes near the sea floor (low average current velocity co-existed with high particle fluxes). Winds storm deepen the upper mixed layer and exert influence on the particle fluxes. Apparently, atmospheric forcing also influences the pulses of primary production exports to the benthic communities (chlorophyll depth maximum, caused by the passing of a storm in the area, coincided with high sedimentation rates). The data in the top left-hand corner of the bottom three panels are the mean values for the data-set. mab, Metres above bottom. Modified from Isla *et al.* (2006).

the high primary production generated by melting ice is not immediately exploited by planktonic grazers and settles on the shelf seabed in a time span of hours to days. The seabed sediments in these regions store high-quality food and grain sizes suitable for the anatomic structures of benthic suspension feeders and other benthic groups. On average, the measured concentrations of protein ($3\,mg\,g^{-1}$) and lipids ($2\,mg\,g^{-1}$) were higher than on other continental shelves and similar to the contents found in settling particles (Isla *et al.* 2006). Sediment inventories of algal pigments and amino acids in Antarctic shelf sediments are additional evidence of the high nutritional value of sedimented materials (Mincks *et al* 2005).

In polar regions hydrodynamic features, such as mesoscale and near-bottom currents, water intrusions from open seas, and the effects of shelves on barotropic eddies (Fahrbach *et al.* 1992), all facilitate the continuous renewal of water close to the sea floor. These physical processes may greatly contribute to explain the dense communities on the shelf and slope (Fig. 8.12). It is not necessary that the water renewal is caused by strong bottom currents but it must be at a relatively constant rate. Tides also act as a mechanism by which resuspension of the food banks takes place and supplies particles to benthic communties throughout the year (Smith *et al.* 2006).

The fact that benthic communities are controlled by vertical flux and that ice cover severely reduces primary production have been discussed extensively. However, benthic communities must also rely on horizontal advection of organic material produced further away in ice-free or littoral areas. The high sedimentation rates observed in the eastern and western McMurdo Sound (Dunbar *et al.* 1989) are associated with meso-scale water-circulation patterns and corroborate the importance of horizontal advection. They observed that advective transport from the Ross Sea supplied biogenic sediment to the eastern McMurdo Sound. In this area, dense assemblages of sponges and coelenterates were observed ,one order of magnitude greater than in the western part where the assemblages are dominated by soft-bottom infaunal species (polychaetes and nemerteans; Dayton, in Smith 1990). Interestingly, the sediments in these areas contain high concentrations of benthic microalgae (300–$900\,mg$ chlorophyll $a\,m^{-2}$).

The link between general water circulation and benthic distribution and abundance is a general phenomenon in the Arctic: Feder *et al.* (1994) observed that higher numbers of benthic organisms in the Chukchi Bight and Kotzebue Sound were attributed to complex current patterns where organic carbon of high quality was advected into the area associated with the Alaska Coastal Water. This POM provided a food source for the fauna when it fluxed to the bottom while POM in the benthic boundary layer

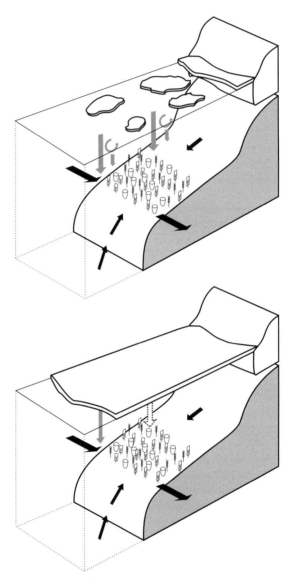

Fig. 8.12   Idealized diagram of food-supply processes to the Antarctic benthos in summer (top) and winter (bottom). Grey arrows mean microplankton and faecal pellet (summer) and pico- and nanoplankton (winter) vertical flux; white arrows show zooplankton vertical migration (winter); black arrows show lateral water flow from the shelf, slope, and shallow areas (all year). Modified from Gili et al. (2001).

supported interface- and suspension-feeding taxa in areas with strong bottom currents.

The food reaching the sea floor via vertical flux can strongly affect the benthos of adjacent areas if it is advected laterally once it has sunk from the surface waters (Feder *et al.* 2005): Vertical flux of organic matter in areas where low phytoplankton production does not supply adequate food to the benthos (under Alaska Coastal Water in the Bight and Sound), but resuspension of bottom material from productive areas via the Bering Shelf–Anadyr Water together with horizontal advective forces can increase availability of organic matter to the benthos. A different situation is described for the the northern Bering and south-eastern Chukchi Seas where almost 70% of water-column primary production escapes zooplankton grazing and settles to the bottom (Grebmeier *et al.* 2006). In these areas, organic-matter flux to the bottom is supplemented by organic material derived from ice algae and also with materials advected from nearby regions.

Orejas *et al.* (2000, 2003) postulate that the resuspension of POM and microalgae deposited during summer months may contribute substantially to food availability for Antarctic benthic organisms over the winter. At the same time, they proposed two strategies for suspension feeders, which are closely related with physical processes:

1. species with a high renewal rate, high ingestion rate of zooplankton, and low maintenance energy requirements can accumulate biomass and energy reserves rapidly during summer months;
2. species with low ingestion rates, less able to use seasonal food supplies when (and if) they appear have adaptations to feed and meet low metabolic costs at very low food concentrations throughout most (or all) of the Antarctic winter (Fig. 8.13).

The first group of species are associated with spring and summer blooms, and these organisms benefit from the short-term high production level. The second group would be related to the continuous food availability in which the role of sediment and resuspension processes also have a major impact on the feeding strategies. This model is in accordance with other studies than demonstrate both the continuous activity of sessile organisms (Barnes and Clarke 1995) and the availability of small particles in near bottom layers all through the year (Ahn *et al.* 1997).

Therefore the new evidence of the physicochemical conditions at the High Antarctic shelf seabed have made it necessary to reconsider the paradox that has served as a cornerstone for the understanding of polar marine benthic ecosystems for many years. Resuspension by tidal currents and the high nutritional quality of the seabed sediment after the summer permit benthic trophic conditions to remain almost constant throughout the year, setting the basis for a new model of Antarctic seasonality.

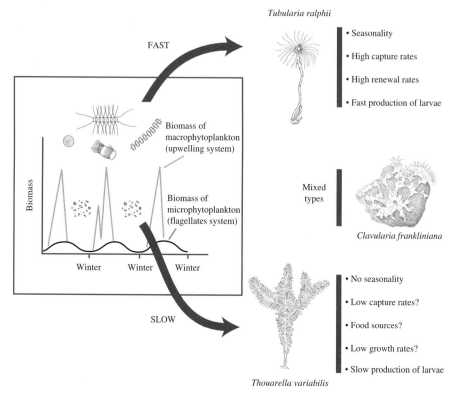

Fig. 8.13   Schematic of a trophic model for Antarctic benthos, the pelagic Antarctic recycling system. The fast system shows periodical summer blooms with a typical phytoplankton community, represented here by *Tubularia ralphii*. The slow system shows a stable continuous system with a typical microbial planktonic community, represented here by *Thouarella variabilis*. In between are the mixed types, with *Clavularia frankliniana* as a representative. See text for details. Modified from Orejas *et al.* (2000).

A model including these characteristics could help explain the diversity and high biomass of benthic communities in the High Antarctic, even when food input from the euphotic zone becomes scarce during the long winter (Fig. 8.14).

8.7 Comparisons and conclusions

Below the reach of floe ice, the sea bottom provides a stable environment in which desiccation, the major limiting factor for life in polar land habitats, is inoperative. Light is a limiting factor for the plants but in shelf waters the benthos also receives organic material produced by photosynthesis in

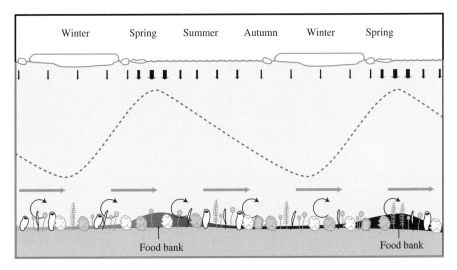

| Winter | Spring | Summer | Autumn | Winter | Spring |

Food bank   Food bank

Fig. 8.14   Synoptic view of the processes described in the text showing the seasonal vertical flux of new organic matter originated mainly at the beginning of spring (wavy line), the seasonal variation of food banks and the lateral and resuspension transport just above the seabed (arrows close the bottom).

the water column. The biomass per unit area in the polar benthos is usually greater, often much greater, than that on nearby terrestrial sites, although rather less than that of benthos in temperate or tropical regions.

Predictable environments with an intermediate level of stability generally have greater species diversity than unstable ones and this is evident in the polar benthos. Diversity is also promoted by the variety of substrates available. This high diversity is coupled with a tendency to even distribution in the abundance of the different species. In one area of McMurdo Sound, 11 species were found to have population densities of more than 2000 m^{-2}.

The idea of a consistently very poor Arctic benthos in contrast to the rich Antarctic bottom fauna is now questioned because the benthic communities vary broadly in faunistic diversity between both Arctic and Antarctic. On a global scale, in terms of biogeographical diversity, both Arctic and Antarctic benthic communities seem to be characterized by intermediate species richness (Piepenburg 2005). Mesoscale hydrography and sea-ice cover events form 'hot spots' of tight pelago–benthic coupling and, hence, high benthic biomass occurs in both regions. The factors that major affect the benthic diversity and structure are more diverse in the Arctic that in the Antarctic but, the recent discovery of the great disturbance by the iceberg scouring in Antarctic shelf ecosystems indicates that both regions are subject to similar impacts.

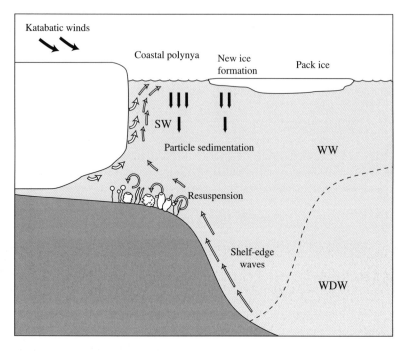

Fig. 8.15   Thermohaline circulation under the open polynya. Resuspension processes and their influence on the benthic communities in the Weddell Sea Shelf. SW, surface water; WDW, warm deep water; WW, winter water. Modified from Scharek and Nöthig (1995).

As a broad generalization, predation was lower and community structure more archaic in offshore, deep-water, and shelf habitats compared to nearshore, shallow-water habitats in the Antarctic (Aronson and Blake 2001). The list of biological and physical disturbance in the Arctic is much higher. In comparison to the Antarctic, the benthic Arctic communities experience grazing pressure from a long list of predators, from crabs to whales, which feed mainly on epifauna and endofauna but also on macrofauna (Feder and Jewett 1981). For example, 25 species of mammals in the Bering Sea consume around 9×10^6–10×10^6 t of nekton and benthic species.

River run-offs in the Arctic deposit a huge quantity of fine sediments on the seafloor, which favours the development of a rich sedimentivorous fauna rather than sessile filter feeders. These also increase the level of perturbation because of their bioturbation effects. This is more similar to the benthic communities in all other oceans, except the Antarctic (Gili *et al.* 2006).

The Antarctic pelagic zone is a highly efficient retention system since losses due to sinking particles are exceptionally low (Fig. 8.15; Scharek and Nöthig 1995). In general vertical flux rates are much higher from new production (bloom) systems than from those dominated by regenerated primary production. New production can be channeled directly and efficiently into a system that is a classical regenerating community and also behaves as a retention system (Smetacek *et al.* 1990). This explains why such large stocks of zooplankton, such as krill, can be maintained in a low-production system, as well as the high benthic diversity and biomass in the High Antarctic. Well-developed benthic communities may exert an important predatory impact on plankton populations and on the abundance of suspended organic matter in the water mass adjacent to the

Fig. 8.16   Seabed photographs of Antarctic and Arctic benthic communities. Assemblages dominated by suspension feeders, gorgonians (a) and hexactinellid sponges (b) from the shelf of the southern Weddell Sea, Antarctic, at 250 m depth. Assemblages dominated by suspension feeders from the waters off north-west Spitsbergen, Arctic, at 100 m depth (c) and from the north-western Barents Sea, Arctic, at 80 m depth (d). Photographs by (a, b) Julian Gutt, Alfred Wegener Institut for Polar and Marine Research, Bremerhaven and (c, d) Dieter Piepenburg, Institute for Polar Ecology, University of Kiel (see colour plate).

bottom. Moreover their activity may increase remineralization of organic matter close to the bottom. Since suspension and sediment feeders are extremely efficient organisms in terms of energy transfer from the pelagic system to the benthic system, they will build patchy communities to mirror the patchiness of food distribution (Fig. 8.16).

9 Birds and mammals in polar regions

9.1 Introduction

The birds and mammals are in the main the predators in the open seas and coastal regions of the polar regions, and again they are mainly of different species north and south. Birds and mammals are visible on top of the sea ice (Fig. 9.1), and they have evolved foraging techniques adapted to the physical nature of the ice and also make use of it as a comparatively safe breeding ground. The emperor penguin in the Antarctic and the polar bear in the Arctic are some of the most iconic marine species on Earth and, although they are as geographically distinct as is possible, sea ice provides a platform for, and dictates the timing of, their entire ways of life. The similarities in the way bird and mammal life histories have adapted to the sea-ice environment the Arctic and Antarctic mean they are dealt with together here, despite the different species involved (Table 9.1).

9.2 Seabirds

The diversity of birds frequenting the open waters of the Arctic is low, about 95% of those breeding in the Arctic belonging to four species, the northern fulmar (*Fulmarus glacialis*), the kittiwake (*Rissa tridactyla*), Brünnich's guillemot or thick-billed murre (*Uria lomvia*), and the Dovekie or little Auk (*Alle alle*; Sage 1986, Pielou 1994). These breed on land in colonies which are often huge (Fig. 9.2). Arrival at their breeding grounds is synchronized with the break-up of ice so that zooplankton and young fish are available when young are hatched. A colony of Dovekies containing 100 000 pairs transports some 71 t of zooplankton from seas to the colony during the 4 weeks of summer.

The distribution of colonies and of the birds at sea is related to regional abundance and accessibility of prey. Polynyas (see Chapter 7), being areas

Fig. 9.1   Large numbers of penguins and seals using the sea ice as a platform, a substitute for land (photograph by David N. Thomas).

of upwelling, are productive and particularly important for overwintering, when they remain open. Fronts (see Chapter 6) provide zones of spatially predictable, high productivity and are also marked by concentrations of seabirds. Lancaster Sound (approximately 74°N 85°W) is an area where several million birds congregate in the summer. Of the eight species found there, three, the northern fulmar (Fig. 9.3), kittiwake, and Brünnich's guillemot, feed largely on the Arctic cod. Of these three species the kittiwake is a surface feeder, the fulmar has limited diving capability, and the Brünnich's guillemot undertakes wing-propelled dives to depths of 50–75 m for 2–3 min. The cod is sustained by algae growing on the undersurface of the sea ice. This, then, is a simple food web with the flow of energy channelled mainly through one species of fish.

The marginal ice zones (see Chapter 7), where melting can create high localized productivity, are important for regional biological activity. Many species of seabirds and mammals use the ice edges as migration routes in the spring where they are dependent on the reliable food supply they offer (Fig. 9.3). Availability of prey is reflected, too, in the density of seabirds, which in the marginal ice zone of the Bering Sea is estimated as 500 individuals km^{-2} as compared with 0.1 km^{-2} in the ice to the north and 10 km^{-2} in adjacent open water. Large numbers of seabirds are regularly found in the summer pack ice, feeding on Arctic cod and crustaceans, but only Ross's gull (*Rhodostethia rosea*), ivory gull (*Pagophila eburnea*), and the black guillemot (*Cepphus grylle*) are characteristic of the pack and dependent on ice-associated fauna for the bulk of their food. The two gulls are rarely seen over the open sea although some races of the black guillemot have distributions going into temperate regions. Just south of the ice edge, however, a variety of species is to be found. In the Barents Sea, for example, glaucous gulls (*Larus hyperboreus*), herring gulls (*L. argentatus*),

Table 9.1 A list of marine mammals and birds whose presence is characteristic of sea-ice covered waters.

Common name	Genus and species
Mammals	
Antarctic	
Crabeater seal	*Lobodon carcinophagus*
Leopard seal	*Hydrurga leptonyx*
Weddell seal	*Leptonychotes weddellii*
Ross seal	*Ommatophoca rossi*
Antarctic fur seal	*Arctocephalus gazella*
Common Minke whale	*Balaenoptera bonaerensis*
Killer whale	*Orcinus orca*
Sperm whale	*Physeter macrocephalus*
Southern bottlenose whale or Antarctic bottlenosed whale	*Hyperoodon planifroms*
Arctic	
Ringed seal	*Phoca hispida*
Harp seal	*Phoca groenlandica*
Hooded seal	*Cystophora cristata*
Bearded seal	*Erignathus barbatus*
Walrus	*Odobenus rosmarus*
Polar bear	*Ursus maritimus*
Bowhead whale	*Balaena mysticetus*
Minke whale	*Balaenopterus acutorostrata*
Gray whale	*Eschrichtius robustus*
Narwhal	*Monodon monoceros*
Beluga or white whale	*Delphinapterus leucas*
Killer whale	*Orcinus orca*
Birds	
Antarctic	
Emperor penguin	*Aptenodytes forsteri*
Adélie penguin	*Pygoscelis adeliae*
Southern giant fulmar	*Macronectes giganteus*
Antarctic fulmar	*Fulmarus glacialoides*
Snow petrel	*Pagodroma nivea*
Antarctic petrel	*Thalassoica antarctica*
Blue petrel	*Halobaena caerulea*
Wilson's storm-petrel	*Oceanites oceanicus*
South Polar skua	*Catharacta maccormicki*

(Continued)

Table 9.1 (Continued)

Common name	Genus and species
Arctic	
Northern fulmar	*Fulmarus glacialis*
Ivory gull	*Pagophila eburnea*
Ross's gull	*Rhodostethia rosea*
Eider duck	Somateria spp.
Long tailed duck	*Clangula hyemalis*
Thick-billed murre or Brünnich's guillemot	*Uria lomvia*
Black guillemot	*Cepphus grylle*
Dovekie (little Auk)	*Alle alle*

From Ainley *et al.*, in Thomas and Dieckmann (2003).

Brünnich's guillemot, and little auks move in from open water to congregate at the ice edge. The ice zone around the coast is thus a barrier to them and breeding colonies are perforce located to the south of the summer ice margin. In the Davis Strait and the Labrador Sea ivory gulls are absent only during the short ice-free period in late summer. During March the largest numbers are at the ice edge near the breeding areas of the hooded seal, where they feed on the afterbirth. In April and May they prey particularly on lantern fish near the ice edge.

Since the great auk (*Alca impennis*), which nested on offshore skerries, became extinct on 4 June 1844 there has been no penguin-like, flightless, diving seabird in the Arctic. Presumably, the great auk foraged at the ice edge as far away as possible from polar bears. As in the Antarctic, flighted birds, such as the existing auks, use the pack ice as feeding ground but do not breed on it. Access to open water through predictable polynyas seem particularly important for the wintering of birds in high latitudes. Most of the population of Ross's gulls seem to winter around them, for example, at the Velikaya Sibirskaya polynya in the Russian Arctic. Regular wintering populations of Brünnich's guillemot, Dovekies (*Alle alle*), and long-tailed ducks (*Clangula hyemalis*) are found in the vicinity of polynyas north of the edge of the pack ice. One of the most astounding examples of the importance of areas of open water in pack ice regions is the recently discovered wintering areas of the spectacled eider (*Somateria fischeri*), where satellite tracking and subsequent aerial surveys discovered over 300 000 birds in the pack-ice region of the Bering Sea, accounting for the majority of the global population.

In the Antarctic, there are two main groups of seabirds, the procellariforms (albatrosses, petrels, etc.) and the penguins; these are two very different groups that are both very highly adapted to marine conditions (Croxall, in Laws 1984, Knox 1994, Williams 1995). The wandering albatross (*Diomeda*

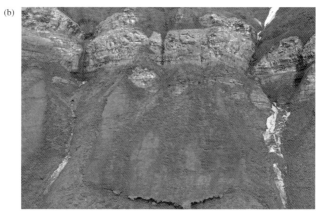

Fig. 9.2 (a) Colonies of sea birds such as these Brünnich's guillemots (*Uria lomvia*) produce copious amounts of guano that runs off, fertilizing surrounding ground and coastal waters (photograph by Rupert Krapp). (b) Stripes of more luxuriant vegetation below areas with dense concentrations of nesting birds on the coast of Svalbard (photograph by Marcel Nicolaus) see chapter 3.

exulans; Fig. 9.4) navigates and finds prey in immense areas of ocean. It is a large bird, weighing up to 10 kg, with a wing span of up to 3.5 m, which spends most of its adult life at sea and may live to be 70–80 years old. Its flight is seemingly effortless, using updraughts to give it height to glide with minimum movement of its stiffly held, narrow wings. It is mostly at latitudes between 40 and 50°S that albatrosses can find sufficiently strong and constant winds for this mode of flight; however, the closely related southern giant petrel has colonies below 60°S. Using satellite telemetry to track birds equipped with small transmitters has revealed foraging trips from nests on Îles Crozet (45°S 52°W) that may cover between 3600 and 15 000 km, over

Fig. 9.3 Northern fulmar (*Fulmaris glacialis*), photographed off the coast of Svalbard (photograph by Marcel Nicolaus).

distances of up to 900 km day^{-1}. The recent development of miniature light-level recorders has allowed the movements of albatrosses to be recorded over longer periods and have revealed the regular circumnavigations of the Southern Ocean undertaken by these birds. The flight path and foraging strategy are determined by the wind. When following the wind they fly at a slight angle to the left, which, in the southern hemisphere, leads them away from cyclonic lows and towards high pressure. However, even with these advances in knowledge of where albatrosses travel understanding exactly how they capture squid, their principal prey, remains uncertain, although some species of albatross can dive up to 10 m.

As is obvious to the seafarer, these birds are also opportunistic scavengers. It is this habit that has brought albatrosses into conflict with humans as the areas in which albatrosses overlap with long-line fishing have increased. As its name suggests, long-line fishing involves setting long lines of baited hooks which prove irresistible to the albatrosses, which become hooked and then drown. The resultant declines in the populations of albatross species has made them one of the most threatened bird families on Earth.

A number of petrels species are clearly associated with sea ice, although no flighted birds breed on sea ice and visit it only on transit between areas of open water. The snow petrel (*Pagodroma nivea*) is even mentioned in maritime guides for sailors as a good indicator for sea-ice conditions, and like the two penguins discussed below this is an obligate ice species (Fig. 9.5). The entirely white plumage of the snow petrel allows them to sit on the edges of ice floes waiting for fish and crustaceans to venture from under the floe edge, when the petrel will dive to catch its prey. However, snow petrels will take advantage of any available food source and they have been observed picking at Weddell seal faecal matter around the holes used by the seals to haul out.

Fig. 9.4   A wandering albatross (*Diomedea exulans*; photograph by Finlo Cottier).

Fig. 9.5   The snow petrel (*Pagodroma nivea*) is an obligate ice species well camouflaged against white snow covered ice flues (photograph by David N. Thomas).

Snow petrels do not nest near the pack, but instead nest on rocky ledges or cliff faces. Snow petrels have been found nesting more than 180 km from the coast on mountain summits (nunataks) on the Antarctic continent. This is even more remarkable considering that the adults must fly to and from the sea ice to collect food for their chicks.

Ross and ivory gulls in the Arctic use similar strategies to the snow petrel, where their white coloration enables them to feed from the edges of ice floes on Arctic cod, other fish, and amphipods grazing on the underside of ice floes. Just like the snow petrels ivory gulls also nest in snow-free mountain outcrops. It is thought that snow petrels and ivory gulls seek out these nesting sites to have a huge expanse of ice around them to protect them from predation from skuas and foxes/polar bears respectively.

Another common petrel in the Antarctic pack ice, especially in marginal ice zones, is the black and white Antarctic petrel. In contrast to the snow petrel this species flies quickly along water edges and dives into the water to pursue prey underwater, swimming using their wings for propulsion. Likewise the Arctic black guillemot and thick-billed murre feed by diving under ice floes to search for crustaceans and fish.

Of the 17 species of penguins there are six—the emperor, king, Adélie, chinstrap, gentoo, and macaroni—that are found in the Southern Ocean (Fig. 9.6). For all of these species their characteristic feeding technique is pursuit diving. Miniature depth recorders, logging the number of dives and their depth and duration, have been used to study several species in

(a)

(b)

(*Continued*)

Fig. 9.6   (a) King, (b) emperor, and (c) Adélie (on left of emperor) penguins (photographs by David N. Thomas) (see colour plate).

several locations. King penguins (*Aptenodytes patagonica*) dive to depths of 236–265 m after fish and squid. Chinstrap and Adélie penguins, both krill eaters, dive mostly to shallow depths up to 70 and 170 m respectively. Radiotelemetry, movement sensors, and even miniature cameras have been used to monitor behaviour at sea. This has revealed a range of foraging behaviors related to local distribution of prey including benthic diving in chinstrap penguin, although apparently still feeding on krill. Foraging ranges and chick feeding intervals are different for the various species; gentoo and chinstrap penguins range within about 20–30 km of the nesting site whereas Adélie penguins go further, up to 100 km, although these distances differ from location to location. Of the six penguin species mentioned the king and emperor lay a single egg, whereas the others lay two eggs, although the macaroni penguin always discards the first egg once the second has been laid (reasons for this are the subject of much, unresolved, debate).

The emperor penguin (Williams 1995, Ancel *et al.* 1997) is superbly adapted to existence on sea ice and has had perhaps 40 million years to evolve to this state. The minimum total population is about 195 400, distributed in 42 breeding colonies around the continent. It feeds by pursuit diving for fish, squid, and krill, going to depths of as much as 500 m. Living entirely on the ice when not at sea, it gets round the difficulty of nesting by incubating its single egg balanced on top of its feet and enveloped in a feathered fold of abdominal skin. This needs a flat surface and breeding sites are on level sea ice in sheltered situations (Fig. 9.7). Eggs are laid at the end of the the summer and the females then go off to sea to recuperate. Incubation is carried out over winter by the males, so that the chicks eventually fledge by midsummer and adults have sufficient pause to moult and before the onset of the next winter. It was thought initially that young birds drifted

Fig. 9.7   Young emperor penguins, preparing for their first swim from the fast ice edge in Drescher Inlet, Weddell Sea (photograph by David N. Thomas).

north on ice floes after fledging; however, recent satellite tracking of young birds immediately after fledging suggests that they may travel well away from the sea ice, even north of 60°S.

The emperor penguin has adaptations for survival through the Antarctic winter when temperatures on the ice may fall to −48°C and winds reach 180 km h^{-1}. First, the bird is large, weighing up to 46 kg, and thus has a small surface area/volume ratio which is further minimized by having flippers and bill about 25% smaller in proportion than other penguins. Insulation is provided by double-layered, high-density feathers and a 2–3-cm-thick layer of subcutaneous fat. The thermal conductivity of fat is a quarter of that of water but the feathers, with the air layer which they entrap, provide more than 80% of the thermal insulation. The feet have a vascular counter-current heat exchange system that reduces heat loss and avoids them becoming frozen to the ice. There is also a nasal heat-recovery system that retains the warmth of the outgoing air to heat the incoming breath, thus avoiding the drawing in of freezing-cold air into the lungs on each breath. These features allow the metabolic rate to remain at normal level down to a critical temperature of −10°C, below which it must be increased if body temperature is to be maintained. A remarkable behavioural characteristic also enables heat and energy reserves to be conserved below this temperature. Unlike other penguins, which in common with other birds have strong territorial instincts, the male emperor does not object to being in close proximity to his fellows and under adverse conditions joins a huddle which may contain as many as 5000 birds, packed 10 to the square metre. The whole huddle moves slowly downwind as birds on the windward side move along the flanks and re-enter it on the lee side. If the emperors are startled and raise their heads, steam can be seen rising from the huddle. It is estimated that by this means the daily loss of body weight is cut by 25–50%. At the end of the winter the male emperor, which by then has lost up to 40% of his summer body weight, still retains enough reserves to produce a nutritious secretion which keeps its chick going for its first few days until the female arrives from the sea with food.

Having a life history so intrinsically linked to sea ice means that emperor penguins are inevitably sensitive to changes in both the distribution and duration of sea-ice. Clearly the early break-out of ice before chicks have moulted into their adult plumage can cause substantially losses. Large-scale climatic changes have been attributed to the reduction by almost 50% of an emperor penguin colony in Adélie Land, although there is little evidence of large-scale changes in the population. However, the remote and inhospitable location of many colonies means that regular monitoring numbers is restricted to a relatively small number of the known colonies.

The Adélie penguin, *Pygoscelis adeliae* (Williams 1995), is the most numerous of Antarctic penguins and although it is frequently seen in the pack ice it nests on land during the Antarctic summer. Arrival at the breeding colonies in November often entails a march across extensive sea ice in the expectation that the ice will be replaced by open water during the chick-rearing period in January, when demand for prey is greatest (Fig. 9.8). The consequences of a failure of the ice to be replaced by open water was demonstrated in the Ross Sea when a 1000-km-long iceberg prevented the normal pattern of ice break-up and many penguin chicks starved as their parents could not gain access to open water to feed. Once the iceberg moved away the normal pattern of ice formation and break-up resumed and the fortunes of the Adélie penguins were restored.

A tentative estimate of krill consumption by seabirds around South Georgia and the Scotia Arc is 10.9×10^6 t per annum. There is no doubt that, quantitatively, they are an important element, comparable with other predators in the pelagic food web. Their predation must also play an important

Fig. 9.8   A young Adélie penguin shedding juvenile plumage (photograph by David N. Thomas) (see colour plate).

qualitative role in structuring pelagic communities. As has already been pointed out they have a profound effect locally on terrestrial habitats. Broadly, the ecological impact of seabirds seems similar in Arctic and Antarctic waters. An attempt to make quantitative comparisons around islands at roughly comparable latitudes north and south—in the Bering Sea in the area 55–63°N 169–173°W, and the South Orkneys 61°S 45°W—bears this out. Numbers of birds per unit sea area tended to be higher in the Arctic but since Antarctic birds have a larger mean size the biomasses are similar. Because smaller birds require proportionately more food than larger ones, consumption of pelagic prey is probably greater in the Arctic than in the Antarctic.

9.3 Seals

Seals, being accomplished swimmers amply insulated with dense fur and blubber, are well adapted to polar waters. Nevertheless, they are visibly most associated with their breeding sites on land or ice. Studies in the open sea and precise information on their lives in the pelagic environment is increasing with the use of satellite-linked behavioural recording devices; however, for many species such information remains sparse.

The 25 or more species of marine mammals in the Bering Sea are estimated to consume between 9×10^6 and 10×10^6 t of pelagic and benthic organisms per year. Most seals eat fin-fish but some such as the bearded seal (*Erignathus barbatus*), prey on benthic invertebrates, and the walrus (*Odobenus rosmarus*) feeds on bivalve molluscs. The walrus therefore prefers shallow waters and although it may congregate in large numbers on beaches, it characteristically lives in the pack and at ice edges. It mates around midwinter among the ice, copulating under water with one bull serving perhaps 15 females. Walruses also use sea ice for hauling out, often using their distinctive tusks as ice axes, giving them one of their popular names, the *ice walkers*. They can also break through ice up to 20 cm by ramming the ice with their heads. Their prey is made up mainly from molluscs such as clams, cockles, and welks growing in the benthos, and so they are restricted to hauling out on ice covering shallow (<70 m) coastal regions. However, walruses are also known to eat ringed seals as well. Walruses do not keep open breathing holes, but instead rely on polynyas and open leads between ice floes for their access to the water.

The walrus (Fig. 9.9) is widespread and, since a Soviet ban on its hunting, has multiplied considerably. It does not use its tusks for digging, as one might suppose, but supports its head on them while it extracts the soft-shelled clam (*Mya truncata*) by sucking or hydraulic jetting to excavate a pit up to 30 cm deep. It may find these clams, which have conspicuous siphons, visually but if visibility is poor it furrows the upper few centimetres

Fig. 9.9   Walruses crammed on to the remnants of a sea-ice floe in Arctic coastal waters off Svalbard (photograph by Rupert Krapp).

of sediment and identifies its prey by touch. One such furrow was found to be more than 60 m long and yielded 34 clams from depths of over 30 cm. The amount of sediment resuspended by walruses in the Chirikov Basin of the north-east Bering Sea is about 100 million t per year.

Pits left by walruses offer a habitat to many kinds of benthic invertebrate. A remarkable number and variety of agencies disturbing the sediments of Arctic seas can be listed. Another large mammal, excavating pits 1–2 m long and up to 0.5 m deep is the grey whale (*Esrichtius robustus*), for which major foods are the amphipods *Ampelisca* and *Bybles* spp. This animal is estimated to resuspend 172 million t of sediment per year in the Cherikov Basin, three times the amount of sediment deposited by the Yukon River (King 1983, Ainley and DeMaster, in Smith 1990, Ainley *et al.*, in Thomas and Dieckmann 2003).

There are seven species of Arctic seal that use sea ice as a habitat (King 1983, Sage 1986, Pielou 1994). The spotted (*Phoca largha*) and ribbon (*Phoca fasciata*) seals are found in the Pacific/Bering sea region, the hooded (*Cystophora cristata*) and harp (*Phoca groenlandica*) seals are found in the north Atlantic/Russian Arctic region, whereas the ringed (*Phoca hispida*) and bearded (*Eringnathus barbatus*) seals have a circumpolar distribution. The bearded seal prefers shallow waters free of fast ice, with moving floes and open leads, but it can keep open breathing holes by means of the strong claws on its fore flippers. The ringed seal is the commonest seal in the Arctic Ocean, with a total population of between 3 and 4 million. It is found in open water in the fast ice but rarely on floating pack ice or in the open sea. In winter, adults stay under the ice in bays and fjords. The younger ones mostly stay further out at the edge of the fast ice. In summer most lie out on the ice basking in the sun, moulting, and fasting. Like the bearded seal the ringed seal keeps open breathing holes, which may extend through ice 2 m thick, by abrasion with its flipper claws. A low dome of ice, about 4 cm high, marks the hole at the surface and the snow which accumulates round this, if it is sufficiently deep, may be hollowed out and used as a lair. Larger lairs are constructed by pregnant females and the pups are born in them in spring. These lairs, with their adjacent breathing holes giving access to the sea, not only give some protection against polar bears and Arctic foxes but provide shelter from cold and wind. The pup, which has not got the insulating layer of blubber possessed by adults, must derive appreciable warmth from its mother in the confines of the lair.

In the suite of 'ice-breeding' seals in the North Atlantic region there are two quite distinct lactation strategies that may well reflect differences in the types of ice habitat used by the different species (Lydersen and Kovacs 1999). The grey (which breeds in the ice but is also found in more temperate regions), the harp, and the hooded seals all frequent the outer regions of the pack ice and have a short lactation period; once weaned pups remain relatively inactive while converting blubber into useful body tissues. In contrast, the bearded and ringed seals are found in nearshore fast ice and have a much longer lactation period. During this time the pups enter the water, and even feed for themselves, so that there is a less abrupt transition to nutritional independence.

Several of the Arctic species frequent shallow coastal waters associated with pack ice, and although some appear to remain resident the distribution of others changes in response to the seasonal movements of the pack ice. The harp seal is the most abundant seal in the northern hemisphere and is found in three discrete populations. The largest, of between 4 and 6.5 million animals, is in the north-west Atlantic, and the others are in the Barent Sea and east Greenland. It breeds in the spring on the margins of large ice fields. Pups feed on planktonic crustaceans and adults on shoaling fish such as capelin, herring, and polar cod. The total population is believed to be between 6 and 8 million, a considerable reduction from what it was before commercial exploitation. Like the harp seal the hooded seal moves seasonally north and south with the movement of the ice although it is not distributed quite so extensively. It is solitary at sea, feeding on Greenland halibut, capelin, cod, and squid, for which it can dive to 1000 m. The total population is of the order of half a million animals. In the North Pacific, Chukchi, and Bering Seas, the predominant pelagic seal is the ribbon seal. Other than the periods of breeding on ice in the spring and the summer moult, scarcely anything is known of its movements thereafter. Fish and cephalopods appear to be its main food.

The total population is now about a quarter of a million (Sage 1986, Ainley and DeMaster, in Smith 1990).

In the Antarctic, Weddell, Ross, crabeater, and leopard seals are associated with ice and only the Antarctic fur seal (*Arctocepthalus gazella*) and Southern elephant seal (*Mirounga leonina*) are more characteristic of open water (Fig. 9.10). The elephant seal is circumpolar in distribution, although divided into three breeding stocks. The fur seal was almost exterminated on most circum-Antarctic islands by the fur trade; however, from near extinction in the nineteenth century it has made a spectacular recovery, with the population now totalling about 3–4 million animals. At South

(a)

(b)

Fig. 9.10   (a) Antarctic fur seal and (b) Southern elephant seal on South Georgia (photographs by David N. Thomas) (see colour plate).

Georgia, home to 95% of its population, its diet is dominated by krill, although fish are also important. Diving behaviour has been monitored by attached instrument packages which show a diurnal pattern with most diving taking place at night when krill are near the surface. The depth of dives is mostly between 20 and 50 m with a maximum of 200 m.

Between May and October, when they are absent from the breeding grounds, female southern elephant seals disperse widely, with some individuals reaching the pack ice while others winter on the Patagonian shelf north of the Falkland Islands. What southern elephant seals do between going to sea after moulting in March to April and returning to shore to breed in September has also been investigated using satellite tracking and time–depth recorders. This has revealed that when at sea, they spend 80–90% of their time underwater, most of it at depths of 200–400 m but sometimes going down to 1500 m. It seems that they feed at the pycnocline, where particulate matter tends to accumulate and attracts squid. The time needed to take breath at the surface is astonishingly short and their muscles must be able to work without oxygen for much of the time when they are submerged. How they navigate, evidently with great precision, when travelling in deep water over the long distances between their breeding and feeding grounds, is not known. Such are the movements of elephant seals that they are now used to carry oceanographic equipment and are actually providing a better understanding of the physical characteristics of the Southern Ocean, as well as of their own foraging strategies.

During the austral winter, some southern elephant seals occur in close association with sea ice, and adult female elephant seals from King George Island fitted with satellite transmitters have revealed that the seals ranged in the sea-ice zone along the Antarctic Peninsula. During winter, some of the female elephant seals also spent several months in heavy pack ice. The availability of the Antarctic silverfish, *Pleurogramma antarcticum*, may explain why elephant seals are attracted to the pack ice to forage. In contrast to the adults, juvenile elephant seals from King George Island appear to avoid the sea ice and range in deeper, open water (Bornemann *et al.* 2000).

There are four species of ice-breeding seals in the Antarctic: Weddell (*Leptonychotes weddelli*), crabeater (*Lobodon carcinophagus*), Ross (*Ommatophoca rossii*), and leopard (*Hydrurga leptonyx*) (King 1983, Knox 1994, Laws, in Laws 1984) seals. The crabeater is the most abundant seal in the world, although the total population is probably much less than the early estimates of 30 million (Fig. 9.11). The Weddell and crabeater seals afford a contrast in use of the ice. The Weddell breeds on the nearshore ice while outside the breeding season they move to the outer limit of the fast ice and the inner zones of the pack ice. Here they are usually seen singly and their mean density has been estimated as 0.14 km^{-2}. The Weddell seal extends as far north as there is reliable nearshore fast ice in the winter.

(a)

(b)

(c)

Fig. 9.11   (a) Weddell, (b) crabeater, and (c) leopard seals (photographs by David N. Thomas).

Weddell seals do not feed on krill, but instead mainly feed on squid and fish, in particular the Antarctic silverfish, of which they can get a hundred or more in a single dive. Weddell seals can dive as deep as 700 m and can stay underwater for over an hour. It appears that they have two main diving habits; shallow dives within the top 100 m of the water column and deeper dives between 200 and 400 m (Plötz *et al.* 2001). The shallow dives tend to be the longer and the seals can travel distances in excess of 10 km from the breathing hole on a single shallow dive. Deep dives of short duration probably reflect feeding on benthic fish, whereas long shallow dives go to depths that do not take the seal out of sight of the surface and probably reflect feeding on fish on the underside of the ice. During these under-ice dives seals have been recorded blowing bubbles into fissures in the ice to flush out and capture fish.

It also appears that the seals preferentially haul out during daylight and spend the night in the water, and also that the deepest dives tend to happen during the day. One reason for these diurnal differences in diving behaviour may be related to the different distribution of the Antarctic silverfish at different times of day. The fish move into shallower waters under the ice at night, possibly to feed on krill that are feeding on ice algae on the undersides of the ice, and investigations into the stomach contents of silverfish that have been vomited up by Weddell seals have been shown to be full of krill.

During the winter the Weddell seal remains in the water and uses breathing holes that are kept open by rasping away the ice, using the well-developed incisor and canine teeth. The eventual wearing away of these teeth seems to be a common cause of death. In spring, pupping colonies assemble around breathing holes and predictable tide cracks and most pups are born, on the ice, by mid-November. Pups are suckled for a month or two but enter the water in their second week under their mothers' care. Mating takes place underwater around midsummer. The Weddell seal is polygynous and a male is able to exert his authority over about 10 the females which frequent one particular breathing hole.

Crabeater seals (Southwell *et al.* 2004) are similar in size to Weddell seals but occupy a different niche in the sea-ice habitat, preferring pack to fast ice. They are more gregarious than Weddell seals and are sometimes found in summer in aggregations of as many as 600 within a radius of 5 km. Counting animals which are spread out over an enormous area of sea ice requires the combined use of aerial photography, ice-breakers, and helicopters to sample strips orientated north–south to penetrate the ice to a maximum distance. The data obtained indicate mean densities of about 4.8 km^{-2} in summer and 0.5 km^{-2} in winter and spring. Their usual food, somewhat contrary to their name, is krill and their molar teeth have prominent cusps well adapted for filtering out these shrimp-sized organisms and they seem to dive only to the relatively shallow depths where krill are most abundant. Pupping takes place on the ice from late September until early November but, because of their inaccessibility in the pack at this time, information about breeding behaviour is sparse. Unlike the Weddell seal and those seals which breed on land, the crabeater is monogynous. Satellite tracking studies, although somewhat limited in number, have shown that the male and female crabeater seal stay together on the ice after the pup has been born and separate once the female and pup enter the water.

A major predator on both of these seals is the killer whale (*Orcinus orca*). Another reknowned Antarctic predator, often labelled the polar bear of the South, is the leopard seal, which like the crabeater seals tend to live in the marginal ice zones traveling many thousands of kilometres on drifting ice floes. It is regularly recorded on sub-Antarctic islands during the winter. Forty-five per cent of the diet of leopard seals is made up from krill; however, they do also feed on penguins and in particular crabeater seal pups, and they frequently swim at ice floe edges or in waters close to penguin colonies waiting for there prey to enter the water, where the aggressive and agile leopard seals are seen to be superb hunters (Hiruki *et al.* 1999).

Not too much is known about the seasonal distribution of the estimated 200 000 Ross seals. This species is thought to feed predominantly on squid and fish, and it is rarely seen because it hauls out on the thickest ice in fields of pack ice where ships and researchers seldom venture. This species is highly dependent on sea ice for giving birth to its young and for hauling out when it moults. Most recent observations show that they can make up to 100 dives a day to depths between 100 and 300m, although the deepest dive was 762m (Blix and Nordøy, 2007).

9.4 Whales

The prerequisite for all whales in ice-covered waters is that there are sufficient areas of open water for the animals to surface and breathe. This is why so few whales are found deep within the pack ice during winter, and why typically whales migrate to sea-ice-covered regions from lower latitudes in summer months. However, where polynyas and areas of open water persist it is possible that whales can survive.

Whales, cetaceans, are of two kinds, the whalebone or baleen whales, Mysticetes, and the toothed whales, Odontocetes. Baleen whales feed by sieving out zooplankton, such as krill and copepods (Fig. 9.12), by means of rows of hairy triangular plates carried on each side of the palate. They include the blue (*Balaenoptera musculus*), fin (*Balaenoptera physalus*), which are found in both the Arctic and the Antarctic, and the common minke (*Balaenoptera acutorostrata*) and Antarctic minke (*Balaenoptera bonaerensis*) that are found in the north and south respectively. The toothed

Fig. 9.12   Whales coming to surface of the water in turn attracting numerous birds to feed on zooplankton (photograph by Finlo Cottier).

whales, which feed on larger prey, include the sperm (*Physeter catodon*) and killer (*O. orca*) whales. Many large cetacean species are found in both Arctic and Antarctic seas but, because their life pattern is to feed in polar waters and breed in equatorial waters and there is therefore a 6-month phase difference in breeding between the hemispheres, it seems that there is limited mixing of the stocks across the equator. In the Antarctic there is latitudinal zonation in the distribution of whale species.

The baleen whales feed near the surface, rarely diving to any great depth, and probably finding their prey by both sight and echolocation. The technique of feeding varies; the right whale skims the sea surface, swimming slowly with jaws agape, filtering water as it goes. The blue, fin, and minke whales take gulps of water plus krill then, with mouth closed, the water is forced through the baleen plates by expansion and pushing forward of the tongue with contraction of the ventral grooves which run backwards from the chin. Whales may concentrate krill by encircling a patch then diving to come up vertically beneath it with open mouth, or, by swimming beneath the surface and releasing a trail of air bubbles in which the kill become collected. Balleen whales caught off South Georgia had stomachs full, or nearly full, in 70% of cases as compared with less than 25% for those caught off South Africa. When feeding maximally, up to 4% of body weight is consumed daily.

Sperm whales feed largely on squid, which they evidently find by echolocation and it has been suggested that they can immobilize prey by a projected beam of sound. They can stay below the surface for as long as an hour. The maximum quantity of food found in their stomachs is about 200 kg but in one instance this consisted of one 12-m giant specimen of the squid *Architeuthis*. Again, daily consumption is about 3% of body weight.

Little is known of the feeding of the smaller Odontocetes—killer whales, pilot whales, and dolphins—in the polar regions. Killer whales have been classified according to whether they exhibit different behavioural traits and those that feed on fish tend to be resident in a certain area while those that feed on marine mammals are transients. In high-latitude pack-ice regions orcas often hunt in packs and co-ordinate attacks on large prey such as seals, penguins, and even the large baleen whales. They do not dive deeply but take their victims at the surface. Estimates of food consumption are 4% of body weight for killer whales, up to 6% for pilot whales, and up to 11% for dolphins. As with most animals, the relative food requirement increases as the size of the animal decreases.

It is generally thought that killer whales move out from pack-ice regions during winter, or at most keep to marginal ice zones. However, there have been several reported sightings of killer whales within winter pack ice (Gill and Thiele 1997). In the Antarctic on one occasion in August, 60 killer whales were spotted together with 120 minke whales in pools of open water that were cut off from the open sea by 65 km of compacted sea ice. On another occasion a group of 40 killer whales of mixed ages were spotted in leads of water, 400 km south of the ice edge. The presence of a calf within this group would indicate that the whales may have given birth in these sea-ice-covered waters. There have also been sightings of killer whales in Arctic winter sea ice off west Greenland and western Alaska.

In the Antarctic, two types of killer whale have been described, although it is unclear whether this distinction is strictly apropriate: The *white form*, that feeds on mammals (penguins, seals, and other whales), is found in more open waters and loose pack ice. In contrast the *yellow form* feeds mainly on fish and found deeper in the pack ice. The 'yellow' coloration comes from these whales being covered in thick biofilms of diatoms.

A small toothed whale in the Arctic are the Belugas (*Delphinapterus leucas*), also called white whales, that migrate over the Arctic often traveling far into the permanent pack ice (Richard *et al.* 2001). Populations of belugas have characteristic migration patterns: for example, Belugas summering along the north coast of Alaska, move great distances well offshore, in deep (>3000 m) water and beneath areas where there is almost complete ice cover to reach their spring feeding areas (Suydam *et al.* 2001). The whales can cover between 50 and 80 km a day through waters with more than a 90% ice cover. Presumably Belugas select their migratory routes in relation to the availability of their prey, such as Arctic cod. They are capable of deep (>300 m) dives during which they are thought to forage on benthic organisms (Martin *et al.* 2001).

Narwhals (*Monodon monoceros*) are probably among the most unusual looking of the toothed whales. All narwhals have two teeth in their upper jaw. After the first year of a male narwhal's life, its left tooth grows outward, spirally. This long, single tooth projects from its upper jaw and can grow to be a maximum of 3 m long. The hollow tusk is usually twisted in a counterclockwise direction. The tusk's function is uncertain, although it is not used to catch the narwhal's prey of fish and squid. It is thought that narwhals can dive to depths greater than 1500 m which makes them the deepest divers of all mammals, and during winter they tend to move from coastal shelf regions where they spend the summer to living over deep Arctic basin waters where they are thought to feed on deep-sea squid (Heide-Jørgensen *et al.* 2002). They are closely associated with the Arctic pack ice throughout the year and utilize leads and polynyas for moving throughout the pack ice in winter. Neither Beluga whales or narwhals have dorsal fins. This enables them to swim close to the underside of ice floes, and this also enables them to break thin ice with their backs.

Whales have a considerable impact on the pelagic ecosystems of polar regions. Before whaling reduced stocks so drastically, baleen whales in the Antarctic took an estimated 190 million t of krill per annum and the sperm whale 10 million t of squid, the corresponding recent figures being 43 and 4.6 million. Possibly because of a greater availability of food for those which have survived, baleen whales have recently shown increased growth rates, earlier maturity, and higher pregnancy rates than formerly. There is now evidence of population increases of baleen whales in many regions; however, the large nature of the ecosystem changes brought about by whaling mean that they may not necessarily reach their pre-exploitation population sizes. Because the large whales are migratory they export the biomass accumulated in polar waters to equatorial regions where they undergo almost total fast. This is an estimated loss of some 18 million t per annum of whale material from the Antarctic, and a corresponding enrichment in energy and nutrients of their breeding grounds (Brown and Lockyer, in Laws 1984, Ainley and DeMaster, in Smith 1990, Knox 1994, Pielou 1994).

9.5 Bears and foxes

Although they produce their cubs on land and are closely related to the terrestrial grizzly bear (*Ursus arctos*), polar bears (*Ursus maritimus*; Sage 1986, Pielou 1994) are essentially sea-ice animals and are strong swimmers (Fig. 9.13), aided by well-developed fat layers for buoyancy. Their large forepaws are used as effective paddles to swim with, whereas they use their rear paws as stabilizers or rudders. Apart from occasionally eating berries, seaweeds, or grasses, and a recently developed habit of raiding rubbish dumps, they are dependent on the sea for food, which is mostly seals. They are well adapted to survive low temperatures by virtue of their large size and fur, which has six or seven times the insulating capacity of

Fig. 9.13   Polar bears are powerful swimmers (photograph by Kit Kovacs and Christian Lydersen).

the clothing normally worn by humans at rest in an ambient temperature of 20°C. The fur (made from hollow fibers) is colourless and transparent to short-wave radiation, which is absorbed by their black skin, but opaque to infrared: hence polar bears are not detected by infrared imaging equipment. The thick undercoat of fur is a highly efficient insulator, and so good that after vigorous exercise or in warm weather bears are likely to overheat.

Polar bears are found throughout the ice-covered waters of the Arctic Ocean. The greatest majority of bears roam near pack ice that is thinner or breaks open on a regular basis. Generally bears avoid heavily ridged, rough sea ice and thick multi-year ice, mostly because the densities of seals are low in these ice types. The southern limits of polar bears is basically governed by the southernmost extent of sea ice, and a few have been reported close to the north pole, although generally it is thought that very few stray further than 80°N since the ice generally comproses thick multi-year ice floes.

It is a difficult task to estimate how many polar bears there are, but reliable estimates place the number at about 40 000 in the whole of the Arctic basin, although more cautious researchers would say that it lies somewhere between 20 000 and 40 000. These bears are not part of one large population, but rather divided up into several subpopulations. Bear have been tagged and recaptured since the 1970s, and more lately bears have been equipped with radio transmitters and their positions logged using satellite tracking systems. The results of these studies show that from year to year individual bears remain in the same geographic region.

However, on shorter time scales bears have been shown to travel in excess of 30 km a day for several days in a row. Therefore, within a year many polar bears may travel many hundreds, if not thousands, of kilometres. These large roaming distances are of course extended because the bears are travelling on a moving platform of ice that is blown by wind and carried on ocean currents, transporting the ice over many thousands of kilometres. Tracking studies have demonstrated clearly that the bears are not roaming aimlessly but that they know exactly where they want to go. Although much of these distances are completed on top of the ice, because polar bears are such strong swimmers, they can swim distances in excess of 100 km at a time. This results in polar bears sometimes being observed several hundreds of kilometres offshore, probably because the ice floes they were travelling on had melted from underneath them (Fig. 9.14).

Polar bears from Svalbard and the Barents Sea that roam huge territories (up to 270 000 km^2) accumulate significantly higher concentrations of polychlorinated biphenyls (PCBs) in their fat, blood, and milk compared with bears confined to smaller coastal territories. PCBs are pollutants that used to be used in electrical equipment and are particularly resistant to breakdown when released into the environment. The bears that are covering large distances have to consume significantly greater food reserves than the bears with smaller ranges. The PCBs are contained within the prey, and so since they need to consume more prey they consequentially consume more of the pollutants (Haave *et al.* 2003).

Polar bears largely lead a solitary existence, and bears may have to travel considerable distances in April to find a mate. After mating females start to feed in excessive amounts (mostly on seals) to gain weight as quickly as possible. Typically female bears weigh 150–175 kg, but before they give birth to their cubs they may have laid down fat stores so that weigh up

Fig. 9.14   Polar bears are dependent on sea ice for hunting their prey (photograph by Finlo Cottier) (see colour plate).

to 450 kg. In late August to October the females move to land where they make maternity dens, and once they enter they have no opportunity to feed again until the following spring. These dens are made in the hard snow of snow drifts or snowbanks, and some dens may be as far as 100 km inland from the coast, although typically they are within 50 km. Maternity dens can also be dug in earth banks in the permafrost. The dens tend to be 2–3 m^3 and may have an exit hole 2 m away from the den itself. Cubs (normally twins) are born in November and are nursed by the mothers with fat-rich milk until late February or March. When born the cubs are less than 1 kg in weight, and when they emerge from the den they weigh between 10 and 15 kg. Cubs remain with their mothers up to 2.5 years and so females normally only have cubs once every 2 or 3 years. The mortality rate of cubs can be as high as 40% in the first year, but if the bears make it to adulthood they can be long lived, reaching 30 years of age, because they have no natural enemies (except humans, of course).

Polar bears feed predominantly on ringed seals and bearded seals as well as other species of seal and, occasionally, small whales. They catch seals by lying quietly in wait by their breathing holes and also attack female ringed seals and pups in their lairs. The number and survival of cubs shows great variation, as it is dependent on the level of fat storage in the females prior to giving birth. This in turn is dependent on the availability of ringed seal pups which is linked to fluctuations in key/climatic factors (Stirling and Lunn, in Woodin and Marquiss 1997). Polar bears have been hunted by humans, probably causing population declines in some areas, but they are now protected and hunting is managed closely.

The Arctic fox (*Alopex lagopus*; Sage 1986, Pielou 1994), although chiefly an animal of the land, is a scavenger of the leftovers of polar bears and is itself an important predator on pups of the ringed seal, catching them, like the polar bear, in their birth lairs. The Arctic fox does not migrate seasonally, and its fur is the most effective insulator known for any mammal, meaning that its thermoregulation is not challenged even under the most extreme conditions faced in the Arctic winter. From a 3-year study in the Canadian Arctic an average seal-pup predation level of 26% by Arctic foxes was estimated on nearshore ice. Stable isotope analysis of diet of Arctic foxes have shown that the marine component of their diet increases during periods when lemming populations are low. Dalén *et al.* (2005, 2006) used molecular techniques, among others, to look at Arctic foxes, in particular Scandinavian populations. They have reported large decreases of fox populations despite over 65 years of protection, and that the Scandinavian populations is threatened due to dwindling habitat, a decrease in food availability, and an increase in the numbers of a competing species, the red fox (*Vulpes vulpes*).

10 Climate change in polar regions

10.1 Introduction

To understand the present situation it is well to consider what has happened in the past. Over many millions of years the polar regions have undergone major changes in configuration of their land masses and climate with corresponding transformation of habitats. However, it is the geologically brief period beginning around 120000 years before present (BP), including a warm interglacial followed by the Würm glaciation and then the interglacial in which we now live, which is of most relevance here. Study of fossils, of beetles in particular, in deposits laid down in glaciated areas have shown that conditions have never been settled for very long and that change can be rapid. Physicochemical examination of cores of polar ice and sediments confirms this picture and yields detailed records of the chronology and magnitude of glaciations and global warmings. It appears that the last ice age ended abruptly, perhaps within the span of a human life. These changes cannot be put down to any human agency. Now, human impact on the environment is mounting. We have no reason to suppose that the present state which we take to be normal is any more stable than those in the past and by our activities we are destabilizing the global system and making change more likely. The totally unexpected development within a decade of the *ozone hole* over Antarctica makes this abundantly clear.

There is no doubt that global warming is taking place and despite the debate about the consequences of this and the international political wrangling and debate it is indisputable is that the current rate of warming is 10–100 times faster than the rise of approximately 5°C which took place between 15000 and 10000 years ago (IPCC. 2007a, b, c). It is argued that polar ecosystems are more sensitive and reliable indicators of change than elsewhere, and changes in climate are predicted to be greater and more rapid around the Poles. Polar species are particularly vulnerable to change because of their slow growth and low fecundity and they are especially responsive to rise in temperature since they exist so close to the lower temperature limit for survival. However, information on responses of Arctic plants and animals in their natural habitats to climate changes is still sparse. In the Antarctic the predominantly cryptogamic vegetation is not markedly responsive to temperature change and barriers to population shifts are formidable. But there is some suggestion that recent rapid population expansions of the two indigenous flowering plants are related to climatic amelioration.

10.2 Changes during geological time: the ice ages

It is incorrect to thick of the polar regions as having always been as they are now. In both the Arctic and Antarctic, sedimentary rocks, still in more or less the same geographical position as when they were laid down, contain fossils of plants and animals characteristic of warmer climes and unlikely to have existed under conditions similar to the present ones (Cocks 1981). Existing habitats may not contain all the types of organism that one might expect and some give the impression of supporting immature communities which are still developing towards their climaxes as plants and animals migrate into them. It is necessary to look back through geological time if we are properly to understand the present ecology of polar habitats.

The scene was set by the drifting of continents according to the theory, now generally accepted, put forward by Wegener in 1912 (Cocks 1981, Walton 1987). Some 250 million years ago the Atlantic and Indian Oceans did not exist and a single land mass, Pangaea, was surrounded by the precursor of the present-day Pacific Ocean. Pangaea broke in two about 220 million years ago, the northern part fragmenting, with the pieces separating so that what are now the Siberian coast, Greenland, and the North American archipelago enclosed a sea area containing the North Pole. The northern edges of the continental plates on which these lands are carried form the physiographic borderlines of the Arctic Ocean. The movement of land masses into the vicinity of the North Pole had a profound effect on climate. The heat-storage effect of the sea was reduced and winter cooling of the land brought about a fall in mean annual temperature, leading to extensive glaciation by the beginning of the Pleistocene, 3 million years BP.

The situation of the Arctic Ocean in a polar basin communicating with the world ocean by narrow channels has given rise to remarkable climatic fluctuations with a periodicity of about 100 000 years. This is explained simplistically by a theory put forward by Ewing and Donn (1956). Ice-free, the Arctic Ocean is the site of a permanent cyclone which moves moist air northwards, resulting in heavy precipitation and the build-up of ice caps on circumpolar land masses. This transfer of water from sea to land causes a fall in sea level, throttling the channels through which warm Atlantic water reaches the Polar Basin. Freezing of the Arctic Ocean follows and on the extended cold polar mass an anticyclone develops, driving dry easterly winds. The ice cover then begins to ablate and the release of water brings about a rise in sea level, allowing warm water into the Polar Basin with progressive melting of ice and eventually an open Arctic Ocean again.

The southern part of Pangaea, known as Gondwana, drifted south and between 180 and 200 million years ago itself split up to give the pieces which now form South America, South Africa, India, Australia, and Antarctica. This left Antarctica in about the position which it still occupies, in the vicinity of the South Pole. The presence of substantial coal seams of Permian age, some 248 million years, in the Transantarctic Mountains is evidence that the new continent originally had a temperate climate. However, the gap between Antarctica and its neighbouring land masses widened as these drifted north, until around 25 million years ago it had opened sufficiently for the Circumpolar Current to become established and isolate the southern continent. This had a profound effect on the Antarctic climate and from this time its ice cap began to expand. Nevertheless, as shown by discoveries of fossil invertebrates and wood as far south as the Beardmore Glacier at 85°S, Antarctica was warmer at various times between the Miocene and possibly the Pliocene, between 14 and 2 million years ago, than at present.

The succession of glaciations and interglacial periods that is still with us began about 3.2 million years ago. Before this, high alpine regions had most of the world's ice. The fluctuations of the ice have been mapped and dated from a variety of different kinds of evidence. Periglacial features such as moraines, eskers, drumlins, ice-scratched rocks, and drainage patterns, provide basic information. Cores from ice and from sediments from lakes and the seabed yield data which is often surprisingly detailed and precise; laminations in cores allow years to be counted; oxygen isotope ratios in 'fossil' water or in remains of foraminiferan shells give measures of temperatures at the time of deposition; diatoms, pollen grains, and other microfossils give a picture of flora and fauna and help in dating (Barrett 1991, Lorius 1991, Thiede, in Wadhams *et al.* 1995). Rock fragments from icebergs mark the occurrence and extent of periods of glaciation. Accompanying these glaciations there have been falls in sea level as water became locked up in ice, with corresponding rises in interglacials.

The cause of the onset of these glaciations is obscure. During the Pleistocene the glacial/interglacial fluctuations have shown a relation with the Milankovich cycle, the cycle of variations, with a period of 120 000–150 000 years, in incident solar radiation resulting from perturbations in the Earth's motion. Of the five or six major glaciations which the Earth has experienced the last one is abnormal in affecting both poles and, at least latterly, the succession of glacials and interglacials have been synchronous in the Arctic and Antarctic. Greenland ice cores show 22 interstadial (warm) events in the period 105 000–20 000 years BP. Cores from the Antarctic show nine interstadials in this period, these occurring whenever those in Greenland lasted more than 2000 years. This linkage between events at the two poles seems to have been mediated by partial deglaciation, causing changes in sea level and ocean circulation.

10.3 Biological responses to long-term changes

Implicit in the foregoing descriptions and discussion is the assumption that many, if not most, of the biota described are relatively recent residents of the areas included in their contemporary polar distributions. This is driven, albeit somewhat simplistically, by the generalization that, as recently as the Pleistocene glacial maximum only approximately 20 000 years ago and also previous glaciations in the Miocene, vast areas of what is currently exposed land were covered by extensive and deep ice sheets. In North America, these extended to cover much of what are now the northern states of the USA, in Europe most land north of the Alps, with centres also radiating from other major mountain ranges, and in Asia far into what is now taiga.

A simple view would be that marine plants, algae, and animals withdrew before the advancing ice and that, when it retreated, recolonization of habitats would not present any major problem to species which are mobile or have drifting propagules. On land, some plants and animals seem to have survived in favoured spots in both polar regions (Convey and Stevens, 2007).

10.3.1 Survival in refugia

In the Arctic there is geological evidence, as well as that from the distribution of species, for numerous refugia remaining ice-free within glaciated areas at the time of the Last Glacial Maximum. As the Pleistocene ice sheets spread south a zone of low precipitation developed on their northern edges with a consequent shrinking of the extent of the ice. This, together with a lowering of sea level, created extensive potential refugia. The most important were in the interior of Alaska and Yukon, eastern Siberia, and a region known as Beringia between the two, which was exposed intermittently as sea level oscillated by 100 m or so. Beringia provided a corridor for migration of plants and animals from the unglaciated forest and mountain areas of central Asia until it was finally inundated about 10 000 years ago. The vegetation which covered it consisted mainly of steppe-tundra, dominated by grasses and sagebrush (*Artemisia*), which supported large herds of grazing mammals such as bison, horse, and mammoth. Smaller and less hospitable refugia, provided by peaks protruding through the ice sheet—*nunataks*—enabled mosses, lichens, and a microflora of algae, yeasts, and bacteria to survive (Sage 1986). The Arctic poppy,

Papaver radicatum, evidently survived the Würm glaciation in a refugium in northern Norway (Crawford 1995).

On continental Antarctica there would presumably be few refugia during the glacial maximum. The dry valleys remained unglaciated at this time but were more arid even than they are today and are unlikely to have supported a flora and fauna other than microorganisms. Nunataks remained above the ice and, since today they support mosses as far south as 84°S and lichens and cyanobacteria beyond 86°S (Kappen, in Friedmann 1993), it seems reasonable to assume that they did so in the past. Among the less obvious organisms are bacteria and yeasts, recovered in a viable state from 87°21'S. The Polar Front must have been further north at the height of the glaciation and some of the peri-Antarctic islands seem to have been covered by ice caps extending out to sea. They may, of course, have had nunataks but Kerguelen was probably the only major refugium for pre-glacial fauna and flora.

Populations isolated in refugia evolve along their own lines. Both the Arctic and Antarctic have endemic species of algae, lichens, and mosses and invertebrates. In the latter the proportion of these relative to cosmopolitan species is striking, leading to a reinterpretation of their evolutionary history on the Continent (Convey and Stevens, 2007). Endemic species, of course, can also arise by genetic change in relatively recent invaders. Bipolar species are also few. The two polar regions seem to have recruited most of their floras and faunas independently from their respective hemispheres, presumably mainly from alpine localities. Arctic species of flowering plants and animals frequently seem different from their counterparts in temperate climes but it is not always clear whether this is due to phenotypic plasticity or genetic change. The circumpolar Arctic-alpine bitter cress (*Cardamine bellidifolia*), transplanted from alpine habitats in Alaska, British Columbia, or Yukon, into temperate conditions, grows into a tall plant. Seeds of the same species collected on Ellesmere Island (approximately 80°N 80°W) give rise to dwarf plants under the same temperate conditions, showing that they come from an ecotype genetically different from the others. Greenland freshwater cladoceran crustacea appear to afford another instance of evolution in refugia. One of the most widely distributed of these today in the Arctic is *Chydorus arcticus*. Remains of a form similar to, but not identical with, this species and the closely related *Chydorus sphaericus* have been found in 1.5-million-year-old Pliocene/Pleistocene deposits in northern Greenland. Thus, species having Arctic distributions at the present time seem to have evolved in ice-free refugia and become better adapted to polar conditions (Røen 1994).

10.3.2 Colonization of recently deglaciated areas

Retreating glaciers, of which there is an abundance at the present time, leave behind them debris which is invaded by microorganisms, plants, and animals and eventually develops into a soil. As one moves away from the glacier snout one passes along a *chronosequence*, in which the stages of colonization are laid out, approximately at least, in order of age. Propagules are derived mainly from the immediate surroundings, being brought in by wind and run-off, and large numbers, rather than specialized adaptations, are most important in assuring rapid establishment. Bacteria, fungi, algae, and cyanobacteria establish themselves in and on the ground surface and stabilize it. Mosses and lichens follow. On Signy Island, vegetative fragments of mosses, such as leaves and stem apices, act as propagules, the ice cap evidently being an important reservoir supplying them. Far more show initial growth than eventually survive. In favourable sub-Arctic and alpine situations, shrubs and trees are among the earliest arrivals and relatively stable forest may be established in 100 years. In the Eurasian tundra zone, stable communities take 800–1000 years to develop. For Svalbard and harsher climates the times required may be between 1000 and 9000 years. Recently glaciated terrain provides a unique opportunity for the study of colonization and succession but it has to be remembered that it differs from the late glacial situation in that the areas are small and surrounded by established vegetation (Matthews 1992).

10.3.3 Colonization

In the Antarctic, which today remains almost 99.7% under permanent ice and effectively still in an ice age, analogous ice expansions occurred. Ice sheets were deeper than those of today (and the contemporary average depth of ice across Antarctica approaches 3 km) and around much of the continent they extended offshore to the point of continental shelf drop-off. As the Antarctic continental shelf is somewhat deeper than that of other continents, through the sheer weight of ice pressing down on the continent, this also does not appear to allow for continued exposure of coastal ice-free ground as a result of the lower sea levels (maximum approximately 120 m below those of today) that were associated with glacial maxima. With this glacial background there is a general, but untested, assumption that most contemporary biota are recent in origin, a view that is now changing (Convey and Stevens, 2007).

10.3.4 Stages in the colonization process

The colonization process can usefully be separated into three elements with reference to the polar regions (Clarke 2003b, Hughes *et al.*, in Bergstrom *et al.* 2006), as follows.

1. A transfer stage, in which survival of the various stresses faced by propagules is prerequisite. This stage effectively creates a biodiversity filter, as only a subset of the potentially transferable biota from the source region will possess the necessary features (pre-adaptations) allowing them to survive this transfer.

2. An establishment stage, which contains a stochastic element whereby the propagules must be deposited into a suitable habitat at a suitable time of year, and a biological element in which they, again, must possess the appropriate physiological and biochemical capacities.
3. A population-development stage, whereby not only survival but also the establishment of a long-term and viable reproducing population sustainable over subsequent years is ensured.

10.3.5 Mechanisms of transfer

With the above background, it is clear that on the one hand dispersal (transfer) challenges are much greater to the Antarctic than the Arctic through its isolation, while establishment challenges are also much greater because of both the generally more severe environmental stresses and the magnified difference in these between southern hemisphere source and destination locations. However, the principles applying in both polar regions are the same, and several modes of dispersal exist (Hughes *et al.*, in Bergstrom *et al.* 2006): organisms and propagules can be transported in air or water currents (Marshall 1996, Coulson *et al.* 2002, but see Pugh 2003), in zoochoric association with other biota, often migratory birds and mammals (Schlichting *et al.* 1978, Pugh 1997), or in water attached to natural or human-made flotsam (Barnes and Fraser 2003).

Finally, as already highlighted, human movements around the globe are a biologically recent but highly effective means of dispersal, with considerable impacts already created in the sub-Antarctic in particular over as little as the last two centuries (Frenot *et al.* 2005). It remains the case that there are very few explicit demonstrations of specific colonization events associated with any of these mechanisms, although the circumstantial evidence for their occurrence is overwhelming.

10.3.6 Invasion of the Arctic

At first sight there seems little obvious restriction on any of the colonization stages in the Arctic. Across the region, land masses project northwards and provide avenues along which microorganisms, plants, and invertebrates can be carried in short or longer steps by wind or water. Larger animals and birds may migrate under their own power, while also carrying with them a cargo of inadvertent passengers that may be deposited along the way. Nevertheless, some biogeographical features of the Arctic are difficult to explain without recognition of a refugial element (Halliday 2002, Jensen and Christensen 2003). Given recent results obtained in the Antarctic (described below), it is likely that the application of molecular biological techniques may also reveal a more ancient history than is currently generally assumed.

10.3.7 Invasion of the Antarctic

The fundamental stages of colonization face far greater obstacles in the Antarctic than in the Arctic, as is clear from the much lower species diversity present today (Table 3.3). The circumpolar ocean is today about 900 km wide at its narrowest (the Drake Passage separating the Antarctic Peninsula from South America) and the set of the prevailing winds and surface currents in the latitudes into which the southern extremities of America, Africa, and Australasia protrude have little southerly component. These barriers act to isolate the continent from colonization by both terrestrial and marine taxa, although they are to some extent *leaky* over the long time scales concerned (Barnes *et al.* 2006).

Dispersal of small propagules into and around the Antarctic by wind is plausible, especially as many of the characteristic biota (lichens, mosses, tardigrades, nematodes, various microbiota) are well known to have small propagules or specific life stages highly resistant to desiccation and/or cold with the capacity to survive the stresses of intercontinental transfer in the air column. Following storms, samples of air taken over Signy Island (South Orkney Islands) carry 13–24 times greater concentrations of biological particles than usual, although even these are much less, at around 3.3 particles m^{-3}, than is seen over southern England, where 12 500 particles m^{-3} is typical in summer (Marshall 1996).

Meteorological track analyses can be used to trace the origin of occasional arrivals of migratory insects (such as moths arriving at the sub-Antarctic islands of South Georgia and Marion; Greenslade *et al.* 1999, Convey 2005), illustrating the possibility of transfer in small numbers of days, well within the insects' capacities. Direct demonstration of establishment of species carried by wind has not yet been achieved but there is convincing circumstantial evidence that it happens (Muñoz *et al.* 2004, Chown and Convey 2007). Viable algae have been recovered from air over Antarctica (Vincent 1988) and the presence in moss cushions on the South Shetlands (approximately 62°S 60°W) of pollen grains of South American plants suggests that viable spores of similar size may arrive too (Kappen and Straka 1988). As described above, algae and bryophytes, not of local provenance, have appeared spontaneously around fumaroles on Deception Island (South Shetland Islands), at various locations in the South Sandwich Islands, and on Continental Antarctica.

It is also becoming clear that contemporary Antarctic terrestrial biota carry a stronger signal of ancient origin than has generally been thought to be the case. In some specific instances, the long-term existence of ice-free ground in parts of the Continental Antarctic even through glacial maxima is accepted. This is particularly the case with reference to the Victoria Land Dry Valleys and some analogous habitats in the Transantarctic Mountains where, despite considerable fluctuations in glacial extent, some areas are thought to have been ice-free for at least 5 million and possibly up to 10–12 million years. Much smaller nunatak refuges are also very likely to have persisted through glacial maxima.

There is evidence from classical biogeographic (Marshall and Pugh 1996) and modern molecular phylogeographic (Stevens *et al.* 2006) studies that elements of their endemic micro-arthropod faunas have a regional evolutionary history on the same multi-million year timescale. The identification of potential refuge regions is far more problematic for the Antarctic Peninsula and much of the continental coastline (Convey 2003). However, again, different lines of biological evidence, such as high levels of endemism and diversity hotspots (Maslen and Convey 2006), and indications using molecular-clock analyses of regional biota divergence times of at least 40 million years (Allegrucci *et al.* 2006), point to patterns of biodiversity and biogeography that are not possible to reconcile with recent post-Pleistocene colonization.

10.3.8 Polar marine benthos

Historically, the surprising diversity of polar benthos may be attributable to the circumstances that expansion of the ice caps would not necessarily obliterate benthic biota as completely as those on land, that transport of propagules by water movements is less hazardous than by air, and that the more uniform conditions of the seabed are more favourable to establishment than those on land.

The Antarctic fauna is found in areas with low-nutrient water, low content of suspended sediment, a low intensity of grazing and predation, and relatively few endobenthic bioturbators; these are conditions similar to those postulated for Palaeozoic seas (Thayer 1983), and several aspects Antarctic system are comparable with what we postulate about the Palaeozoic. These include the lack of riverine sediment input and oligotrophic waters (outside the period of the short summer phytoplankton bloom). The Antarctic benthic community also has a Palaeozoic characteristic with the lack of durophagous predators, reduced bioturbation, and a dominance of sessile or slow-moving taxa. However, there are also aspects that are clearly different from those in the Palaeozoic. These include the rich zooplankton community that mediates transfer of phytoplankton to the benthos, severe iceberg disturbance, a significant mobile fauna (which includes amphipods and isopods), and taxa that were apparently absent from Palaeozoic seas (for example ascidians, Clarke and Crame 1989, 1992). At the same time, it is generally accepted that the lack of several key predator groups, coupled with the dominance of suspension feeders and the presence of dense ophiuroid beds and crinoid populations, give the modern Antarctic benthic marine fauna a more than passing resemblance to Palaeozoic assemblages (Aronson *et al.* 1997). In contrast in the Arctic benthos the structure and ecology of these communities is similar to those living in the post-Cambrian Palaeozoic, although of course the species composition is entirely modern (Gili *et al.* 2006).

There remain to be explained striking differences in the composition and diversity of the benthos in the two areas. There are a few bipolar species. Some species of the seaweeds, *Ulva*, *Enteromorpha*, and *Ceramium*, are cosmopolitan in distribution and the green alga, *Acrosiphonia arcta*, and the brown seaweed, *Desmarestia viridis*, have disjunct bipolar distributions. The genetic affinities of these disjunct populations have been established by sequencing of nucleotides in rDNA. Comparing northern and southern samples, *Desmarestia* showed only one base change among 1073 nucleotide positions and *Acrosiphonia* exhibited 17 variable sites among 626 nucleotides. This indicates that for both species the disjunction has been recent. The temperature requirements of strains of *Acrosiphonia* show that those from polar habitats have growth optima between 0 and 10°C and upper survival temperatures of around 22°C, whereas for cold-temperate strains these temperatures are respectively 15 and 23–25°C. The upper survival temperatures suggest a possibility of transfer across the equator during the Pleistocene lowering of seawater temperature in the tropics, 18 000 years BP. Growth would not have been possible during this passage. The presence of *Acrosiphonia* on the east coast of Chile and its absence from the Atlantic coasts of South America and South Africa indicate that the route of transfer may have been through the East Pacific, where at present, the tropical belt is relatively narrow (Bischoff and Wiencke 1995).

A contrasting situation is presented by priapulid worms, relicts of a phylum which evolved in the early Cambrian. The bipolar disjunct pairs of species *Priapulus caudatus*/*Priapulus tuberculatospinosus* and *Priapulopsis bicaudatus*/*Priapulopsis australis* are morphologically similar but have genetically distinct enzyme patterns. In the long period since the populations were separated, evolution at the molecular level has been faster than that in anatomy (Schreiber *et al.* 1996).

Apart from the few cosmopolitan and bipolar species, the marine benthic floras and faunas are quite different. Although individual numbers are questionable there is a consistent tendency for the numbers of species in the various groups of animals to be greater by a factor of between 1.5 and 6 in the Antarctic than in the Arctic (White, in Laws 1984). The corresponding data for algae are ambiguous. Another feature is that whereas Arctic species show little endemism and have close affinities with cold-temperate Atlantic or Pacific forms, a high proportion, ranging between 57 and 95%, of endemic species occurs in Antarctic animal groups. Endemism at the generic level is less, between 5 and 70%. Despite its richness in species the Antarctic fauna is marked by the absence of certain major groups: many forms that are common in the Arctic, such as crabs, flatfish, and balanomorph barnacles, are not represented in the Antarctic.

There are differences between the present-day benthic habitats in the Arctic and Antarctic. The Antarctic, lacking appreciable freshwater run-off from the land and the massive bioturbation seen in the Arctic, is the more stable and this has been favourable to diversification. The Arctic benthos is more accessible to invading species whereas the Polar Front and lack of shallow sea approaches are impediments for forms migrating into the Antarctic. However, it is necessary to look to the past for the full explanation of the differences in benthic floras and faunas.

The habitats in the Arctic are mainly the outcome of events in the Quarternary period; that is, the last 2 million years. An Arctic estuarine water mass was formed by the inputs from the Siberian rivers and was colonized by a mixture of freshwater and euryhaline immigrants which distributed themselves around the Polar Basin during the last sea-level transgression. These assemblages seem, however, to have been nearly eradicated during the last major glaciation. In post-glacial times there has been invasion by low-temperature-tolerant species from the Atlantic and, to a lesser extent, from the Pacific. Thus, the present flora and fauna comprise relatively young assemblages with notably few endemics. An odd thing is that, whereas animal species with Pacific affinities are numerous in the inshore regions between 145°E and 120°W, algal species with the same origin are scarce. This might arise from the greater vulnerability of seaweeds requiring hard rock and shallow water (Dunton 1992). The highest diversity occurs in areas of hydrographic mixing. Areas of lowest species richness in the Arctic benthos are those influenced by brackish waters and the deep sea. The benthos is subject to much the same range of low temperatures in the Arctic and Antarctic but nevertheless there are differences in the degree to which organisms are adapted to this. Endemic brown algae of the genus *Desmarestia* in the Antarctic have growth optima around 0°C, an upper limit for growth at around 5°C, and a survival limit at 13°C. In the Arctic, the endemic *Laminaria solidungula* grows up to 16°C and survives at 18°C.

A few Arctic species have temperature requirements similar to those of Antarctic species but on the whole the situation is consonant with the idea that the southern species have had much longer to adapt to the polar environment. The Antarctic became cut off by the Circumantarctic Current about 25 million years ago and then began to cool, although for a time it continued to enjoy a mild climate. There is fossil evidence of a focus for the appearance of cool water taxa at high southern latitudes around this time. It seems that much of the present flora and fauna of Antarctic waters is descended from forms which originated then. This would explain the large numbers of endemics, a tendency which would be reinforced by the hydrological barriers and the propensity of the dominant benthic invertebrate groups not to have pelagic larval stages. In addition to the relict autochthonous flora and fauna there are some species to be distinguished (e.g. among the sponges), which appear to have emerged from deep water into shallower coastal waters. There are also cool-temperate species, mostly from South America, which have invaded via the chain of islands in the Scotia Arc. This has involved negotiating numerous gaps and pitfalls that eliminate whole families as well as individual species. Apart from those that appear to have followed this path, most Antarctic species have circumpolar distributions (Dell 1972, Dayton, in Smith 1990).

The longer time which has been available for evolution of polar biota in the Antarctic as compared with the Arctic is evident in the physiology of organisms as well as in their taxonomic features. Endemic brown seaweeds, for example, have lower requirements for light and their growth is inhibited by rise in temperature to a greater extent than it is in corresponding endemic Arctic species (Wiencke 1996).

10.4 Present-day global climate change and polar regions

There can be no doubt that one of the major environmental debates facing the human population is the potential effects of global climate change, sometimes referred to as global warming. It is well beyond the scope of this book to cover this topic in anything but the most cursory overview, although the wealth of information and synthesis of current understanding is readily available from scientific and political organizations who have recognized the importance making up to date information as widely accessible as possible (see the following in the recommended Further reading section: Arctic Climate Impact Assessment 2005, European Science Foundation 2007a, Intergovernmental Panel on Climate Change 2007a–2007c).

The analyses of long ice cores from the Antarctic and Greenland ice sheets have been fundamental for our understanding of climate history over the past 800000 years (EPICA Community Members 2004, 2006). The high resolution of temperature and gas changes deduced from an ice core in Droning Maud Land, in Antarctica, shows similar trends to the findings deduced for Greenland ice cores and are indicative that there is the clear link between the climate in the northern and southern polar regions over 150000 years or so, governed by ocean circulation. The climate variation being discussed is on the millennial scale and is thought to be linked to meridional overturning circulation (or MOC) in the Atlantic ocean: strong MOC brings heat to the North Atlantic at the expense of the southern Ocean and so increases and decreases in the strength of the MOC results in a climatic 'sea-saw' between the southern and northern hemispheres (Steig 2006).

But rather than these climate changes over hundreds of thousands of years it is the global climate changes since the 1850s that are foremost in the general public's thinking and that are driving global environmental policy agendas. *Greenhouse gases* include a cocktail of many different gases, but the main contributors are carbon dioxide, methane, nitrous oxide, hydrofluorocarbons (HFCs), perfluorocarbons (PFCs), and sulphur hexafluoride. When an object is warmed up by absorbing solar energy, thermal energy is then lost by emission of infrared radiation. Clouds, atmospheric water vapor, and the greenhouse gases absorb this radiation, the greenhouse gases alone trapping about 40% of the radiation being emitted from the Earth. This trapping of energy is the *greenhouse effect*, and without it the Earth would be 30–40°C cooler than it is. Therefore in itself this effect is vital to life as we know it. Water vapor is the greatest contributor to the effect, but carbon dioxide can contribute up to 25% of the effect. It is the increased loading into the greenhouse gases through human activities in the past two centuries that is attributed to much of the global tempoerature increase beyond previously observed ranges. The Antarctic and Greenland ice cores are the best evidence we have that current levels of greenhouse gases, in particular methane and carbon dioxide, are at their highest for 650 000 years.

The huge post-industrialization increases in greenhouse gases are attributable to many sources, but the major ones are energy supply and transportation and subsequent burning of fossil fuels. However, some greenhouse gases, such as the HFCs, PFCs, and sulphur hexafluoride, although minor in comparison to carbon dioxide and methane, have long lifetimes in the atmosphere: HFCs 10–100s of years and PFCs and sulphur hexafluoride up to 1000 years. HFCs were introduced as replacements for chlorofluorocarbons (CFCs) in the 1990s when the latter were implicated in atmospheric ozone depletions.

We now have over three decades of satellite data about the extent of ice sheets, sea ice, sea-surface temperature, and other factors pertinent to recording environmental change. Comiso and Parkinson (2004) and Serreze *et al.* (2007) have described how latitudes higher than 60°N between 1981 and 2003 increased by 0.5°C per decade over sea ice, 0.85°C per decade over Greenland, and 0.79°C over North America. During the same time there was a cooling of about 0.14°C over Eurasia. In general springs were warmer in the 1990s compared with the 1980s and came earlier in the year. A direct consequence of this warming trend is that some areas at the periphery of the Greenland ice sheet have melted by up to 1 m, and in general there was an increase of 17% in the melt region of the Greenland ice sheet between 1992 and 2002. This increase in ice-sheet melt has been mirrored in increased melting of large glaciers in other Arctic regions, leading in part to the observed 10–20-cm increase in sea level since 1900, or the currently estimated rise of 3 mm year^{-1} (Shepard and Wingham 2007).

Some of the most dramatic changes to be recorded in the past 30 years in the Arctic are the changes in perennial ice cover, which is decreasing at a rate of about 9.2% each decade (Comiso 2002, Stroeve *et al.* 2007). These reductions in sea ice are greatest in the Beaufort Sea. Haas (2004) reports on a series of transects made in 1991, 1996, 1998, and 2001 between Svalbard and the North Pole where sea ice thickness was measured routinely. During the whole period there was a 22.5% reduction in the average sea ice thickness from 3.11 to 2.41 m. Rothrock and Zhang (2005) using a larger data-set also report a greater 2-m decrease in sea ice thickness in the central Arctic between the 1960s and 1990s. It is not just the summer sea ice extent that is decreasing, but the 2006 and 2007 winter sea ice maximum extents were about 1 million km^2 less than the average mean maximum extent for all years between 1979 and 2000 (Comiso 2006). These sorts of major findings have led several researchers to conclude that there will be no perennial sea ice in the Arctic summers after the 2050s (Comiso 2003, Johannessen *et al.* 2004, Serreze *et al.* 2007). This would mean that although sea ice would form in winter, it would all melt in the following spring/summer and there would be little thick ice which is characteristic of the Arctic (Nghiem *et al.* 2006). In fact the seasonal dynamics of ice formation, consolidation, and melt would be similar to that we know of the Southern Ocean pack ice today.

Naturally such reductions in ice and snow cover will have great impact on the albedo (Chapter 1) with consequences for heat budgets of the regions. When ice melts, the albedo of the remaining ice is reduced and therefore more energy can be absorbed. This in turn will increase the rate of melting. This is termed a *positive albedo feedback loop* where the absorption of heat energy leads eventually to an even greater absorption of energy. This happens on a seasonal basis during the melting of sea ice, but more dramatically it is thought that such positive feedbacks may play a large role in changes brought about by global climate change. Potentially if there is a increased melting of sea ice in Polar regions, as in the Arctic, the reduced albedo will induce further warming of the surface waters and thinner ice resulting in accelerated ice melt. There are other proposed related loops that will also enhance melting, but some that will also slow it down (Comiso and Parkinson 2004).

Clearly such dramatic changes in the seasonal dynamics of the physical structure of a whole ocean basin will have profound implications on the biology living there. These are well documented in the Arctic Climate Impact Assessment (2005) report that comprehensively discusses the consequences of these warming trends on marine, freshwater, and terrestrial systems.

A brief and obvious example of the potential and complicated consequences of the predicted decreases in summer sea ice and decreasing concentrations of multi-year sea ice is that polar bears will need to migrate further north to forage for food than their present territories. However, increasingly the Arctic pack ice will be cut off from land for longer periods of time. These effects have been highlighted by many researchers (Derocher *et al.* 2004, Stirling and Parkinson 2006, Fischbach *et al.* 2007) and have bought the effects of climate change in polar regions to a wider-public audience: the combined effects of observed decreases in sea-ice thickness and the progressively early sea ice break-up are linked to decreasing polar bear populations in Western Hudson Bay and Baffin Bay. A consequence of this is that the surviving bears have to search for alternative food sources close to towns and communities. This is because earlier break-up of sea ice is forcing bears on to land earlier in the year, thereby preventing them feeding fully on seals and building up sufficient fat stores with which to survive before freeze-up, when they can return to the sea ice to hunt seals. Naturally as ice further retreats the bears will become increasingly stressed and their numbers will decline significantly. As the populations decline, problem interactions between bears and humans will continue to increase, as the bears seek alternative food sources. Polar bears usually return to the same area for overwintering and can utilize the same maternity den several times. However, warming of some permafrost regions has already resulted in such dens collapsing, and making the suitability of dens more precarious than has been the case up to now (see Lunn *et al.* 2004).

In contrast, in the case of the whale species that seasonally move into ice-free waters, presumably a decline of Arctic sea ice, especially thick multi-year ice, will encourage them to move into regions of the Arctic basin where they have not frequented before. Therefore whale migration routes and consequent effects on their prey may be greatly affected. However, seal species that need the ice to haul out on to will be concentrated into ever-decreasing regions of ice as it decreases in extent. This will not only affect the behavioural patterns of the seals, but also those animals that predate them, namely the polar bears.

Schiermeier (2007) gives a very thought-provoking synopsis of how the Arctic region may change over the next 20 years as a result of warming climate. Among the examples of possible developments the following stand out:

1. by 2012 the polar bear may be listed as a threatened species;
2. by 2017 vector-borne diseases may become more widespread among the Inuit and mosquitoes may migrate to the Arctic, and the Arctic cod could be replaced by temperate fish;
3. by 2027 the Arctic Ocean may remain ice-free in September and oil tankers and freight ships may be able to sail the Northeast Passage. Already the consequences for territorial rights are being challenged.

The effects of global climate change on Antarctic regions are less obvious than those in the Arctic regions, and there is very much debate about interpreting trends, and conclusions are difficult to draw. This is exemplified by the synopsis of Turner *et al.* (2005) who investigated near-surface temperatures recorded over the last 50 years at 19 Antarctic bases. Eleven of these had warming trends and seven had cooling trends in the annual data (one station had too few data to allow an annual trend to be computed).

Certainly there have been no massive increases in melting on the Antarctic ice sheets, and Monaghan *et al.* (2006) indicate that there has been no statistically significant change in Antarctic snowfall over the past 50 years, although this is debated by Shepard and Wingham (2007). Measurements show that the West Antarctic ice sheet is getting thinner, while at the same time the East Antarctic ice sheet is thickening (Davis *et al.* 2005, Oppenheimer and Alley 2005). In contrast there can be no doubt that the Antarctic Peninsula region is one of the regions of the world that is warming the fastest, where average air temperatures have risen by 2°C per decade (Vaughan *et al.* 2001, 2003). This warming has been linked to increased glacial melt and retreat on the peninsula, and Cook *et al.* (2005), by examining aerial images, report that 87% of 244 investigated glaciers show evidence of significant retreat since the 1950s. Some glaciers towards the north of the peninsula retreated by up to 13 m in a 10-year period. There is considerable debate as to whether or not collapses of large ice shelves such as the Larsen B Ice Shelf on the eastern side of the Antarctic Peninsula are the result of warming, although a collapse of such magnitude seems to have not occurred since the end of the last ice age (Domack *et al.* 2005).

In contrast to that in the Arctic, sea ice extent in the Southern Ocean does not show trends of decreasing. There are reports of slightly shortened sea-ice season for some of the regions (Parkinson 2004). There is actually a slight overall increase in the Antarctic sea ice extent, rising at approximately 0.4–1% per decade since satellite observations started (Zwally *et al.* 2002). In the Weddell Sea, Indian Ocean, and Pacific sectors the changes in extent are rather minor, but there are significant decreases of 8% per decade of ice extent in the Bellingshausen and Amundsen Seas and a corresponding increase of 7% per decade in the Ross Sea. These two sectors are adjacent to each other and it would appear that there is a movement of ice from the Amundsen Sea into the Ross Sea, as well as increased ice production in the latter. These generalizations mask significant regional differences. There are measurements to show that the ice seasons in the eastern Ross Sea, far western Weddell Sea, far eastern Weddell Sea, and coastal regions of East Antarctic have shortened. In contrast, sea ice seasons have lengthened in the western Ross Sea, Bellingshausen Sea, and central Weddell Sea (Comiso, in Thomas and Dieckmann 2003, Parkinson 2004).

The most well-reported effect of the warming at the peninsula region is that of the penguin populations living there. Since the 1970s the Adélie penguin population has declined, whereas the chinstrap and gentoo penguin populations have increased. Adélie penguins are dependent on sea ice (Chapter 9). The localized decline in sea ice has resulted in there being 85 days less sea ice than 25 years ago (Stokstad 2007).

Curren *et al.* (2003) provided indirect evidence that at least in some regions of the Southern Ocean, that instead of this increasing, or at least static, sea ice extent, there have been decreases since 1950. They measured the concentration of methane-sulphonic acid (MSA) in ice cores. This is a breakdown product of dimethylsulphide (DMS; Chapters 6 and 7), which in turn is a breakdown product of dimethylsulphoniopropionate (DMSP) produced in large quantities by ice alage. By measuring concentrations of MSA in the glacial ice core, researchers have been able to estimate sea ice extent in a particular sector (80–140°E) off the East Antarctic coast back to the 1841. They estimate that sea ice extent in this region was rather constant between 1841 and 1950 and thereafter there was a sharp decrease of about 20%. Dixon *et al.* (2005), also by measuring MSA concentrations in several ice sheet ice cores, have estimated sea ice extents over a 200-year period. MSA records from South Pole cores have been related to the extent of sea ice in the Amundsen and Ross Seas. The interpretation was that the sea ice extent in these regions were generally higher from 1800 to 1992 compared with the period 1487–1800. They also demonstrated that El Niño Southern Oscillation-related atmospheric circulation systems have affected Antarctica for at least the last 500 years.

Interestingly the results of Curren *et al.* (2003) were not too dissimilar from conclusions drawn from the early whaling records. de la Mare (1997) estimated the extent of the Antarctic summer ice edge from positions where whales were caught, or recorded by scientific expeditions, from the early 1920s up to the 1960s. This is because the whaling for blue, fin, and minke whales was focused on productive ice-edge regions and there was an international requirement to record the position of whale catches. Therefore it was possible to make estimations of previous ice-edge extents before satellites were used, and compare these with the satellite data collected since the 1970s. de la Mare concluded that there had been a 25% drop in the extent of the summer sea ice in the Southern Ocean in just 20 years. However, other analyses of the data indicated that the whale-catch data may not as reliable as first thought (Vaughan 2000, Ackley *et al.* 2003). The debate has continued and most recently Cotté and Guinet (2007), accounting for the bias in the whaling data, also highlight a significant reduction in the sea-ice extent happening in the 1960s, especially in the Weddell Sea region.

11 Human impact on polar regions

11.1 The first invasions by humans

Only relatively recently has mankind invaded the areas around the Poles: the Arctic, being nearer centres of population and with a continuous land connection, was the first to be colonized (Hoffecker 2004). Wandering north via the unglaciated lowlands of north-east Siberia and the Beringia Peninsula tribes established themselves at least 20 000 years BP, hunting mammoths, bison, and reindeer (caribou) with stone-tipped spears and moving across the land bridge to North America. This Palaeo-Arctic culture persisted until around 7000 years BP, although around 11 000 years BP as the climate warmed, and melted the ice caps, the land bridge disappeared under the sea. The population was thus fragmented and the groups that remained inhabited the coastal areas, living on sea mammals, muskox, and reindeer (also known as caribou). One group moved to northern Greenland while another moved to the eastern Canadian Arctic, giving rise to the Dorset culture. This latter group developed the building of ice houses (igloos), skin-covered boats (umiaks and kayaks), and an efficient harpoon technology for hunting from the ice.

Exploitation of sea mammals was the key to successful occupation of all the high-Arctic areas. Another wave of migration, probably from the area of east Siberia/west Alaska occurred around 1000 AD, establishing the Thule culture, which extended as far as west Greenland and adapted the Dorset culture techniques to hunting of the bowhead whale. It was around this time that the Thule culture first came into contact with Europeans when Norsemen discovered Greenland and established colonies on its west coast. The Thule culture ended during the sixteenth century, probably through a combination of climatic deterioration, disease, and European whaling activities.

The present-day Inuit are a generally homogenous people of North America and Greenland speaking variants of the same language and clearly demarcated from the forest Indians of North America. During the twentieth century their isolation made them especially prone to death from many common diseases, for example measles, carried into their settlements by Europeans. The subsistence hunting culture of the Inuit has always made them especially sensitive to climate changes because of its effects on the key species of seals, whales, and reindeer. Their insecurity is clearly reflected in their religion—Shamanism—which believes that all living creatures have human and animal, spiritual and physical qualities. Meanwhile in Europe several different groups—the Saami in Scandinavia, and the Nganasans, Chuckchi, and other groups in Siberia—have colonized principally the sub-Arctic, using the High Arctic only for summer grazing of their herds. They have evolved both different languages and a migratory culture based mainly on reindeer herding.

In contrast the Antarctic has been protected from invasion by its remoteness and the stormy seas of the Southern Ocean. The first recorded landing south of the Antarctic Polar Front was when Captain James Cook set foot on South Georgia on 17 January 1775. As for the continent itself, the first landing was probably in the Hughes Bay area of the peninsula from the US sealing tender *Cecilia*, captained by John Davis, in February 1821. It may be that the Polynesians had penetrated into Antarctic waters before this: a Rarotongan legend recounts a voyage into a region of fogs, monstrous seas, and what may have been icebergs, but they evidently did not establish settlements or leave any traces of their visit. Two projectile heads of a type used until 1500 by the Indians of central Chile have been dredged up at King George Island but there is no other evidence supporting such an early visit this far south.

11.2 The ecology of pre-industrial humans in the Arctic

At present there are approximately 150 000 native people living in a variety of environments across the vast Arctic areas that surround the Arctic Ocean (Couzin 2007). This population is considerably in excess of the pre-industrial levels, which may have been less than 100 000, as all the communities had then to be self sufficient, dependent on renewable resources. If food became scarce locally the people moved elsewhere. Mobility required a minimum of possessions since travel was either on foot or sledge on land, or in kayak or umiak on water. The boats, constructed with a frame of driftwood held together with treenails, whalebone lashings or seal sinews, and covered with stretched seal skin coated with boiled seal oil, are a remarkable combination of lightness with robustness. Their construction requires great skill and even an expert might take up to a year to complete one (Arima 1988). Only men were allowed to use kayaks with women restricted to the much larger and open-decked umiak.

To live in the sub-zero temperatures it is essential to have clothing that is windproof and warm, does not trap moisture, and is light and durable.

Throughout the Arctic reindeer skins offered the warmest material for garments because the hairs were hollow and insulating. The Inuit also understood the principle of layering, wearing an inner skin with the hairs facing inward and an outer skin with the hairs facing outwards. The air space between the two layers insulated the wearer who could regulate air flow with a belt or a hood. Skins from seals, polar bears, and small fur-bearers were also used, all clothing being sewn together with sinews as thread. Clothing was embroidered and decorated as a celebration of both group or tribal identity and personal declaration.

Whereas summer habitations were tents made from skins, in the winter many groups in the High Arctic lived in snow houses (igloos) constructed from blocks of firn. The snow must be of the right consistency and some experience is needed to get the blocks to spiral inwards evenly. Using dogs to pull sledges is thought to have begun around 7000 years ago, probably in Siberia. Dog sleds allowed people to move around more efficiently in winter. Sledges have been constructed from driftwood, whalebones, and even antlers, with runners made of bone, ivory, or even frozen fish wrapped in skins. Whips and harnesses were made from seal skin and whale skin. Just two or three dogs seem to have been used at first, but the modern way is for up to 10 dogs, often in a fan trace.

The ancient hunters of reindeer in the Eurasian Arctic perhaps numbered no more than 11 000, assuming around 3 million wild reindeer with an annual increase of about 7%. Exploitation of the reindeer required a nomadic existence, following them south in the winter to the protection of the forest tundra and northern taiga. Reindeer grazing on lichens needed to be sustainable so that reindeer numbers also imposed a limit on the human population it could support. With the development of reindeer husbandry people were able to settle in the sub-Arctic, hunting of other animals began as well as fishing, and the human population increased. Norsemen settled on the west coast of Greenland in 986 AD and at the height of its development the colony contained about 3000 people on 2080 farms. The main livelihoods were trade in walrus skins and ivory, and cattle rearing. The settlements persisted until the early fifteenth century and then died out, possibly because of diseases introduced by Europeans or because of a deterioration in climate. The ecological impact of this colonization was small other than in the introduction of around 50 European species to the local flora. Denmark began to colonize Greenland in 1721, the start of over 230 years of Danish colonial rule.

11.3 Sealing and whaling

Humans have headed the list of predators of Arctic seals for several thousand years and hunting in the sea ice for seals developed particularly with palaeo-Eskimos, who moved into the north-west Canadian Arctic and western Greenland 2000 years BP and became the present-day Inuit. In winter, hunters moved over the inshore ice by dog sledge and, where the ice was clear, breathing holes were approached with feet muffled in polar bear skin shoes and watched until a seal appeared and could be harpooned.

When snow obscured the holes, dogs were used to sniff them out. Winter sealing was combined with fishing through holes in the ice. In spring, when the seals bask on the ice, a hunter might creep up, lying down to imitate a seal if sighted by his prey. In summer, the sealskin kayak provided a swift and silent means of approach through the pack. Walruses and, especially, ringed seals were taken around Greenland and the Northwest Territories.

Now canoes with outboard motors are used and the seals shot with high-powered rifles. Many thousands of ringed seals are caught each year but the walrus is now protected, apart from small numbers which local populations are allowed to take. The bearded seal is of great importance to coastal natives in Alaska and around the Bering, Chukchi, and Beaufort Seas. In 18 months in 1977–1978 rather more than 8000 were taken around Alaska. In the Bering and Chukchi Sea areas they are caught not only by the natives but by Russian commercial sealers (Sugden 1982, McGhee, in Ives and Sugden 1995).

The taking of seals and small cetaceans for food by local people began in Neolithic times but it was not until the seventeenth century that a commercial whale fishery was established (Tønesson and Johnsen 1982). Willem Barents, who discovered Svalbard in 1596, reported abundant whales in its vicinity and this attracted the attention of Basque fishermen who had been hunting the northern right whale in European waters. Soon English whalers, followed by Dutch and Danish, were operating in the Svalbard area and the interactions (including trading) between the native people and the Europeans intensified (McGhee 2006). The whales were pursued in small boats and harpooned by hand. Catching mainly the bowhead whale and the northern right whale for whale oil used in lamps, only the blubber was utilized. Although many of the ships boiled out the oil on board, Dutch whalers established a whaling station on Svalbard, called Smeerenburg, to process the blubber.

Interestingly William Scoresby Jr, the most successful English whaling captain of the time, had a strong scientific interest in the Arctic and his book on the Arctic regions published in 1820 became one of the foundations for Arctic science. Whales became scarce around Svalbard in the early eighteenth century and whaling shifted to the Davis Strait off the west coast of Greenland. An increasing number of ships with no controls inevitably resulted in over-hunting. The invention of the harpoon gun in the mid-nineteenth century followed by the development of steam-powered whale catchers allowed a wider range of species to be caught and hastened the decline in numbers of all the whales until by the end of the century Arctic whaling had become uneconomical. Whalers began looking southwards for new grounds in Antarctic waters.

Seals, and especially the walrus, were taken for blubber and ivory in the early years of commercial whaling in the North Atlantic. However, it was not until whales became scarce that seals were recognized as an economically important part of the total catch. Throughout this period harp, hooded, and bearded seals were taken on a regular basis and in large numbers for their skins by professional sealers.

The northern fur seal (*Callorhinus ursinus*) was discovered by Pribilof in 1786 with a population estimated at 2.5 million. The skins were highly prized, especially by the Chinese who used the fur to make felt, so Russian and American sealers soon reduced the population to around 300 000 by the early twentieth century. The report by Captain Cook of extensive populations of fur seals in southern waters sounded the death knell for hundreds of thousands in the Southern Ocean. The slaughter that began on South Georgia, lead again by British and American vessels, soon moved southwards to the South Shetland Islands and over several decades almost wiped out both species of fur seal. By 1822 over 1.25 million animals had been killed. Although fur seals were the main target elephant seals were also taken for the oil in their blubber. Sealing continued intermittently throughout the nineteenth century but never again achieved the remarkable returns of the early commercial expeditions. Only a small number of fur seals survived and, from a population of probably a few hundred, began to recover in the mid-twentieth century under protection at South Georgia. The population is now estimated to exceed 4 million, with animals colonizing southwards almost to 70°S. The elephant seals were never reduced to such small numbers and a profitable, controlled annual cull of around 6000 was established on South Georgia (Fig. 11.1) by the Compañia Argentina de Pesca with a scientific management plan to promote sustainability.

The whales that were abundant in the Southern Ocean at the turn of the century were blue and fin. Captain H.A. Larsen, while south with the Swedish South Polar Expedition, recognized a commercial opportunity. With Argentinean finance he established a whaling station at Grytviken on South Georgia in 1904. This flourished and several other companies began operating from South Georgia. The number of whales of all species caught from South Georgian stations rose to over 7000 in 1916.

The early methods wasted much of the carcass and the British Government quickly became concerned that uncontrolled hunting would result in the collapse of the industry. Establishing the Falkland Islands Dependencies in 1908, which included not only South Georgia but also the potentially

Fig. 11.1   Whale catchers at Leith Harbour, South Georgia, at the start of the whaling season in 1957 (photograph by W.N. Bonner) (see colour plate).

profitable whaling grounds further south towards the Antarctic Peninsula, the government began to regulate the industry (Hart 2006). By controlling the factories it was possible to prohibit the taking of females with calves and insist on fuller utilization of the whole whale. In addition there was concern that not enough was known about the biology of the species so a research programme called the Discovery Investigations began in 1925 into the factors controlling whale abundance.

These measures might have saved the whale stocks if it had not been for the development of factory ships capable of processing the whales at sea, beyond the jurisdiction of the government. The first of these ships began operating in the Southern Ocean in 1925 and by 1930 there were 41 of them. Between 1925 and 1931 the number of whales killed per year rose from 14 219 to 40 207. International concern was manifested by an agreement called the Convention for the Regulation of Whaling, which came into effect in 1935, but since only Norway and the UK made any effort to observe its principles it was of little use in slowing the slaughter. In 1938 a whale sanctuary was designated in the area south of 40°S between 70 and 160°W, and complete protection was agreed for the humpback whale, after which the Second World War began and whaling effectively stopped. The International Whaling Commission (IWC) was established in 1946 to regulate the industry but has failed to live up to expectations. It was systematically undermined by companies that failed to report catch data and by countries, like the Soviet Union, that deliberately provided false data, making it impossible to set scientifically sensible quotas.

As the whale populations continued to spiral downwards most countries eventually found whaling uneconomical and by the 1970s there was little active interest except from Japan, Iceland, and Norway. With some species apparently likely to become extinct a major international effort by non-governmental organizations (NGOs) like the Cousteau Foundation and Greenpeace drove international public sentiment against whaling and allowed the IWC to agree a moratorium on whaling in 1984. Japan was unhappy with this and began to conduct so-called "scientific whaling" under a special clause in the moratorium. This has allowed it to kill a small number of whales each year for supposed scientific studies and then sell the meat for public consumption. With the recovery in stocks of some smaller whales, for instance the minke whale, it is now scientifically possible to undertake limited whaling again, although the majority of IWC countries do not wish to do this for political reasons. Japan is now trying to get the moratorium reversed, with support from a wide range of countries that have no historical interest in whaling but have joined the IWC in recent years for political purposes.

Sealing has suffered a similar fate at the hands of public opinion. The first international agreement for controlling exploitation was the North Pacific Fur Seal Convention of 1911. Hunting of other seals has been regulated around the coasts of Britain, Norway, Sweden, USA, and Russia for some time so that none of the culls endangered the populations and in the far north the income generated for the Inuit was a key element of the local economy. However, the NGOs have been adept at exploiting the brutality of the hunting of harp seal pups to generate a backlash against the sale of the fur and the world market has largely collapsed with significant repercussions for the aboriginal people. In the Antarctic a pilot sealing expedition in 1964 from Norway rang alarm bells and led to the development of the Convention for the Conservation of Antarctic Seals, ratified in 1972. The Convention has never had to be used because of the collapse in the market but it does provide a sound basis for conservation and rational use of this resource.

11.4 Hunting

The first human colonizers of the Arctic were hunters and, although they were few in number and operated over an enormous area, their impact on some animals was probably significant. On the American continent fossils show no decline in diversity or territorial range of large mammals until the spread of human invaders began around 11 000 years BP. The extinctions happened suddenly. Similarly, some of the ancient animals of the Eurasian tundra seem to have been exterminated by small numbers of hunters. Some palaeolithic sites in northern Eurasia contain astonishing quantities of the remains of slaughtered animals. More recently the great auk was hunted to extinction (the last reported sighting was in 1852) and today the walrus and muskox are seriously threatened.

Whereas the aboriginal people killed animals at a subsistence level the expansion of European exploration led to hunting on a much larger scale. The centuries-old Russian fur trade became a state-supported monopoly and by the end of the seventeenth century operated throughout nearly all of northern Siberia. In Canada the Hudson's Bay Company, granted a charter in 1670, gradually extended its activities across the Northwest Territories and into all the Arctic and sub-Arctic areas, encouraging unregulated slaughter of fur-bearing animals. Steel knives, guns, and patent traps enabled the Inuit to enter into the trade, which greatly reduced the numbers of foxes, but no species was hunted to extinction. Again public sentiment was turned against fur as a fashion accessory and the enactment of conservation legislation and the establishment of reserves have safeguarded the future of these species. More recently demand has developed for trophy hunting, especially of polar bears, which are now closely protected throughout their range in all circumpolar countries. A quota system has now been introduced and a small number can be shot annually on payment of a high licence fee.

11.5 Fishing

Fishing has always been part of the subsistence economy in the Arctic and at that level there was never any threat to stocks. The Arctic seas host a rich and diverse range of fish species, with around 150 species of fish in the Barents, White, and Kara Seas of which the most important are the large numbers of cod, herring, capelin, and salmon. There are as many species in the Bering and Chukchi Seas, which also includes the heavily exploited pollock. In fact the Barents and the Bering Seas are two of the most commercially productive fisheries in the world and the Arctic fisheries together supply a significant part of the world's total fish (Hoel and Viljamsson 2005). The Bering Sea fisheries alone comprise half the US catches. But increasing demand and exhaustion of other stocks, as well as greater accessibility due to sea ice retreat, has allowed commercial fishing to steadily expand northwards. A major expansion took place in the 1950s, the Barents Sea and the coasts of Greenland and Iceland being fished intensively, especially for Atlantic cod. Competition led to the "Cod War" between Britain and Iceland in the 1960s and 1970s after which Iceland adopted the Exclusive Economic Zone principle to keep out foreign vessels. In the last 50 years there have been some spectacular crashes of populations of commercially important species, such as the cod and Atlantic salmon off the coasts of Canada and Greenland and herring in the Norwegian and Icelandic waters. Strict conservation measures including quotas, net mesh sizes, and no-catch zones were put in place. However, even with these, some recovery has been slow and not a certainty.

Other populations such as the haddock between northern Norway and Svalbard have also seen a gradual but steady decline. The Icelandic fishing ban on Atlantic herring between 1972 and 1975 made a difference, with stocks gradually recovering and now considered to be within safe biological limits. These declining stocks put pressure on the Arctic indigenous people who often depend on fish catches as a key part of their diet. Catches in the North Atlantic reached a plateau around 3.5 million t in 1974 with declining totals since. This may also be important in the decline in seal populations in recent years.

There are major fisheries, dominated by Japan and Russia, for halibut and Alaska pollock in the Bering Sea. Here too there have been impacts on top predators with numbers of guillemots in Norton Sound, Alaska, declining as the pollock are fished out in the areas where the birds spend the winter. The indications are that fish production in the Arctic Ocean is low because of slow growth in near-zero temperatures and heavy predation by birds and marine mammals. There has also been a sport fishery in Canada and Alaska, especially in inland lakes and rivers, for trophy specimens of lake trout, Arctic grayling, Arctic char, and pike. This has been developing over the past few decades as part of a broader tourist initiative that has important economic implications, especially for the more remote regions. More recently there have been projects to develop fish farming of Arctic char off the northern coast of Norway and the establishment of a new fishery for Kamchatka king crab.

The Barents Sea cod fishery is the single most important fishery for Norway, both commercially and in terms of maintaining viable communities along the northern coasts. Since 1976 Norway and the Soviet Union/Russian Federation have managed this fishery bilaterally through the Joint Norwegian-Soviet/Russian Fisheries Commission. Although generally considered to be an example of successful international collaboration, the management regime has met new challenges since the late 1990s: massive over-fishing by Russian vessels, difficulties for Norwegian research vessels in getting access to the Russian economic zone, a tougher stance by the Russians in the Fishery Protection Zone around Svalbard, and pressure from the Russian side to set quotas far above the precautionary reference points. A continuing problem for all the Arctic fisheries that use gill netting is the large numbers of birds that are drowned: over 300 000 a year at present.

In Antarctica fishing got off to a much later start. Commercial exploitation did not begin until the 1960s when the imposition of the 200-nautical-mile (370-km) fishing zones elsewhere caused the redeployment of the distant-water fleets, especially the Soviet ones. The Soviet trawlers and factory ships focused on the area around South Georgia (Agnew 2004). The total annual catch, mostly of *Notothenia rossii*, rose rapidly to a peak of 400 000 t in 1969–1970 before declining rapidly, causing the fleets to move to other sub-Antarctic islands, like Kerguelen, or further south to the South Orkneys and South Shetland islands. Other species were now included with growing catches of ice fish *Champsocephalus gunnari* and krill *Euphausia superba* and a number of other countries (Poland, Bulgaria, and German Democratic Republic) competed for stocks.

Krill had been caught experimentally by the Soviets in 1962. It posed difficult problems in terms of processing as its chitinous exoskeleton contains high levels of fluoride, and once caught it must be shelled and frozen within 3 h to stop enzymic degradation of the protein-rich flesh. By 1970 the Soviets had developed processing equipment and began harvesting, followed by the Japanese in 1972. At this stage the emphasis was on krill-based foods and with catch rates of up to 40 t h^{-1} and an annual catch of 600 000 t international concern was suddenly aroused that uncontrolled exploitation could wreck the entire Southern Ocean food web.

The Scientific Committee for Antarctic Research (SCAR; www.scar.org) established a major international research programme called Biological Investigation of Marine Antarctic Systems and Stocks (BIOMASS), which provided the scientific basis for international management. Using these data the Antarctic Treaty Parties negotiated the Convention for the Conservation of Antarctic Marine Living Resources (CCAMLR), which came into force in 1984 to manage all marine biological resources on a sustainable ecosystem-wide basis (Hempel 2007). Continued research on krill has shown that there are a variety of other markets for krill products with interests in biochemical products (especially enzymes) and in the chitin from the shell. At present almost all krill is processed for use as a protein supplement, either for cattle food or for fish farms.

The efforts of CCAMLR over the past 20 years have allowed the recovery of some over-fished stocks, controlled legal access to all other fish, squid, and krill stocks, and established a long-term monitoring system to assess the effects of fishing on key components of the food web. The increasing use of baited long-lines to catch fish and squid had profound effects on many Antarctic seabirds. Populations of albatrosses, especially wandering albatross, began to fall dramatically. Observers on the fishing vessels as well as data from satellite tracking of birds showed that they were getting their beaks caught on the hooks when diving for bait as the long-line was let out and being drowned. CCAMLR has found ways of mitigating this unfortunate by-catch within the legal fisheries. However, the growth of the fishery for the very valuable Patagonian toothfish (*Dissostichus eleginoides*) has encouraged the rapid development of illegal, unreported, and unregulated (IUU) fishing which now threatens the continuing existence of these long-lived, slow growing fish and is now largely responsible for the present albatross and petrel deaths as by-catch (Fig. 11.2). At least 11

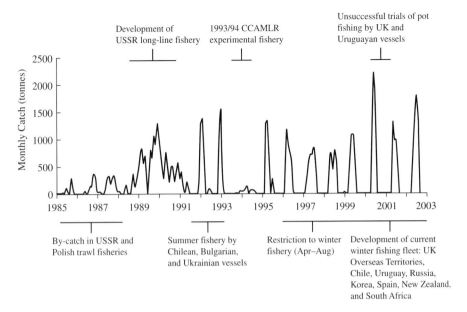

Fig. 11.2   Development of South Georgia toothfish fishery. From Agnew (2004).

countries are involved in this IUU activity, which now accounts for half of all the toothfish sold worldwide. CCAMLR methods to counter this include satellite tracking of fishing vessels and catch certification at the port of landing.

11.6 Pastoral and agriculture development

The pastoral way of life in the sub-Arctic centres on the reindeer and is essentially a European development. In Eurasia hunting of the wild animals changed into husbandry as the societies developed. Initially Arctic peoples probably used their herded reindeer for transport, for raw materials, and for food. The change in human–animal relationships inherent in this meant that now pastoralists were protecting the animals from predators, ownership of the animals changed the structure of their previously egalitarian society, and the eventual development of commercial trading in skins and meat forced economic stratification as some herders became much richer than others (Ingold 1980).

Until the middle of the twentieth century the Lapps followed traditional patterns; during the summer entire families lived with their herds in the tundra and then migrated south for the winter. Now reindeer herding is motorized using skidoos, motorcycles, jeeps, boats, and even helicopters.

In the Russian Arctic there are now around 2.5 million domesticated reindeer as against 600 000 wild animals. The latter occupy marginal terrain that will not support the densities of the domesticated herds without permanent damage. The explosion at the Chernobyl nuclear power station in the Ukraine had a serious effect on Scandinavian reindeer herds. Fallout clouds carried large amounts of the radioactive isotope ^{137}Cs with a half-life of around 30 years. This was absorbed by the vegetation and accumulated in the reindeer, especially those herds in central Sweden and Norway, making the meat dangerous to eat. Large numbers of reindeer were slaughtered and buried in both countries and, although the governments paid compensation, the Saami cultural system was changed by a break in its traditional practices. Contamination of some pastures continues to be a problem today.

In North America the Inuit hunted wild reindeer in summer and the Indians hunted them in winter. The introduction of guns made hunting much easier and had a dramatic effect on the reindeer with numbers falling from 1.75 million in 1900 to around 200 000 in 1950. In contrast reindeer husbandry was founded on stock imported from Chukotka in the late nineteenth and early twentieth centuries, using Saami herders to demonstrate the techniques in use then. Herding was seen both in the Northwest Territories and in Alaska as a suitable occupation for the Inuit and one with commercial potential. However, the stocking was too high and in the 1940s and 1950s herd numbers crashed as poor management allowed serious overgrazing. In the post-war period management has followed the pattern in Eurasia and become mechanized but in Alaska, in contrast to everywhere else, the majority of reindeer keepers are not now indigenous people.

Arctic agriculture is a small activity in global terms, although some nations, such as Iceland, produce more than enough meat and dairy products to sustain their populations. The emphasis is on cool-season forage crops, cool-season vegetables, small grains, and some raising of cattle, sheep, goats, pigs, and poultry. In some townships there are horticultural facilities to produce vegetables but at high energy cost. While agriculture is limited by climate principally to the Low Arctic and sub-Arctic, it is also limited by the lack of infrastructure, small population base, remoteness from markets, and land-ownership issues.

The Arctic now offers an important service to agriculture in the rest of the world through the establishment on Svalbard of an international seed bank for crop species. This underground rock vault is meant to be a resource of last resort and uses the permafrost to keep the seeds frozen. More than 100 nations have endorsed its construction and many are already making arrangements both to fund its endowment and to contribute seeds from the estimated 2 million crop varieties currently used.

Fig. 11.3   Boreal forest in Alaska where melting permafrost is undermining the stability of many trees (photograph by David W.H. Walton) (see colour plate).

The most important biological resource from an economic point of view is forestry. The boreal forest, stretching from the Bering Straits through Canada, Northern Europe, and Siberia to the coast of Kamchatka, is the world's second largest terrestrial biome (Fig. 11.3). Understanding this habitat for sustainable long-term management must be a priority for research (Chapin *et al.* 2006). Commercial forestry has already fragmented and depleted the boreal forests in northern Arctic Russia and northern Scandinavia and Finland. The increase in the harvesting of timber for pulp, paper, and wood products is eliminating the remaining biodiversity of these once thriving systems. In the middle of the twentieth century the wood-processing industry began to devastate large tracts of Arctic forest with clear cutting, leaving only limited areas of virgin forest. The harsh climate slows natural regeneration and so much of the present reforestation uses non-endemic tree species.

The prognosis for the future of the remaining Arctic forests is mixed. The most important international agreements so far have developed from the UNCED Meeting in Rio de Janeiro in 1992. Initiatives under the Convention for Biological Diversity and Agenda 21 coupled with the UN Forum on Forests should have produced major change by now but heavy government lobbying has made international obstruction the name of the game. The most hopeful developments appear to be private initiatives on forest stewardship and sustainable forest initiatives. In some commercial forestry areas, new and innovative management regimes are being implemented to allow sustainable exploitation of the natural forest systems. Areas in north Sweden and Finland have been given national park or nature reserve status to protect against deforestation. As yet this has not been duplicated in Norway, or the Russian Federation. In these countries, as in Canada and other areas of the Arctic, commercial forestry and infrastructure development continue in their move northwards, extending the area of habitat fragmentation.

11.7 Introduction of non-native organisms by humans

The journeys of humans around the world have been accompanied by both the intentional and unintentional spread of non-native organisms. Of course, natural dispersal of organisms has always occurred, even across vast expanses of sea, but humans have provided a wealth of new opportunities for the spread of plants, animals, and microorganisms, sometimes with devastating results.

11.7.1 The Arctic

Introductions into the Arctic have been more extensive and have occurred over a much longer period than in the Antarctic. However, for both geographic and ethnographic reasons they are unevenly distributed about the Pole. In North America and Greenland the Inuit, hunters travelling light, took little with them although they must have transported a range of microorganisms and seeds from some of the plants. The people going to the North Atlantic islands were Europeans and carried both agricultural species and most certainly a range of fellow travellers of which they were not aware. In Greenland the Norse farmers provided a range of new immigrant plant species, some of which survived and spread after the settlements died out. Greenland now has at least 86 established introduced plant species compared to 427 native species, whereas in Alaska there are 144 compared to 1229 native species (Vitousek *et al.* 1997).

The Eurasian Arctic peoples, having a direct land connection further south, have brought about more introductions. The Chukchi in Siberia have used a variety of plants in their diet and dunghills around their campsites support many species otherwise absent from those localities. Disturbance of the native vegetation can also demonstrate the hidden presence of non-native species. After ploughing of hayfields at Noril'sk (69°21'N, 88°02'E) 19 species of plants appeared, some hundreds to thousands of kilometres away from their native habitats. Some weedy species show remarkable acclimatization potential and species like groundsel (*Senecio vulgaris*), pinapple weed (*Matricaria matricarioides*), and annual meadow grass (*Poa annua*) have been found both within native vegetation and also on disturbed areas where competition is much less.

There are a range of invertebrate introductions associated with man. Some, for example the house fly (*Musca domestica*) are found only in human company whereas others, such as the carrion flies *Protophormia terranovae* and *Cynomyia mortuorum*, are widely distributed throughout the Arctic wherever human refuse is found. Among birds it is more difficult to be certain which species have been unnaturally introduced but it would appear that sparrows, starlings, and swallows, all of which are associated with settlements, are probably contenders. Rats and house mice do not seem to be established in the Arctic (Chernov 1985) outside some urban areas.

11.7.2 The Antarctic

The sub-Antarctic islands that lie around the Antarctic continent have suffered considerably from introduced species. Initially sealers and whalers brought the species accidentally but there were also later attempts to introduce species for food. The diversity of the native flowering plants is not high on any of the islands so the fact that more than 17 alien species are established on South Georgia with only 25 native species is very worrying. Rabbits, cats, pigs, sheep, horses, cattle, reindeer, goats, mouflon, geese, and even trout have been deliberately introduced to these islands, and there are also rats and mice on several islands.

Only Heard and Macdonald Islands remain free of alien animals. Some species such as geese and horses did not persist but cats, rats, rabbits, and reindeer have had dramatic effects on the bird and plant species on many islands. On Kerguelen the rabbits have grazed much of the native vegetation down to the roots, changing the colour of the islands from green to brown. The rabbits on Macquarie Island are undermining the peaty soil, causing landslips, and endangering the survival of the megaherbs. On South Georgia overgrazing by the reindeer herds threatens the long-term future of much of the native vegetation on the north side of the island while rats prey on the eggs and young of most of the smaller birds. On some islands the alien mammals are managed as a fresh food resource, with the French keeping sheep and reindeer on Kerguelen and cattle on Amsterdam Island for meat. The trout persist on Kerguelen with apparently little impact on the local freshwater systems.

Control measures are being undertaken successfully with the eradication of feral cats from Marion Island, cattle and sheep from Campbell, cattle and rabbits from Enderby, goats from Auckland Islands, and rats from St Paul and Campbell Islands, and there are trials for rat eradication at Kerguelen and South Georgia.

The situation with both plants and increasingly with introduced invertebrates is worrying. With marked glacial recession on many of the islands and a general warming of the sub-Antarctic climate introduced species are spreading. Several of the plant species (*Poa annua*, *Sagina procumbens*) are aggressive colonizers and can outcompete the native vegetation, establishing themselves in both disturbed areas and in natural vegetation. *P. annua* is especially good at adapting to grazing and so spreads quickly on those islands with introduced herbivores. More recently an increasing range of introduced invertebrates have been identified, many of which are spreading steadily across the islands, often eating their competitors.

On the Antarctic continent and the associated Maritime Antarctic islands there is considerable concern about the possibility of introduced species establishing as a direct result of the increase in tourism. So far there is little evidence that this is a real problem (Frenot *et al.* 2005). Prior to the Antarctic Treaty there were various minor attempts to grow vegetables or keep chickens at Antarctic stations but none of them were successful. Small patches of alien grasses (*P. annua*, *Poa pratensis*, *Poa trivialis*) have established at various times at Deception Island and near Syowa Station and have flowered but grow very slowly. There were sledge dogs present at several stations throughout the continent until their enforced removal in 1994, but they had no measurable impact on the native flora and fauna. Growth trials at Signy Island of species from South Georgia have left a legacy of two introduced species associated with the trials area: an enchytraeid worm and a chironomid midge (*Eretmoptera murphyi*). Although the flora of the continent is now quite well established the microbial flora is largely unknown so that the conservation of microbial ecosystems is virtually impossible with no baseline against which to measure introductions. An attempt has been made to limit inadvertent introductions to some geothermal areas with especially interesting microbial communities by declaring them Antarctic Specially Protected Areas and requiring that those entering them wear sterile outer garments and use sterilized equipment.

11.8 Mineral and oil extraction

Exploration of the polar regions was originally for fame, personal gain, and the hope that there would be valuable mineral deposits. Even as late as the 1950s many scientists still believed that the polar regions were there to be developed in some way, with little thought of environmental damage or the concept of stewardship. This has certainly changed dramatically for the Antarctic but there are still powerful economic and political drivers for mining the Arctic.

Mining began in Greenland in 1854 for cryolite at Ivittuut at 61°N and the mine closed in 1987 after producing 3.7 million t of ore. Lead and zinc were discovered at Marmorilik during quarrying for marble but mining did not begin until 1973. This was the Black Angel Mine, which remained profitable until 1990. Various other mines were established for copper, graphite, and silver but most were on a small scale and proved unprofitable. There is now interest in developing a gold mine at Nalunak. Other companies are prospecting for diamonds and other gem material while the Greenland Government is promoting investigation of possible hydrocarbon reserves in West Greenland.

This is small scale compared with industrial developments elsewhere in the Arctic (Walton, in Ives and Sugden 1995). Economics should apparently determine which mining is undertaken but political decisions on sovereignty or self-sufficiency have at times encouraged wholly uneconomic and damaging activities, for example in Svalbard where Norway and Russia have been mining coal for many decades (Fig. 11.4). In many cases there is often little economic benefit for the local people from the mining development and every possibility that it will adversely disrupt the local ecosystem on which people may depend for food. An exception is the Red Dog Mine north of Kotzebue, the largest zinc mine in the world, which is owned by the native people. Mines in the Canadian territories include major lead and zinc mines in the High Arctic as well as gold and diamond mines. In Sweden there are very large iron-ore mines and in the Kola Peninsula the Russians mine nickel, copper, platinum, and iron. All this activity generates significant impacts on the environment, not only from the mine itself but from the effects of waste and toxic tailings on rivers and ground water, from the fumes from smelters, from the infrastructure of roads and ports needed to support the mines and from the abandonment of the sites when uneconomic. Clearly, large-scale mineral

Fig. 11.4   Community of Barentsberg on Svalbard (approximately 500 inhabitants) that was established mainly because of coal mining (photograph by David N. Thomas) (see colour plate).

extraction has created a strong need for regulations and impact analysis. Thoughtful management and utilization of mineral resources is imperative so that environmental impacts may be kept at a minimum but not all Arctic governments apparently accept this.

The extensive oilfield at Prudhoe Bay (70°N) on the Arctic Slope of Alaska was discovered in 1968 and began production in less than a decade. The engineering associated with the drilling and extraction was difficult, with much equipment having to be delivered to the sites in winter to avoid damaging the tundra, but to transport the oil to the tankers a pipeline 1289 km long had to be built across Alaska to the port at Valdez. At a cost of US$7.7 billion dollars for pipeline and terminal this was the most expensive civil-engineering project the world had seen (Fig. 11.5). The oil had to be transported warm to make it flow through the pipe, and so to avoid damaging the permafrost the pipe had to be heavily insulated and raised above the ground. To avoid heat conduction through the supporting columns a novel ammonia-based self-regulating refrigeration system was devised. Indeed the environmental regulations surrounding the building of the pipeline and the oil installations made the engineering much more difficult and expensive yet went a long way to protecting this wilderness area. Roads across the tundra and pads under buildings were made of gravel 1.5 m thick to prevent melting of the permafrost, and to allow the reindeer to migrate sections of refrigerated pipeline were buried underground. The reindeer did at first make use of these passing points but once the wolves learnt about them they were rapidly deserted and animals now pass under the pipe anywhere except at the points arranged for them. The political battle over oil extraction from the Arctic National Wildlife

Fig. 11.5   The Aleyaska Trans-Alaska Pipeline from Prudhoe Bay on North Slope down to Valdez on Prince William Sound (photograph by David W.H. Walton).

Refuge continues but the promises from the oil industry of minimal environmental damage are not convincing in the light of the damage already caused in the Arctic.

There are many other oil- and gas-production sites across the Arctic and for some new sites advanced technology using ice roads and ice pads instead of gravel for exploration, three-dimensional imaging of the oil field for better extraction, and directional drilling systems allows greater oil recovery with fewer holes. The extent of the onshore and offshore reserves in Russia, which stretch from the Caucasus to the Pacific, make it certain that Arctic Russia will be a dominant player in future energy supply (Krajick 2007).

The story of the future energy industry in the Arctic is likely however, to be a tale of Russian gas, according to a recent study that found that 80% of the overall hydrocarbon resources in the Arctic are gas, and 69% is specifically Russian natural gas. Subsea technology, which can operate underneath pack ice, will be the key to the economic development of much of this gas while delivery by pipeline from the Yamal Peninsula, over 4000 km from Western Europe, will be a huge challenge. In Canada and Alaska there are trials to extract gas from deposits of gas hydrates which are crystalline solids containing gas molecules, usually methane, each surrounded by a cage of water molecules. It looks very much like water ice and there appears to be immense quantities of it, both in the Arctic and elsewhere, trapped in deep sediments.

It has proved impossible to extract oil anywhere in the world without spilling some of it. In the Arctic these spills can be onshore or offshore and both can have catastrophic and long-lasting effects. There are, for example, 55 major contaminated sites on North Slope with hundreds of old exploration drilling sites with their waste pits that have yet to be cleaned up and restored. With roughly one spill a day there is a considerable area of contaminated land around Prudhoe Bay. In Siberia the situation is much worse, with lax regulation, old pipelines, and poor maintenance, and huge areas of tundra contaminated by spills from broken pipes and leaking valves and little effort being made to clean them up. Not only have they seriously damaged huge areas of tundra vegetation but oil has also leaked into ground water and rivers, to carry the contamination much further afield.

The most serious marine oil spill was in Prince William Sound, Alaska, in March 1989 when the fully loaded 300-m tanker *Exxon Valdez* ran on to a rock. The size of the resultant spill is still disputed but it must have exceeded 30 million l of crude oil. The slick covered 2500 km^2 of the sub-Arctic Sound, an area of exceptional ecological concern where the oil terminal had been built despite the original objections of environmentalists. The response to the spill was disorganized and decisions were delayed.

There was no plan or equipment to deal with a spill this large and the treatment of the shoreline with high-pressure hot-water hoses seems likely to have exacerbated the original damage to birds, fish, otters, and the benthic fauna. Exxon spent US$2 billion cleaning up the area and a further $1 billion settling civil claims for damages, mainly from those whose livelihoods had been ruined. Both the long- and short-term effects of the oil spill have been studied comprehensively. Thousands of animals died immediately; the best estimates include 250 000 seabirds, 2800 sea otters, 300 harbor seals, 250 bald eagles, up to 22 orcas, and billions of salmon and herring eggs. Little visual evidence of the event remained in most areas just 1 year later, but the effects of the spill continue to be felt today. In the long term, reductions in population have been seen in various ocean animals, including stunted growth in pink salmon populations. Sea otters and ducks also showed higher death rates in following years, partly because they ingested contaminated creatures. The animals also were exposed to oil when they dug up their prey in dirty soil. Some shoreline habitats, such as contaminated mussel beds, could take up to 30 years to recover.

In 1988 the Antarctic Treaty parties had agreed a Convention on the Regulation of Antarctic Mineral Resources Activities which might have provided the basis for an eventual mining industry. The environmental groups disliked this greatly and their international campaign to kill the Convention succeeded when several countries refused to ratify it. Now mineral resource development and the extraction of hydrocarbons is completely banned for 50 years under the Protocol on Environmental Protection to the Antarctic Treaty. Since almost all of the land is under ice and the two main marine sedimentary basins are covered by ice it has been difficult to estimate the distribution and extent of any mineral or energy resources. Analogies have been drawn with other parts of Gondwana and on this basis it is expected that there are possibly significant deposits of minerals like gold, platinum, copper, and lead. Traces of oil and gas have also been found during drilling. The only large mineral deposits above the ice are of coal and iron ore. Whatever we might evetually want to extract from the Antarctic will come at immense financial cost and is only likely to be economic when most of the reserves elsewhere in the world are exhausted.

11.9 Pollution

Pollution is a feature of human activities everywhere in the world and the polar regions, although remote, provide object lessons in our damaging effects at the local, regional and global levels. The Arctic was essentially unpolluted when only occupied by its aboriginal people. As industrialization began in Europe and North America and explorers began to move northwards and establish new settlements pollution became a growing problem. Access points to the Arctic are limited and so for some pollutants these increase the impacts.

Much of the Arctic pollution is generated elsewhere and carried northwards by the atmosphere, by migrating animals, or by water. For example, contaminated waste water produced from the nuclear fuel reprocessing plants in the UK (Sellafield) and in France (Cap de la Hague) is continually moved northwards to the Arctic (Fig. 11.6). Atmospheric pollutants generated by industry and urban centres to the south are carried up into the Arctic to join those generated by the northern mining and hydrocarbon industries. Rivers flowing north to the Arctic Ocean are a major artery for pollutants collected from forestry, agriculture, and industry hundreds of kilometres south as well as oil spills within the Arctic. Pesticides and other persistent organic pollutants are ingested by birds and other migratory animals in the more temperate areas and carried north in the summer. A visible sign of this atmospheric pollution, especially in winter, is Arctic haze that can reduce incident solar radiation by as much as 15%. Arctic haze is mostly composed of particles of sulphuric acid and organic compounds formed in the air from the combination of naturally occurring chemicals and pollutant sulphur dioxide or hydrocarbon gases and originates. Called aerosols, these resultant particles are small enough to float in the air but are large enough to reflect sunlight, and hence cause a haze (Law and Stohl 2007).

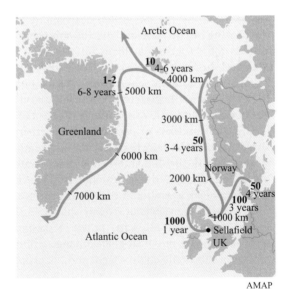

AMAP

Fig. 11.6   Transport of radionuclide-contaminated water to the Arctic from nuclear fuel reprocessing plants in the UK and France. Courtesy of Arctic Monitoring and Assessment Program (AMAP) 2000, from *Arctic Pollution Issues: A State of the Arctic Environment Report.*

The accumulation of persistent organic pollutants through the food chain is a serious health problem in the Arctic where the indigenous people use a variety of species for food. The circumpolar nations became so concerned about the effects on humans by pollutants that they established the Arctic Monitoring and Assessment Program (AMAP) to collect together both baseline data and trends for a wide range of chemical compounds. This has shown that organochlorine pesticides (aldrin, chlordane, deildrin, DDT, etc.) and industrial chemicals like polychlorinated biphenyls (PCBs), dioxins, and furans are accumulating in seals, polar bears, fish, and birds while PCBs are also accumulating in the Inuit with a high intake of traditional foods. This latter is of particular concern as these compounds can be transferred via breast milk and are believed to have their most significant impacts in the pre-natal and early childhood periods. At Broughton Island Inuit mothers eating fish were found to have 47 times the normal levels of PCBs in their breast milk.

Some of the heavy metals also give cause for concern. AMAP has been surveying levels of cadmium, lead, selenium, and mercury but in some areas of industrial activity there are also elevated levels of nickel. A unique combination of photochemical activity involving bromine and ozone leads to enhanced mercury deposition in the Arctic after polar sunrise, providing two to four times the normal dosage of this toxic metal that then accumulates in the food chain (AMAP 2002).

High in the atmosphere at both ends of the world ozone continues to be destroyed by the chlorofluorocarbons (CFCs) still present. Despite the phasing out of their use and the substitution of less-damaging compounds for refrigeration, foam blowing, and aerosol sprays under the Montreal Protocol, there is still a long way to go before the upper atmosphere returns to its previous ozone concentrations. Although the Arctic ozone hole is not at present as persistent and widespread a feature as the one over the Antarctic this could change in the future, possibly exposing hundreds of millions of people in the northern hemisphere to elevated levels of ultraviolet radiation. A new complication has been demonstrated by recent work that has shown that polar stratospheric clouds accelerate ozone destruction and the formation of these clouds is accelerated by the greenhouse-gas warming of the Earth's surface. The recent international assessment of the effects of climate change in the Arctic (Arctic Climate Impact Assessment 2005) has provided the most comprehensive review so far of the way in which global change will increasingly impact on all aspects of the Arctic environment and the people who live there.

The Antarctic has no industries and is a long way from the major centres of population. Although there is some very localized pollution close to major scientific stations, the Antarctic pollutants are principally global pollutants transported there by the atmosphere, the sea, and animals. It therefore provides an ideal location for measuring the baseline levels of global pollution. Persistent organic pollutants have been found in penguins and seals, showing the worldwide distribution of these long-lasting chemicals. Ice cores have provided not only the information on past climates but detailed patterns of global pollution. The changes in global lead levels has been convincingly measured in ice cores with an upwards trend after the Industrial Revolution peaking with the addition of lead to gasoline and then a slow decline after it was withdrawn. It is also possible to detect the fallout from atomic bomb tests in the ice cores.

11.10 Tourism

These days tourism encompasses the whole world, including the polar regions (Hall and Johnston 1995). The Arctic was an early centre of attraction for adventurous visitors with small expeditions aboard steamers and private yachts visiting Alaska and Svalbard in the mid-nineteenth century. As more steamers became available more remote destinations like the Aleutian Islands became possible and now virtually all of the Arctic is open to tourists. Tourist development in the Arctic initially sold itself on its remoteness, an experience of raw and pristine nature. These days it has diversified with specialized adventure tourism (for example skiing across Iceland or Greenland), wildlife and ornithological tours, sport fishing and hunting, dog driving, historical tours of the old Gold Rush areas, and even visits to Santa Claus in Finland, Greenland, or Alaska! The tourist may now arrive by aircraft, by ship, or even by road, especially these days by camper vans and recreational vehicles. As long ago as 1896 there was a hotel on Svalbard catering for tourists and now there are a wide range of facilities across the Arctic as governments have realised that tourism is an important part of the Arctic economy.

Svalbard has been popular as a cruise-ship destination for over 100 years with now over 20 000 visitors coming that way each year. The development of hotels at Longyearbyen and a large number of small companies catering specifically for tourists has pushed the number of bed-nights there to over 20 000 a year. A great deal of tourism, especially that on cruise ships, has limited impact on the environment but it also has limited benefits for the local economy. There is now a growing interest among indigenous communities in Alaska, Canada, and Greenland in closer participation in tourism projects, both to control their impacts and to reap the economic benefits.

In the Antarctic tourism was a comparatively recent development. Although there was some small-scale visits on board Argentine and Chilean government vessels in the late 1950s to the Antarctic Peninsula it was not until

Lars-Eric Linblad built the *Linblad Explorer* in the late 1960s that a market in luxury cruises, with both a taste of adventure and expert lecturers, began to develop.

Growing from a few hundred in 1969 to 8700 in 1992, the total number of tourists landed has now increased to almost 27000 in 2006–2007 season. The ship-based activities now use over 180 sites in most seasons, many of which are only visited once or twice. There are a small number of key sites which attract most of the tour companies and these are all on or near to the Antarctic Peninsula: Whalers Bay on Deception Island, Port Lockroy on Goudier Island (Fig. 11.7), Half Moon Island, Neko Harbour and Dorian Bay, Cuverville Island, and Petermann Island. The majority of the cruises concentrate on South Georgia, the Peninsula, and its associated islands but there are some that visit the sub-Antarctic islands, others that penetrate the Ross Sea to visit the huts of Scott and Shackleton, and every 2 or 3 years there is a cruise aboard a Russian icebreaker that circumnavigates the whole continent.

Antarctic visits are concentrated mainly at ice-free coastal zones over the 5-month period from November to March. Visits ashore are generally of short duration (approx. 3 h), of moderate intensity (fewer than 100 people), and of variable frequency (Fig. 11.8). Typically there are one to three landings per day. Landings are made using Zodiacs (rubber inflatable crafts) or, in the case of Russian icebreakers, by helicopter. Shipboard staff supervise shore visits, with one staff member for each 10–20 passengers. Such staff generally include ornithologists, biologists, geologists, glaciologists, historians, and naturalists. Environmental impact assessments prepared

Fig. 11.7   Over 10000 Antarctic ship-based tourists visit the restored British station at Port Lockroy each summer (photograph by David W.H. Walton).

Fig. 11.8   Tourists photographing walruses from Zodiacs off the coast of Svalbard in the Arctic. The rule that wildlife should not be approached is being closely observed (photograph by Rupert Krapp).

by all International Association of Antarctic Tour Operators (IAATO) members are provided to relevant Antarctic Treaty Parties in advance of each season.

A considerable range of nationalities are represented among the passengers but the most frequent are Americans, British, Germans, Australians, Canadians, and Dutch. The trade association—IAATO, formed in 1991—provides a good example of environmental stewardship in a commercial environment with the companies well aware that many of their passengers are environmentally concerned and expect the highest standards of behaviour. The 80 members in 14 countries currently use around 40 ships for each summer season, many of them also employed during the other half of the year on Arctic cruising. There is also limited land-based tourism run mainly from a summer camp organized by Adventure Network International at the blue-ice runway at Patriot Hills. From here people can climb Mt Vinson (78°2′S 22°00′W), the highest peak in the Antarctic, or ski to the South Pole. Other activities by visitors to Antarctica now include kayaking and scuba diving from tourist vessels. Not all companies belong to IAATO as its guidelines are seen by some as too commercially inhibiting.

Most of the tourist activity is still concentrated in the Antarctic Peninsula region, but the operators are becoming more adventurous, and the customers willing to pay for more: trips to inland sites are on the increase and over time will become more affordable. To manage environmental impacts, the Antarctic Treaty signatories have recently implemented a scheme for site guidelines to encourage sustainable management of the most heavily visited sites. In addition where tourism, logistics and science conflict Antarctic Specially Managed Areas (ASMAs) can be designated. With areas at Deception Island, the Dry Valleys and Admiralty Bay already agreed these offer new opportunities for collaborative management. There can be no doubt that tourism in the Antarctic is a positive thing provided it is done in a way that is ecologically sustainable and socially responsible.

The first sightseeing tourist aircraft flew over the Antarctic Peninsula in 1956 and there were irregular flights in later years. There are now flights from Chile every summer out of Punta Arenas and, on the other side of the continent, there are Quantas overflights from Australia. From 1976 to 1979 there were also overflights by Air New Zealand but these ceased after a DC-10 crashed into Mt Erebus, killing all 257 passengers and crew.

Overflights have a minimal impact on the Antarctic environment but the potential for environmental damage from cruise ships is greater. There have so far only been two vessels (*Bahia Paraiso* in 1989 and the MS Explorer in 2007) wrecked while carrying tourists but the possibility of a tourist ship running aground in these poorly charted waters is of great concern to the Treaty Parties. Not only is there the likelihood of loss of life if the ship sinks but considerable potential for ecosystem damage from the loss of fuel oil. Environmentalists also have concerns about the impacts of the numbers of tourists on the wildlife. Some research on this has been undertaken over the past 10 years, on both vegetation and penguins. An early study on an Adélie penguin rookery at Cape Royds showed that regular visits by helicopters resulted in declining numbers of penguins, and this reversed when helicopters were banned. But monitoring for changes in penguin numbers at some of the most heavily visited Peninsula sites has provided contrasting data. At Port Lockroy penguin numbers have increased alongside visitor numbers, despite the fact that the old station is situated in the middle of the rookery and disturbance of individual birds by visitors is inevitable. A second study, comparing penguin numbers at a tourist site and an unvisited control site nearby, apparently showed a decline at the unvisited site, initially suggesting that visitors were good for penguins. In reality the explanation for the decline lay with differences in snow-lie on the breeding sites and this has highlighted the difficulty in measuring effects that are only attributable to the visitors.

11.11 Military uses of the polar regions

Until fairly recently the Arctic was of major strategic military significance and it is only since the fall of the Soviet Union that it has been considered possible to start demolishing some of the extensive military installations scattered around the Arctic. Chief among these were the American Distant Early Warning (DEW) Line stations, established in Alaska, Canada, and

Greenland to provide early warning of a missile attack from the Soviet Union. The chain of 63 radar and communication stations stretched over 4800 km inside the Arctic Circle and was built in less than 3 years, coming into use in 1957. At the time this was probably the most expensive military development in the world. A major part of the system has been upgraded in the last 20 years and is still in use as the North Warning System in Alaska and Canada. To support the stations there were airfield and road developments and extensive training of troops from both the West and the East in Arctic warfare.

From around 1954 the former USSR used the archipelago Novaya Zemla for nuclear weapons testing. Over 220 atomic weapons were exploded there, most but not all below ground with the last explosion as recently as 1990. In the nearby Kara Sea Russia has apparently dumped old nuclear reactors from submarines as well as radioactive waste from civilian and military nuclear reactors. Although the International Atomic Energy Agency has reported that releases are at present low and localized there are clearly major long-term implications.

As part of military activities by both the Soviet Union and the USA there were frequent submarine patrols under the pack ice of the Arctic Ocean. Data collected over many years on ice thickness above the submarines are now proving useful in assessing the rate of warming of the polar seas.

In the Antarctic, while there were some naval skirmishes in the 1950s between the UK, Argentina, and Chile the Antarctic Treaty has stopped all military activities on the continent, although it does allow the use of military resources, like ships and planes, to support scientific activities. The provision in the Treaty for access and inspection of any site was specifically included to ensure that no military infrastructure could be put in place by any country.

11.12 Conservation

Conservation means different things to different people but here it will be taken as the wise use of resources, in keeping with the concept of environmental stewardship. Management for conservation needs to be based on sound science but also needs to have regard to social and cultural values. In reality conservation policy is also driven by economics, politics, and expediency, which means that sensible objectives are often unexpectedly difficult to achieve. Approaches to conservation differ between the polar regions.

11.12.1 The influence of politics on conservation in the polar regions

Ownership of the Arctic territories has been settled for a century with every portion of land north of the Arctic Circle allocated to one of eight nations: Canada, USA, Iceland, Denmark/Greenland, Norway, Sweden, Finland, and Russia. Each country uses its land as it sees fit and legislates for resource management in different ways. Offshore the situation is slightly different with several points of conflict. There is a continuing dispute between the USA and Canada about the status of the waters between Canada's Arctic islands, with the USA declaring international rights for passage and Canada declaring them national waters under the control of the federal government. Most recently a dispute, begun in 1973 between Canada and Denmark over the ownership of Hans Island (1.3 km^2 in area) in Nares Strait, has re-opened with the argument being seen as an important determinant over control of offshore resources around the island. While Svalbard is governed by Norway under the 1920 Treaty of Svalbard all the other signatories to the Treaty have certain reserved rights that Russia, for example, chose to exercise by mining for coal. The discovery and proposed exploitation by Norway of newly discovered oil and gas fields offshore from Svalbard has precipitated arguments as Norway claims that the Treaty does not cover continental shelf resources which are exclusively Norwegian while other signatories insist that they should be treated in the same way as land resources. At stake are a great deal of valuable energy reserves and their control.

Comparisons between the laws relating to environmental management in the circumpolar countries show quite different governance systems. Despite the so-called democratization of Russia it is still the federal government in Moscow rather than the regional or local government that determines resource development and environmental management in Siberia. The local people have little say in decision-making, unlike the situation in parts of Canada (Nunavut) and Greenland where the native people have taken over control of the government. In Alaska both state and federal governments are closely involved in conservation and resource management and in Iceland and the Scandinavian countries there are democratic systems of control and oversight. There has been little general cross-border agreement although there were specific agreements to deal with reindeer herding, polar bears, and with joint management of fisheries in the Barents Sea. The need for a more cooperative approach emerged in the 1980s and environmental issues were seen as the first step. The adoption of the Arctic Environmental Protection Strategy in 1991 by all eight countries allowed the development of the Arctic Council in 1996, a regional forum for sustainable development, mandated to address any environmental, social, and economic questions important for the Arctic as a whole. This, in turn, established several groups—Conservation of the Arctic Flora and Fauna (CAFF), Protection of the Arctic Marine Environment, Sustainable Development Working Group, and AMAP—that now provide forums for both gathering scientific data and discussing its policy implications. The Council is further assisted by the International Arctic Science Committee, a body analogous to the SCAR, which provides independent scientific advice on topical problems.

Many people remember that there have been various claims to part of the Antarctic. First, in 1908, Britain claimed a sector slice between 20 and 80°W in order to legalize its regulation of whaling. Further claims followed with Chile and Argentina counter-claiming much of the same territory as Britain and most of the remainder of the Antarctic being claimed by Norway, Australia, New Zealand, and France almost always on the grounds of *discovery* by the early explorers. Despite having similar grounds for a claim the USA refused to claim or recognize claims of others, an approach also followed by the Soviet Union. In the early 1950s naval activities around the Peninsula by Argentina, Chile, and the UK heightened the tension but this was defused by the agreement to collaborate for the International Geophysical Year of 1957–1958. Apart from being scientifically successful this demonstrated that nations of opposing views could collaborate effectively in the pursuit of a common purpose in a hostile environment. Pressure from the scientists to build on these achievements allowed negotiation of the Antarctic Treaty. Putting claims of sovereignty aside and dedicating the region to peace and science, it was signed in 1959 and ratified in 1961 by all 12 nations then active in the Antarctic (Fogg 1992). Since then more countries have joined the Treaty and with 45 countries now signatories the Treaty represents over 70% of the global population. The ratification of the Protocol on Environmental Protection to the Antarctic Treaty established a new advisory group, the Committee for Environmental Protection (CEP), which now provides the forum for the discussion of all conservation and environmental management issues. All decisions at the Treaty are by consensus and are legally binding on citizens of each of the parties when passed into national law. The SCAR provides independent scientific advice to the CEP as does the Council of Managers for National Antarctic Programs (COMNAP) on logistic matters.

Nationalism has not gone away despite the provisions in the Treaty. Argentina and Chile have both established settlements in their claimed areas, issuing birth certificates and passports to their "Antarctic citizens" whereas Australia has registered a claim with UNCLOS for future rights over the exploitation of the mineral rights on the continental shelf of Australian Antarctic Territory. Issuing stamps is seen as a governance act and here Australia, New Zealand, the UK, and France all issue special Antarctic stamps whereas the USA, Russia, Argentina, Chile, and many other countries use their normal stamps with an Antarctic cancel.

Various NGOs (and especially members of the Atlantic Southern Ocean Coalition, ASOC) have for many years promoted the suggestion that the Treaty is an exclusive club and the Antarctic would be better governed as a World Park under the United Nations. This would seem a very poor move and certain to result in total paralysis of Antarctic governance as uninformed political agendas ran out of control. If science is to be the main business of the Antarctic, and if sustainable management is to be achieved, those countries with the most direct interest and experience are those most likely to understand the problems and come to practical solutions. The Treaty Parties already represent the majority of the world's population and have demonstrated their commitment and concern for the continent over the last 50 years. The present system is not without its flaws but overall works well.

11.12.2 The Arctic

There are many books and papers on Arctic conservation but the best overviews for the whole of the Arctic are found in publications by CAFF (Conservation of the Arctic Flora and Fauna 2001). The Arctic is resource-rich and especially so in oil and gas, commodities of great value and great political significance. The national and multinational companies, which exploit the mineral resources, naturally put economic considerations first, and must be compelled to recognize the environmental implications of their activities. On the positive side Russia has announced many Siberian nature reserves, national parks, scientific and scenic reserves and wilderness areas amounting to around 19.7 million ha. Although they are staffed, funding is almost totally inadequate so that real management is difficult. Five of the areas are designated strict nature reserves which, by definition, excludes any economic development. It seems unlikely in present-day Russia that such reserves will either be given the resources that they need for protection or be able to resist being re-zoned if economic pressures demand it. A similar situation exists at present in Alaska with the Arctic Wildlife Refuge that the current US administration is trying hard to open up to oil exploitation.

11.12.3 The Antarctic

The early explorers had no concern for conservation and even in the Antarctic Treaty there is little indication that conservation or environmental stewardship were issues of concern. However, any sort of military activity, weapon testing, nuclear explosions, and disposal of nuclear waste were prohibited and these prohibitions, which have been strictly observed, disposed of many possible sources of environmental damage at a stroke. The SCAR, however, recognized the omission on conservation and provided the Treaty with the text that became the Agreed Measures for the Conservation of the Antarctic Flora and Fauna in 1964. These enjoined governments to prohibit the killing of animals save for scientific purposes or in cases of necessity, to minimize the disturbance of birds and seal colonies, to prohibit the collection of plants except for scientific purposes and to alleviate pollution. Building on this basis the Treaty Parties, with advice from the SCAR, went on to designate Sites of Special Scientific Interest and Specially Protected Areas, agree that some species required special protection, and develop the Convention for the Conservation of Antarctic Seals to manage any attempt to establish commercial harvesting. Recognizing an increasing harvesting of marine biological resources the Parties, again on the basis of data provided by the SCAR, negotiated the Convention for the Conservation of Marine Living Resources which spread the authority of the Parties into the high seas, an area not covered by the original Treaty. Finally the Protocol subsumed all these elements as well as waste disposal, marine pollution, heritage sites, and liability to provide a comprehensive framework for conservation and environmental management. Key Treaty documents are now available on the website of the Antarctic Treaty Secretariat.

The progress over the last 50 years has been remarkable, both in utilizing scientific advice as a principal determinant of policy and developing ways of reporting on progress without having a legal basis for sanctioning any country for failure. Of course not everything has worked as well as it should. There has been much criticism of the multiple stations established on King George Island rather than having them spread more widely to advantage the scientific possibilities. CCAMLR was agreed rather late, allowing the Soviet Union to badly damage some key fish stocks, and the present situation of pirate fishing of toothfish is almost certainly being undertaken in part with connivance from some of the CCAMLR countries. The failure of the parties to agree to a State of the Antarctic Environment report is a clear dereliction of duty but a consequence of the intransigence of the USA in matters relating to climate change. There is still no sensible framework for protected areas nor was the opportunity of preparing management plans for all the existing ones used to review the substantive need for some of them. Finally, the CEP still has some way to go to provide a convincing expertly staffed forum for the range of questions it is addressing. The Antarctic is now clearly a key part of the global change models, an increasingly important part of the international science picture, and a unique area for international and interdisciplinary science. Its long-term conservation is of crucial importance for everyone.

12 Some conclusions

The quintessential polar habitat, the ice cap, supports low vital activity but life is not extinguished altogether, and some microorganisms can lie dormant in it for seemingly indefinite periods. Lack of moisture, not low temperature itself, is the ultimate constraint and wherever liquid water appears, active life becomes possible and is usually to be found. Around the ice caps such situations are transient and shifting, being colonized only by such opportunistic organisms as are already present in dormant form or whose propagules arrive via air, water, or human visitors.

Organisms establishing themselves in terrestrial habitats at the ice margin must be able to grow and reproduce in the short intervals when conditions are favourable and to survive desiccation, cold, and starvation when they are not. Among them there need to be autotrophs, usually algae or cyanobacteria, but often there is some form of structured association with heterotrophs. The associations found in lichens, algal mats, endolithic communities, and in the brine channels in sea ice are more successful in utilizing the resources available in these extreme habitats than are individual components alone. In the endolithic and brine-channel communities this is evidently because the combination of phototrophs and heterotrophs allows existence in an almost self-contained microcosm, within which materials can be cycled and conserved. Lichens are more effective in absorbing and retaining water than are the algae they contain. Such associations may support a limited microfauna of detritivores but their productivity is minimal and they do not provide for much secondary production.

Thus far, the Arctic and Antarctic are similar in the habitats they provide and in the species which occupy them. In peri-glacial terrestrial areas, however, there are differences. These stem from the respective degrees of isolation from the rest of the biosphere, the times for which they have been open to colonization, the extents of the land surfaces, and the different climates. In both regions it seems that none of the organisms found have any unique adaptations to polar life but have migrated from milder climates already genetically equipped with physiology and life cycles which can be modified for survival in the harsher environment. Accumulation of biomass, as in Arctic tundra, is sometimes sufficient to ameliorate the environment. Despite the longer time that has been available for colonization, the greater distances and lack of bridges facing migrants into the Antarctic have resulted in floras and faunas much poorer in species than those in the Arctic and, conspicuously, lacking in mammals. With the Arctic mammals there is, again, no fundamentally different feature of physiological adaptation, survival being largely a matter of modification of life cycle and behaviour. This is particularly evident in the Inuit, who have no special structural or metabolic adaptations beyond the level of acclimation for dealing with cold. They create their own microenvironments by effective clothing and housing and avoidance of exposure. The secret of success of polar organisms has been neatly summed up as a combination of an ability to *sit it out* when conditions are adverse with a *get-up-and-go* strategy in the short favourable periods. Neither of these are exclusively polar characteristics. Nevertheless, the way in which structural, physiological, biochemical, and behavioural features are combined, as in the emperor penguin, to enable a species to be successful in a quite atrocious environment is something which one cannot dismiss as fully explained.

The contrasts between terrestrial and marine polar habitats could scarcely be greater (Peck *et al.* 2006). In place of desiccation, temperatures which can fluctuate between extremely low and uncomfortably hot, and exposure to destructive winds, the open sea offers unlimited access to liquid water, stable temperatures never falling below freezing point, and currents which are not usually life-threatening. Habitats in ice-free polar seas or on their shores show no features differentiating them in any fundamental way from temperate waters. Although the Southern Ocean has often been described, superficially and mistakingly, as having the simplest of food chains— diatoms, krill, and whales—the basic trophic structure is the same in polar seas as in temperate ones. The complexities of the food web, particularly at the level where the ultraplankton meshes with the higher trophic system, seem just as great. The sea has fronts, adverse currents, and deep basins as barriers to invading organisms. However, in the Arctic, convoluted shorelines and current patterns, together with extensive and shallow continental shelves, facilitate invasion. Organisms from temperate waters may penetrate far north and it is not always easy to distinguish between Arctic and sub-Arctic flora and fauna. In contrast, the northward edge of the Southern Ocean is marked clearly, biologically as well as physically, by the Antarctic Polar Front and its currents deflect water-borne invasions. For some organisms the front is an almost impermeable barrier. The 25 million years which have elapsed since it was established have allowed the evolution of distinctive biotas with high proportions of endemic species. Among these, the fishes have gone furthest in developing physiological and biochemical mechanisms for survival under polar conditions.

Ice gives polar oceans and seas a unique habitat. Its immediate effects are mostly physical—abrasion, modification of radiation flux and temperature, stability of the water column, refuge from predators, and a solid substratum for birds and seals above and microalgae below. The microhabitats provided by the brine channels in sea ice are inhabited by a microbial community essentially similar in trophic organization to the endolithic communities of the polar desert but lasting for a season or two rather than for centuries. Another difference is that whereas the organic production of the endolithic community finds its way into general circulation only slowly, that of the sea-ice community is immediately available to secondary producers. The existence of this niche, depending essentially on the pattern of ice crystallization, in the Southern Ocean but not in the Arctic Ocean, is another instance of the far-reaching effects of the enclosure of the Arctic Ocean by land masses. The ice edge, a distinct and remarkable habitat which shifts hundreds of kilometres polewards each summer, is highly dynamic, both physically and biologically. The surge in primary and secondary production which follows it is adequately accounted for in terms of release of organic matter from the ice and stabilization of the water column providing conditions favourable for phytoplankton. This is an intensification of what happens in the open sea, not anything radically different.

Primary productivity in polar regions is generally less per unit area than that of equivalent vegetation elsewhere, although in favoured situations it sometimes equals or even exceeds them. As everywhere, the primary productivity of submerged aquatic communities is less than that of terrestrial ecosystems with adequate water. The limitation on photosynthesis is not temperature *per se* but usually desiccation or nutrient supply on land and nutrient deficiency or excessive turbulence in the sea. There is no specifically polar factor preventing plants from doing better than they do.

Polar ecosystems are often described as *fragile*. This requires some qualification. Habitats may be obviously fragile. Frost heaving, wind, and ice movement can wreak havoc, and activities of humans, or even of the native fauna, may have devastating effects. Traffic around a scientific or military installation cuts up the ground and its scars persist for decades. On the other hand, human impact on the ice is minimal and disturbance is obliterated within a season. Equally, polar communities show extremes of fragility and robustness. The macroscopic vegetation of polar desert is usually physically fragile, if biologically robust, and a footstep may destroy a lichen a century or more old. Wet tundra does not invite passage either on foot or in a vehicle and its vegetation is more resilient. However, it is sensitive to alteration in drainage or heat balance and an oil spill, by altering albedo, excluding oxygen, and introducing both organic substrates for microbial growth and toxic materials, can have profound effects. The character of fresh waters can be changed radically (e.g. switched from an oligotrophic to eutrophic state), especially if the volume is small, by quite minor events such as interference with an inflow or introduction of small amounts of inorganic or organic substances. At the higher trophic levels, hunting or fishing may upset the balance of predator and prey. Where a single animal occupies a key position in a terrestrial ecosystem, as the lemming may in Arctic tundra, reduction in numbers of its predators can lead to rapid population increase with drastic effects on the structure and composition of the vegetation.

Nevertheless, many plants and algae are robust and recovery may take place even if it is a lengthy process. Marine ecosystems are resilient. The severe reduction in whale numbers in a few decades cannot really be cited as evidence of fragility since the biomass of these top predators is a minute part of the total and their place can be taken by other predators with no perceptible effect on lower trophic levels (cf. Smetacek and Nicol, 2005). Overfishing of krill, which occupies a key position in the Southern Ocean akin to that of the lemming in the tundra, might have far-reaching effects. It seems probable that the repercussions of even catastrophic changes at these top trophic levels would be imperceptible at lower levels. The relations between the ultra- and microplanktonic communities seem much the same in all parts of the oceans, regardless of what goes on up top. Microorganisms are much less susceptible to the toxic effects of oil and organochlorine pollutants than are the higher animals. Zoologists and microbiologists often have quite different notions of fragility or absence of life.

Investigations of polar ecology have made notable contributions to biological science in general, principally because some polar habitats present situations in which particular problems can be studied under unique conditions. The total absence of mammalian herbivores and the small number of species generally in terrestrial habitats in the Antarctic results in food webs, flows of energy, and cycles of materials of greater simplicity than those elsewhere. They are consequently more amenable to mathematical modelling. Large, accessible, undisturbed colonies of seals and seabirds adjacent to sea areas which are under investigation by marine biologists give unrivalled opportunity of investigating the role of these animals in the marine ecosystem. The tolerance to handling of Antarctic seabirds has helped quantitative studies on diet and growth as well as simplifying the deployment of telemetric and other electronic monitoring equipment. Polar ornithologists have led the way in studies of the ecology of seabirds. The habits of the Weddell seal allow its diving patterns and physiology to be studied under nearly natural conditions using techniques usually possible only in the laboratory. As a result, the diving behaviour of this seal is better known than that of any other marine mammal. Apart from allowing innovative long-term studies of microclimate and microbiology, the Dry Valleys of Antarctica have provided the nearest approximation on Earth to the surface of Mars, a terrain for testing ideas and equipment used in searching for evidence of life on that planet.

The study of polar habitats has other wide perspectives. For various reasons—because the explorers and scientists who first ventured into these regions were men with broad interests and because it has always been practical to send naturalists together with astronomers, magneticians, and geologists on expeditions—polar science has always been holistic in outlook. In contrast to the reductionist approach, which studies processes in simplified situations under controlled conditions, the holistic view recognizes that the manifold processes taking place in the natural world interact with one another to varying degrees and that conditions are never constant. Reductionist science is essential but it is not always realistic to apply it unthinkingly in the natural environment, with all its complexity and inconstancy, and to do so may be disastrously misleading. Study of natural systems as a whole has hitherto been largely empirical but is now acquiring rigour by use of experiments *in situ* and mathematical modelling.

From the holistic viewpoint, the polar regions cannot be put aside as remote, separate, and distinct from the rest of the world. The Arctic and Antarctic, because they are our planet's heat sinks, play key roles in propelling flows of air and water, setting the patterns of circulation in atmosphere and ocean, influencing our weather, shipping, and air transport. At the biological level, the polar oceans and seas act as sinks for carbon dioxide. The isolation and physical conditions of the Arctic and Antarctic enable global perturbations of atmospheric chemistry (e.g. in ozone and pollutants) to be detected more certainly there than anywhere else.

The impacts of global warming on polar habitats and communities are complicated and difficult to predict. The response of individual species to continuous and rapid climate change is generally one of migration rather than of evolutionary adaptation (Huntley 1991). This seems to have been true in the Arctic and, discounting notothenioid fish and emperor penguins, which have had many millions of years to adapt to change, in the Antarctic too. Species differ in the timing, rates, and directions of their migrations so that communities must be regarded, not as being of fixed composition, but as temporary assemblages of species, associating and dissociating as conditions alter. Rapid change will thus produce communities different from those arising from slower change. The maximum migration rate of most trees is between 150 and 500 m year^{-1}, this being just about enough to have kept pace with retreating ice at the end of the ice age. The present rate of warming being so much faster than it was then, we can no longer expect forests to follow retreating ice.

Migration rates of pests, pathogens, or herbivores may be different from those of invading plant species and so upset ecological balance. Perhaps

Fig. 12.1   There is always a great feeling of adventure when a research vessel moves through the pack ice of either the Arctic or Antarctic. Sitting on a darkened bridge, crashing through the ice which is highlighted by the ship's powerful navigation spotlights is a never-to-be-forgotten experience (photograph by David N. Thomas).

the greatest immediate effect on the biota will follow the disappearance of permafrost, which may lead either to waterlogging or to drought according to topography and precipitation regimes. Warming will accelerate the decomposition of the substantial organic carbon reserves in tundra, turning them into net producers of greenhouse gases and thus providing positive feedback to global warming. In the Antarctic there are no great areas of tundra underlaid by permafrost to provide feedback. Attempts to manage terrestrial ecosystems to avoid unwanted effects of global warming in the polar regions will tax our ecological expertise to its limits.

Effects on the polar seas will be of a different kind. In open waters it seems unlikely that a rise in temperature will affect total primary productivity and changes in secondary production seem unpredictable. There may, however, be considerable alterations in the species composition and distribution of marine communities, including mammals and seabirds (and fishermen), in response to rise in sea level and shifting patterns of water movement. However, in waters currently covered by sea ice there will of course be a considerable change in the related ecosystems since so much of the biological processes from the pelagic to underlying benthos are intrinsically linked to dynamics of sea-ice formation, consolidation, and melt.

Polar habitats are unique and of great intrinsic interest to ecologists. Their study also promises to help us understand, and to some extent cope with, the damage to the global environment that we have so unthinkingly wrought.

Further reading and web resources

Further reading

This reading list is in no way meant to be an exhaustive list of books about polar regions. It is simply designed to provide a starting point for finding out more about the topics covered in this brief overview.

Ainley, D.G. (2002) *The Adélie Penguin: Bellweather of Climate Change*. Columbia University Press, New York.

Aleksandrova, V.D. (1980) *The Arctic and Antarctic: their Division into Geobotanical Areas* (translated by D. Love). Cambridge University Press, Cambridge.

Arctic Climate Impact Assessment (2005) *Impacts of a Warming Arctic*. Cambridge University Press, Cambridge.

Ball, P. (1999) *H₂O—A Biography of Water*. Wiedenfeld & Nicolson, London.

Battaglia, B., Valencia, J., and Walton, D.W.H. (ed.) (1997) *Antarctic Communities: Species, Structure and Survival*. Cambridge University Press, Cambridge.

Bergstrom, D., Convey, P., and Huiskes, A. (eds) (2006) *Trends in Antarctic Terrestrial and Limnetic Ecosystems. Antarctica as a Global Indicator*. Springer, Berlin.

Bigg, G. (2003) *The Oceans and Climate*. Cambridge University Press, Cambridge.

Bischoff, J. (2000) *Ice Drift, Ocean Circulation and Climate Change*. Springer, Berlin.

Buckland, D., MacGlip, A., and Parkinson, S. (2006) *Burning Ice—Art & Climate Change*. Cape Farewell, London.

Byatt, A., Fothergill, A., and Holmes, M. (2001) *The Blue Planet*. BBC Worldwide, London.

Campbell, D.G. (2002) *The Crystal Desert: Summers in Antarctica*. Mariner Books, New York.

Castello, J.D. and Rogers, S.O. (eds) (2005) *Life in Ancient Ice*. Princeton University Press, Princeton, NJ.

Chernov, Yu.I. (1985) *The Living Tundra* (translated by D. Love). Cambridge University Press, Cambridge.

Committee on Frontiers in Polar Biology (2003) *Frontiers in Polar Biology in the Genomic Era*. The National Academies Press, Washington DC.

Davenport, J. (1992) *Animal Life at Low Temperature*. Chapman & Hall, London.

Davis, L. (2004) *Penguins: Living in Two Worlds*. Yale University Press, New Haven, CT.

Di Prisco, G., Pisano, E., and Clarke, A. (eds) (1998) *Fishes of Antarctica: A Biological Overview*. Springer Verlag, Berlin.

Dowdeswell, J. and Hambrey, M. (2002) *Islands of the Arctic*. Cambridge University Press, Cambridge.

Eastman, J.T. (1993) *Antarctic Fish Biology: Evolution in a Unique Environment*. Academic Press, San Diego, CA.

European Science Foundation (2007a) *Position Paper 9. Impacts of Climate Change on the European Marine and Coastal Environment. Ecosystems Approach*. ESF Publications (can be downloaded as a pdf file from www.esf.org/publications.html).

European Science Foundation (2007b) *Investigating Life in Extreme Environments. A European Perspective*. ESF Publications (can be downloaded as a pdf file from www.esf.org/publications.html).

Everson, I. (2000) *Krill—Biology, Ecology and Fisheries*. Blackwell Science, Oxford.

Faithfull, S. (2006) *Ice Blink: an Antarctic Essay*. Book Works, London.

Falkowski, P.G. and Raven, J.A. (2006) *Aquatic Photosynthesis*. Princeton University Press, Princeton, NJ.

Fogg, G.E. (1992) *A History of Antarctic Science*. Cambridge University Press, Cambridge.

Fogg, G.E. and Smith, D. (1990) *The Explorations of Antarctica: The Last Unspoilt Continent*. Cassell Publishers, London.

Fothergill, A. (1993) *Life in the Freezer: a Natural History of the Antarctic*. BBC Books, London.

Friedmann, E.I. (ed.) (1993) *Antarctic Microbiolgy*. Wiley-Liss, London.

Fukuchi, M. and Marchant, H.J. (2006) *Antarctic Fishes*. Rosenberg Publishing Pty, Kenthurst, NSW.

Green, B. (1995) *Water, Ice and Stone: Science and Memory on the Antarctic Lakes*. Harmony Books, New York.

Hansom, J.D. and Gordon, J.E. (1998) *Antarctic Environments and Resources: a Geographical Perspective*. Addison Wesley Longman, Harlow.

Hardy, A. (1967) *Great Waters*. Harper & Row, London.

Hempel, G. (ed.) (1994) *Antarctic Science: Global Concerns*. Springer, Berlin.

Intergovernmental Panel on Climate Change (2007a) *Climate Change 2007—Mitigation of Climate Change*. Cambridge University Press, Cambridge.

Intergovernmental Panel on Climate Change (2007b) *Climate Change 2007—Impacts, Adaptation and Vulnerability*. Cambridge University Press, Cambridge.

Intergovernmental Panel on Climate Change (2007c) *Climate Change 2007—The Physical Science Basis*. Cambridge University Press, Cambridge.

Ives, J.D. and Barry, R.G. (eds) (1974) *Arctic and Alpine Environments*. Methuen, London.

Ives, J.D. and Sugden, D. (eds) (1995) *Polar Regions*. RD Press, Surrey Hills, NSW.

Kirchman, D.L. (ed.) (2000) *Microbial Ecology of the Oceans*. Wiley-Liss, New York.

Knox, G.A. (2006) *The Biology of the Southern Ocean* 2nd Edition, CRC Press, Boca Raton, Florida, USA.

Lainema, M. and Nurminen, J. (2001) *Ultima Thula: Arctic Explorations.* John Nurminen Foundation, Finland.

Laws, R.M. (ed.) (1984) *Antarctic Ecology*, vols 1 and 2. Academic Press, London.

Laws, R. (1989) *Antarctica: the Last Frontier.* Boxtree, London.

Laws, R.M. and Franks, F. (ed.) (1990) Life at low temperatures. *Philosophical Transactions of the Royal Society of London* B326, 515–697.

Leppäranta, M. (ed.) (2001) *Physics of Ice Covered Seas* (2 vols). Helsinki University Press, Helsinki.

Longton, R.E. (1988) *Biology of Polar Bryophytes and Lichens.* Cambridge University Press, Cambridge.

Lorius, C. (1991) *Glaces de l'Antarctique: une mémoire, des passions.* Éditions Odile Jacob, Paris.

Lunine, J.I. (2005) *Astrobiology: a Multidisciplinary Approach.* Pearson Addison Wesley, San Francisco, CA.

Margesin, R. and Schinner, F. (eds) (1999) *Cold-Adapted Organisms: Ecology, Physiology, Enzymology and Molecular Biology.* Springer, Berlin.

Margesin, R., Schinner, F., Marx, J.C., and Gerday, C. (2007) *Psychrophiles: from Biodiversity to Biotechnology.* Springer, Berlin.

McGonigal, D. and Woodworth, L. (2003) *Antarctica: the Complete Story.* Frances Lincoln, London.

Melnikov, I.A. (1997) *The Arctic Sea Ice Ecosystem.* Gordon & Breach, Amsterdam.

Mills, M.J. (2003) *Exploring Polar Frontiers. A Historical Encyclopedia* (2 vols). ABC Clio Publishers, Oxford.

Nuttall, M. (ed.) (2004) *Encyclopaedia of the Arctic* (3 vols). Routledge, Taylor & Francis Group, , New York.

Pielou, E.C. (1994) *A Naturalist's Guide to the Arctic.* University of Chicago Press, Chicago, IL.

Pörtner, H.-O. and Playle, R.C. (eds) (1998) *Cold Ocean Physiology.* Society for Experimental Biology Seminar Series. Cambridge University Press, Cambridge.

Pyne, S.J. (2003) *The Ice.* Weidenfeld & Nicolson, London.

Rey, L. (1984) *The Challenging and Elusive Arctic Regions.* Significant Issue Series. Center for Strategic and International Studies, Georgetown University, Washington DC.

Reynolds, J.F. and Tenhunen, J.D. (eds) (1996) *Landscape Function and Disturbance in Arctic Tundra.* Springer, Berlin.

Riffenburgh, B. (ed.) (2006) *Encyclopaedia of the Antarctic* (2 vols). Routledge, Taylor & Francis Group, , New York.

Sage, B. (1986) *The Arctic and its Wildlife.* Croom Helm, London.

Sale, R. (2006) *A Complete Guide to Arctic Wildlife.* Christopher Helm, London.

Scoresby, W. (1820) *An Account of the Arctic Regions with a History and Description of the Northern Whale-fishery.* Vol. 1, *The Arctic*; vol. 2, *The Whale Fishery.* Archibald Constable, Edinburgh (reprinted in 1969 by David & Charles, Newton Abbot).

Seckbach, J. (ed.) (2004) *Origins: Genesis, Evolution and Diversity of Life*. Kluwer Academic Publishers, Dordrecht.

Seckbach, J. (ed.) (2007) *Algae and Cyanobacteria in Extreme Environments*. Springer, Berlin.

Selkirk, P.M., Seppelt, R.D., and Selkirk, D.R. (1990) *Subantarctic Macquarie Island: Environment and Biology*. Cambridge University Press, Cambridge.

Shirihai, H. (2002) *A Complete Guide to Antarctic Wildlife*, Princeton University Press, Princeton, NJ.

Smith, Jr, W.O. (ed.) (1990) *Polar Oceanography. Part A, Physical science. Part B, Chemistry, Biology, and Geology*. Academic Press, San Diego, CA.

Stirling, I. (1998) *Polar Bears*. University of Michigan Press, Ann Arbor, MI.

Stonehouse, B. (1989) *Polar Ecology*. Blackie, Glasgow.

Stonehouse, B. (1990) *North Pole South Pole: a Guide to the Ecology and Resources of the Arctic and Antarctic*. Prion, London.

Sugden, D. (1982) *Arctic and Antarctic: a Modern Geographical Synthesis*. Blackwell, Oxford.

Sullivan, W.T. and Baross, J. (2007) *Planets and Life—The Emerging Science of Astrobiology*. Cambridge University Press, Cambridge.

Thomas, D.N. (2004) *Frozen Oceans*. Natural History Museum, London; Firefly Books, Ontario.

Thomas, D.N. (2007) *Surviving Antarctica*. Natural History Museum, London; Firefly Books, Ontario.

Thomas, D.N. and Dieckmann, G.S. (eds) (2003) *Sea Ice—an Introduction to its Physics, Chemistry, Biology and Geology*. Blackwell Publishing, Oxford.

Vincent, W.F. (2004) *Microbial Ecosystems of Antarctica*. Cambridge University Press, Cambridge.

Wadhams, P. (2000) *Ice in the Ocean*. Gordon & Breach Science Publishers, Amsterdam.

Walton, D.W.H. (ed.) (1987) *Antarctic Science*. Cambridge University Press, Cambridge.

Wheeler, S. (1997) *Terra Incognita: Travels in Antarctica*. Vintage, London.

Wiencke, C. and Clayton, M.N. (2002) *Antarctic Seaweeds*. Koeltz Scientific Books, Koenigstein.

Williams, P.J.leB., Thomas, D.N., and Reynolds, C.S. (2002) *Phytoplankton Productivity. Carbon Assimilation in Marine and Freshwater Ecosystems*. Blackwell Science, Oxford.

Woodin, S.J. and Marquiss, M. (eds) (1997) *Ecology of Arctic Environments*. Blackwell Science, Oxford.

Woodworth, L. (2005) *Antarctica: The Blue Continent*. Frances Lincoln, London.

Wu, N. and Mastro, J. (2004) *Under Antarctic Ice, The Photographs of Norbert Wu*. University of California Press, Berkeley, CA.

Zenkevitch, L. (1963) *Biology of the Seas of the U.S.S.R.* Unwin, London.

Scientific journals

There are several academic journals dedicated to issues in polar research. Listed here are a few that cover biological subjects in detail, although polar research is of such importance that scientific findings are reported in a very wide range of journals.

Antarctic Science, Cambridge University Press
Antarctic Research Series, American Geophysical Union
Extremophiles, Springer
Polar Biology, Springer

Web resources

There are many websites that deal with the Arctic and Antarctic. Again, this is not intended to be an exhaustive list, but rather the first stage to help access information on the Internet. Note: website addresses are subject to change.

Alfred Wegener Institute for Polar and Marine Research, http://www.awi.de/en/home/
Antarctic Climate and Ecosystems Cooperative Research Centre, http://www.acecrc.org.au/
Antarctica New Zealand, http://www.antarcticanz.govt.nz/
Antarctic Treaty Secrateriat, http://www.ats.aq/
Arctic Circle, http://www.arcticcircle.uconn.edu/
Arctic Change Indicator, http://www.arctic.noaa.gov/detect/
Arctic Climate Impact Assessment, http://www.acia.uaf.edu/
Arctic Council, http://www.arctic-council.org/
Arctic Monitoring and Assessment Programme, http://www.amap.no/
Australian Antarctic Division, http://www.aad.gov.au/
British Antarctic Survey, http://www.antarctica.ac.uk/
Byrd Polar Research Center, http://www.bprc.mps.ohio-state.edu/
Canadian Ice Service, http://ice-glaces.ec.gc.ca/
Canadian Polar Commission, http://www.polarcom.gc.ca/
Canadian Wildlife Service, http://www.cws-scf.ec.gc.ca/index_e.cfm/
Circumpolar Flaw Lead System Study, http://www.ipy-cfl.ca/
Cold Regions Research and Engineering Laboratory, http://www.crrel.usace.army.mil/
Commission for the Conservation of Antarctic Marine Living Resources, http://www.ccamlr.org/
Conservation of Arctic Flora and Fauna, http://arcticportal.org/en/caff/
Cool Antarctica, http://www.coolantarctica.com/
Council of Managers for National Antarctic Programs, http://www.comnap.aq/
Danish Polar Center, http://www.dpc.dk/
Intergovernmental Panel on Climate Change, http://www.ipcc.ch/
International Arctic Research Center, Alaska, http://www.gi.alaska.edu/IARC/
International Arctic Science Committee, http://www.iasc.se/
International Association of Antarctica Tour Operators, http://www.iaato.org
International Polar Year 2007–2008, http://www.ipy.org/
International Whaling Commission, http://www.iwcoffice.org/
Inuit Circumpolar Conference, http://www.inuitcircumpolar.com/
McMurdo Dry Valleys Long Term Ecological Research, http://www.mcmlter.org/index.html
NASA, http://www.nasa.gov/
NASA Earth Observatory, http://earthobservatory.nasa.gov/
NASA, JPL Oceanography Group, http://oceans-www.jpl.nasa.gov/polar/

National Institute of Polar Research, Japan, http://www.nipr.ac.jp/
National Oceanic and Atmospheric Administration, http://www.noaa.gov/
National Science Foundation Polar Programme, http://www.nsf.gov/dir/index.
   jsp?org=OPP
National Snow and Ice Data Center, http://nsidc.org/
Norwegian Polar Institute, http://npiweb.npolar.no/
Science Poles, http://www.sciencepoles.org/
Scientific Committee on Antarctic Research (SCAR), http://www.scar.org/
Scott Polar Research Institute, http://www.spri.cam.ac.uk/
Spanish National Antarctic programme, http://tierra.rediris.es/antartida/
The Antarctic Circle, http://www.antarctic-circle.org/
The Antarctic Sun, http://antarcticsun.usap.gov/2005–2006/sctn02–12–2006.cfm/
The Ozone Hole, http://www.theozonehole.com/
United States Antarctic Programme, http://www.usap.gov/
World Meteorological Organisation, http://www.wmo.ch/pages/index_en.html

References

It has not been possible to give references for everything mentioned in the text but it is hoped that by using a combination of the Further reading listed above and the following selected references the reader will be able to locate the sources used. For a chapter in a multi-author work the names of the author(s) and editor(s) (e.g. Lizotte, in Thomas and Dieckmann 2003) are given in the text and the reference appears in this list only under the name(s) of the editor(s). Note: if an item is listed in the Further reading list, it is not repeated again here.

Ackley, S., Wadhams, P., Comiso, J.C., and Worby, A.P. (2003) Decadal decrease of Antarctic sea ice extent inferred from whaling records revisited on the basis of historical and modern sea ice records. *Polar Research* **22**, 19–25.

Adams, B.J., Bardgett, R.D., Ayres, E., Wall, D.H., Aislabie. J., Bamforth, S. *et al.* (2006) Diversity and distribution of Victoria Land biota. *Soil Biology and Biochemistry* **38**, 3003–3018.

Agnew, D. (2004) *Fishing South: the History and Management of South Georgia Fisheries*. Penna Press, St Albans.

Ahn, I.-Y., Chung, H., Kang, J.-S., and Kang, S.-H. (1997) Diatom composition and biomass variability in nearshore waters of Maxwell Bay, Antarctica, during the 1992/1993 austral summer. *Polar Biology* **17**, 123–130.

Alexander, V., Stanley, D.W., Daley, R.J., and McRoy, C.P. (1980) Primary producers. In Hobbie, J.E. (ed.), *Limnology of Tundra Ponds, Barrow, Alaska*, pp. 179–248. Dowden, Hutchinson & Ross, Stroudsberg, PA.

Allegrucci, G., Carchini, G., Todisco, V., Convey, P., and Sbordoni, V. (2006) A molecular phylogeny of Antarctic Chironomidae and its implications for biogeographical history. *Polar Biology* **29**, 320–326.

AMAP (2002) *Arctic Pollution 2002*. Arctic Monitoring and Assessment Programme, Oslo.

Ancel, A., Visser, H., Handrich, Y., Masman, D., and Le Maho, Y. (1997) Energy saving in huddling penguins. *Nature* **385**, 304–305.

Archer, S.D., Leakey, R.J.G., Burkill, P., Sleigh, M.A., and Appleby, C.J. (1996) Microbial ecology of sea ice at a coastal Antarctic site: community composition, biomass and temporal change. *Marine Ecology Progress Series* **135**, 179–195.

Arima, E.Y. (1988) *Inuit Kayaks in Canada: a Review of Historical Records and Construction*. Canadian Museum of Civilisation, Quebec.

Arnaud, P.M. (1974) Contribution à la bionomie marine benthique des régions antarctiques et subantarctiques. *Téthys* **6**, 465–656.

Arnold, R.J., Convey, P., Hughes, K.A., and Wynn-Williams, D.D. (2003) Seasonal periodicity of physical and edaphic factors, and microalgae in Antarctic fell-fields. *Polar Biology* **26**, 396–403.

Arntz, W.E. (1997) Investigación antártica en biología marina: situación actual, proyectos internacionales y perspectivas. *Boletin de la Real Sociedad española de Historia Natural (Sección Biología)* **93**, 13–44.

Arntz, W.E. and Gili, J.M. (2001) A case for tolerance in marine ecology: let us put out the baby with the bathwater. *Scientia Marina* **65** (suppl. 2), 283–299.

Arntz, W.E., Brey, T., and Gallardo, V.A. (1994) Antarctic zoobenthos. *Oceanography and Marine Biology: an Annual Review* **32**, 241–304.

Aronson, R.B. and Blake, D.B. (2001) Global climate change and the origin of modern benthic communities in Antarctica. *American Zoology* **41**, 27–39.

Aronson, R.B., Blake, D.B., and Oji, T. (1997) Retrograde community structure in the late Eocene of Antarctica. *Geology* **25**, 903–906.

Arrigo, K.R. (2005) Marine microorganisms and global nutrient cycles. *Nature* **437**, 349–355.

Arrigo, K.R. and Lizotte, M.P. (eds) (1998) *Antarctic Sea Ice: Biological Processes, Interactions and Variability*. American Geophysical Union, Washington DC.

Arrigo, K.R. and Thomas, D.N. (2004) Large scale importance of sea ice biology in the Southern Ocean. *Antarctic Science* **16**, 471–486.

Arrigo, K.R., Dieckmann, G.S., Gosselin, M., Robinson, D.H., Fritsen, C.H., and Sullivan, C.W. (1995) High resolution study of the platelet ice ecosystem in McMurdo Sound, Antarctica: biomass, nutrient, and production profiles within a dense microalgal bloom. *Marine Ecology Progress Series* **127**, 255–268.

Arrigo, K.R., Worthern, D.L., Lizotte, M.P., Dixon, P., and Dieckmann, G.S. (1997) Primary production in Antarctic sea ice. *Science* **276**, 394–397.

Arrigo, K.R., Worthen, D., Schnell, A., and Lizotte, M.P. (1998) Primary production in Southern Ocean waters. *Journal of Geophysical Research* **103**, 15587–15600.

Arrigo, K.R., Worthen, D.L., and Robinson, D.H. (2003) A coupled ocean-ecosystem model of the Ross Sea: 2. Iron regulation of phytoplankton taxonomic variability and primary production. *Journal of Geophysical Research* **108**, doi:10.1029/2001JC000856.

Atkinson, A., Siegel, V., Pakhomov, E., and Rothery, P. (2004) Long-term decline in krill stock and increase in salps within the Southern Ocean. *Nature* **432**, 100–103

Bano, N., Ruffin, S., Ransom, B., and Hollibaugh, J.T. (2004) Phylogenetic composition of Arctic Ocean archaeal assemblages and comparison with Antarctic assemblages. *Applied and Environmental Microbiology* **70**, 781–789.

Bargagli, R., Skotnicki, M.L., Marri, L., Pepi, M., Mackenzie, A., and Agnorelli, C. (2004) New record of moss and thermophilic bacteria species and physico-chemical properties of geothermal soils on the northwest slope of Mt. Melbourne (Antarctica). *Polar Biology* **27**, 423–431.

Barnes, D.K.A. and Clarke, A. (1995) Seasonality of feeding activity in Antarctic suspension feeders. *Polar Biology* **15**, 335–340.

Barnes, D.K.A. and Fraser, K.P.P. (2003) Rafting by five phyla on man-made flotsam in the Southern Ocean. *Marine Ecology Progress Series* **262**, 289–291.

Barnes, D.K.A., Hodgson, D.A., Convey, P., Allen, C., and Clarke, A. (2006) Incursion and excursion of Antarctic biota: past, present and future. *Global Ecology and Biogeography* **15**, 121–142.

Barr, S. (ed.) (1995) *Franz Josef Land*. Norsk Polarinstitutt, Tromsö.

Barrett, P.J. (1991) Antarctica and global climatic change: a geological perspective. In Harris, C. and Stonehouse, B. (eds), *Antarctica and Global Climate Change*, pp. 35–50. Belhaven Press, London.

Battista, J.R. (1997) Against all odds: the survival strategies of *Deinococcus radiodurans*. *Annual Review of Microbiology* **51**, 203–224.

Belt, S.T., Massé, G., Rowland, S.J., Poulin, M., Michel, C., and LeBlanc, B. (2007) A novel chemical fossil of palaeo sea ice: IP25. *Organic Geochemistry* **38**, 16–27.

Bennett, V.A., Kukal, O., and Lee, Jr, R.E. (1999) Metabolic opportunists: feeding and temperature influence the rate and pattern of respiration in the high arctic woollybear caterpillar *Gynaephora groenlandica* (Lymantriidae). *Journal of Experimental Biology* **202**, 47–53.

Berta, A., Sumich, J., and Kovacs, K.M. (2005) *Evolutionary Biology of Marine Mammals*. Elsevier, San Diego, CA.

Beyer, L. and Bölter, M. (eds) (2002) *Geoecology of Antarctic Ice-free Coastal Landscapes*. Ecological Studies vol. 154. Springer, Berlin.

Bhatia, M., Sharp, M., and Foght, J. (2006) Distinct bacterial communities exist beneath a high Arctic polythermal glacier. *Applied and Environmental Microbiology* **72**, 5838–5845.

Bischoff, B. and Wiencke, C. (1995) Temperature ecotypes and biogeography of *Acrosiphonales* (Chlorophyta) with Arctic–Antarctic disjunct and Arctic/cold-temperature distributions. *European Journal of Phycology* **30**, 19–27.

Blix, A.S. and Nordøy, E.S. (2007) Ross seal (Omnatophoca rossii) annual distribution, diving behaviour, breeding and moulting off Queen Maud Land Antarctica. *Polar Biology* **30**, 1449–1458.

Block, W. (1990) Cold tolerance of insects and other arthropods. *Philosophical Transactions of the Royal Society of London Series B* **326**, 613–633.

Block, W. (1994) Terrestrial ecosystems: Antarctica. *Polar Biology* **14**, 293–300.

Block, W. (1996) Cold or drought—the lesser of two evils for terrestrial arthropods? *European Journal of Entomology* **93**, 325–339.

Bluhm, B. and Gradinger, R. (2008) Regional variability in food availability for Arctic marine mammals. *Ecological Applications* (in press).

Bluhm, B., Gradinger, R., and Piraino, S. (2007) First record of sympagic hydroides (Hydrozoa, Cnidaria) in Arctic coastal fast ice. *Polar Biology* **30**, 1557–1564.

Boenigk, J., Pfandl, K., Garstecki, T., Harms, H., Novarino, G., and Chatzinotas, A. (2006) Evidence for geographic isolation and signs of endemism within a protistan morphospecies. *Applied and Environmental Microbiology* **72**, 5159–5164.

Booth, C.B. and Smith, Jr, W.O. (1997) Autotrophic flagellates and diatoms in the Northeast Water Polynya, Greenland: summer 1993. *Journal of Marine Systems* **10**, 241–261.

Bornemann, H., Kreyscher, M., Ramdohr, S., Martin, T., Carlini, A., Sellmann, L., and Plötz, J. (2000) Southern elephant seal movements and Antarctic sea ice. *Antarctic Science* **12**, 3–15.

Borriss, M., Helmke, E., Hanschke, R., and Schweder, T. (2003) Isolation and characterisation of marine psychrophilic phage-host systems from Arctic sea ice. *Extremophiles* **7**, 377–384.

Boyd, P.W.T., Jickells, T., Law, C.S., Blain, S., Boyle, E.A., Buesseler, K.O. *et al.* (2007) Mesoscale iron enrichment experiments 1993–2005: Synthesis and future directions, *Science* **315**, 612–617.

Brandt, A., de Broyer, C., Gooday, A.J., Hilbig, B., and Thomson, M.R.A. (2004) Introduction to ANDEEP (ANtarctic benthic DEEP-sea biodiversity: colonization history and recent community patterns)—a tribute to Howard L. Sanders. *Deep-Sea Reserach Part II* **51**, 1457–1465.

Brandt, A., Gooday, A.J., Brandão, S.N., Brix, S., Brökeland, W., Cedhagen, T. *et al.* (2007) First insights into the biodiversity and biogeography of the Southern Ocean deep sea. *Nature* **447**, 307–311.

Brierley, A.S. and Thomas, D.N. (2002) Ecology of Southern Ocean pack ice. *Advances in Marine Biology* **43**, 171–276.

Brierley, A.S., Demer, D.A., Watkins, J.L., and Hewitt, R. (1999) Concordance of interannual fluctuations in acoustically estimated densities of Antarctic krill around South Georgia and Elephant Islands: biological evidence of same-year teleconnections across the Scotia Sea. *Marine Biology* **134**, 675–681.

Brierley, A.S., Fernandes, P.G., Brandon, M.A., Armstrong, F., Millard, N.W., McPhail, S.D. *et al.* (2002) Antarctic krill under sea ice: elevated abundance in a narrow band just south of ice edge. *Science* **295**, 1890–1892.

British Antarctic Survey (2004) *Antarctica 1:10,000,000 Map.* BAS (misc) 11. British Antarctic Survey, Cambridge.

Broecker, W.S. (1997) Thermohaline circulation, the Achilles heel of our climate system: will man-made CO_2 upset the current balance? *Science* **278**, 1582–1588.

Broecker, W.S., Sutherland, S., and Peng, T.-H. (1999) A possible 20th-century slowdown of Southern Ocean deep-water formation. *Science* **286**, 1132–1135.

Brouwer, P.E.M. (1996) *In situ* photosynthesis and estimated annual production of the red macroalga *Myriogramme mangini* in relation to underwater irradiance at Signy Island (Antarctica). *Antarctic Science* **8**, 245–252.

Brown, M.V. and Bowman, J.P. (2001) A molecular phylogenetic survey of sea-ice microbial communities (SIMCO). *FEMS Microbiological Ecology* **35**, 267–275.

Bowman, J.P., McCammon, S.A., Brown, M.V., Nichols, D.S., and McMeekin, T.A. (1997) Diversity and association of psychrophilic bacteria in Antarctic sea ice. *Applied and Environmental Microbiology* **63**, 3068–3078.

Buesseler, K.O., Andrews, J.E., Pike, S.M., and Charette, M.A. (2004) The effects of iron fertilization on carbon sequestration in the Southern Ocean. *Science* **304**, 414–417.

Buma, A.G.J., de Boer, M.K., and Boelen, P. (2001) Depth distributions of DNA damage in Antarctic marine phyto- and bacterioplankton exposed to summertime UV radiation. *Journal of Phycology* **37**, 200–208.

Buma, A.G.J., Wright, S.W., van den Enden, R., van de Poll, W.H., and Davidson, A.T. (2006) PAR acclimation and UVBR-induced DNA damage in Antarctic marine microalgae. *Marine Ecology Progress Series* **315**, 33–42.

Cadée, G.C., González, H., and Schnack-Schiel, S.B. (1992) Krill diet affects faecal string settling. *Polar Biology* **12**, 75–80.

Callaghan, T.V., Björn, L.O., Chernov, Y., Chapin, T., Christensen, T.R., Huntley, B. *et al.* (2004) Climate change and UV-B impacts on Arctic tundra and polar desert ecosystems. Key findings and extended summaries. *Ambio* **33**, 386–392.

Cameron, R.E. (1969) *Abundance of Microflora in Soils of Desert Regions.* Technical Report 32–1378. Jet Propulsion Laboratory, Pasadena, CA.

Campbell, I.B. and Claridge, G.G.C. (1987) *Antarctica: Soils, Weathering Processes and Environment.* Elsevier, Amsterdam.

Campen, R.K., Sowers, T., and Alley, R.B. (2003) Evidence of microbial consortia metabolizing within a low-latitude mountain glacier. *Geology* **31**, 231–234.

Cannon, R.J.C. and Block, W. (1988) Cold tolerance of microarthropods. *Biological Reviews* **63**, 23–77.

Cannone, N., Guglielmin, M., and Gerdol, R. (2004) Relationships between vegetation patterns and periglacial landforms in northwestern Svalbard. *Polar Biology* **27**, 562–571.

Carpenter, E.J., Lin, S., and Capone, D.G. (2000) Bacterial activity in South Pole snow. *Applied and Environmental Microbiology* **66**, 4514–4517.

Carr, M.-E., Friedrichs, M.A.M., Schmeltz, M., Noguchi, A.M., Antoine, D., Arrigo, K.R. *et al.* (2006) A comparison of global estimates of marine primary production from ocean color. *Deep Sea Research* **53**, 741–770.

Cavicchioli, R. (2002) Extremophiles and the search for extraterrestrial life. *Astrobiology* **2**, 281–292.

Cavicchioli, R. (2006) Cold adapted archaea. *Nature Reviews Microbiology* **4**, 331–343.

Chapin, F.S., Oswood, M.W., van Cleve, J., Viereck, L.A., and Verbyla, D.L. (2006) *Alaska's Changing Boreal Forests*. Oxford University Press, Oxford.

Chapman, V.J. (ed.) (1977) *Wet Coastal Ecosystems*. Elsevier, Amsterdam.

Cheng, C.C. and DeVries, A.L. (1991) The role of antifreeze glycopeptides and peptides in the freezing avoidance of cold-water fish. In di Priscu, G. (ed.), *Life Under Extreme Conditions. Biochemical Adaptation*, pp. 1–15. Springer, Berlin.

Chernov, Y.I. (1985) *The Living Tundra*. Cambridge University Press, Cambridge.

Chiantore, M., Cattaneo-Vietti, R., Povero, P., and Albertelli, G. (2000) The population structure and ecology of the Antarctic scallop *Adamussium colbecki* in Terra Nova Bay. In Faranda, F.M., Guglielmo, L., and Ianora, A. (eds), *Ross Sea Ecology*, pp. 563–573. Springer, Berlin.

Chown, S.L. and Convey, P. (2007) Spatial and temporal variability across life's hierarchies in the terrestrial Antarctic. *Philosophical Transactions of the Royal Society of London Series B* **362**, doi:10.1098/rstb.2006.1949.

Christner, B.C. (2002) Incorporation of DNA and protein precursors into macromolecules by bacteria at -15°C. *Applied and Environmental Microbiology* **68**, 6435–6438.

Christner, B.C., Mosley-Thompson, E., Thompson, L.G., Zagorodnov, V., Sandman, K., and Reeve, J.N. (2000) Recovery and identification of viable bacteria immured in glacial ice. *Icarus* **144**, 479–485.

Chyba, F.F. and Phillips, C.B. (2002) Europa as an abode of life. *Origins of Life and Evolution of the Biosphere* **32**, 47–68.

Clark, P.U., Pisias, N.G., Stocker, T.F., and Weaver, A.J. (2002) The role of the thermohaline circulation in abrupt climate change. *Nature* **415**, 863–869.

Clarke, A. (1983) Life in cold water: the physiological ecology of polar marine ectotherms. *Oceanography and Marine Biology: an Annual Review* **21**, 341–453.

Clarke, A. (1988) Seasonality in the Antarctic marine environment. *Comparative Biochemical and Physiology* **90**(B), 461–473.

Clarke, A. (2003a) The polar deep seas. In Tyler, P.A. (ed.), *Ecosystems of the Deep Oceans*, pp. 239–260. Elsevier, Amsterdam.

Clarke, A. (2003b) Evolution, adaptation and diversity: global ecology in an Antarctic context. In Huiskes, A.H.L., Gieskes, W.W.C., Rozema, J., Schorno, R.M.L., van der Vies, S.S., and Wolff, W.J., (eds), *Antarctic Biology in a Global Context*, pp. 3–17. Backhuys Publishers, Leiden.

Clarke, A. and Crame, J.A. (1989) The origin of the Southern Ocean marine fauna. In Crame, J.A. (ed.), *Origins and Evolution of the Antarctic Biota*, pp. 253–268. Special Publication 47. Geological Society of London, London.

Clarke, A. and Crame, J.A. (1992) The Southern Ocean benthic fauna and climate change: a historical perspective. *Philosophical Transactions of the Royal Society of London Series B* **338**, 299–309.

Clarke, A. and Leakey, R.J.G. (1996) The seasonal cycle of phytoplankton, macro-nutrients and the microbial community in a nearshore Antarctic marine eco-system. *Limnology and Oceanography* **41**, 1281–1299.

Clarke, A. and Johnston, N.M. (2003) Antarctic marine benthic diversity. *Oceanography and Marine Biology: an Annual Review* **41**, 47–114.

Clarke, A., Meredith, M.P., Wallace, M.I., Brandon, M.A., and Thomas, D.N. (2008) Seasonal and interannual variability in temperature, chlorophyll and macronutrients in northern Marguerite Bay, Antarctica. *Deep Sea Research Part II* (in press).

Coale, K.H., Johnson, K.S., Chavez, F.P., Buesseler, K.O., Barber, R.T., Brzezinski, M.A. *et al.* (2004) Southern Ocean iron enrichment experiment: Carbon cycling in high- and low-Si waters. *Science* **304**, 408–414.

Cockell, C.S. and Stokes, M.D. (2004) Widespread colonization by polar hypo-liths. *Nature* **431**, 414.

Cockell, C.S., Rettberg, P., Horneck, G., Wynn-Williams, D.D., Scherer, K., and Gugg-Helminger, A. (2002) Influence of ice and snow covers on the UV exposure of terrestrial microbial communities: dosimetric studies. *Journal of Photochemistry and Photobiology B: Biology* **68**, 23–32.

Cocks, L.R.M. (ed.) (1981) *The Evolving Earth*. British Museum (Natural History) and Cambridge University Press, Cambridge.

Codispoti, L.A., Flagg, C., Kelly, V., and Swift, J.H. (2005) Hydrographic condi-tions during the 2002 SBI process experiments. *Deep Sea Research Part II*, **52**, 3199–3226.

Comiso, J.C. (2002) A rapidly declining Arctic perennial ice cover. *Geophysical Research Letters* **29**, doi:10.1029/2002GL015650.

Comiso, J.C. (2003) Warming trends in the Arctic. *Journal of Climate* **16**, 3498–3510.

Comiso, J.C. (2006) Abrupt decline in the Arctic winter sea ice cover. *Geophysical Research Letters* **33**, L18504.

Comiso, J.C. and Parkinson, C.L. (2004) Satellite observed changes in the Arctic. *Physics Today* **57**, 38–44.

Conovitz, P.A., McKnight, D.M., MacDonald, L.H., Fountain, A.G., and House, H.R. (1998) Hydrologic processes influencing streamflow variation in Fryxell Basin, Antarctica. In Priscu, J.C. (ed.), *Ecosystem Dynamics in a Polar Desert, the McMurdo Dry Valleys, Antarctica*, pp. 93–108. American Geophysical Union, Washington DC.

Conservation of the Arctic Flora and Fauna (2001) *Arctic Flora and Fauna: Status and Conservation*. Edita, Helsinki.

Convey, P. (1996a) The influence of environmental characteristics on life history attributes of Antarctic terrestrial biota. *Biological Reviews* **71**, 191–225.

Convey, P. (1996b) Overwintering strategies of terrestrial invertebrates from Antarctica—the significance of flexibility in extremely seasonal environments. *European Journal of Entomology* **93**, 489–505.

Convey, P. (2003) Maritime Antarctic climate change: signals from terrestrial biol-ogy. *Antarctic Research Series* **79**, 145–158.

Convey, P. (2005) Recent lepidopteran records from sub-Antarctic South Georgia. *Polar Biology* **28**, 108–110.

Convey, P. (2007) Antarctic ecosystems. In Levin, S.A. (ed.), *Encyclopedia of Biodiversity*, 2nd edn, pp. 174–184. Academic Press, San Diego, CA.

Convey, P. and Smith, R.I.L. (1997) The terrestrial arthropod fauna and its habitats in northern Marguerite Bay and Alexander Island, maritime Antarctic. *Antarctic Science* **9**, 12–26.

Convey, P. and Wynn-Williams, D.D. (2002) Antarctic soil nematode response to artificial environmental manipulation. *European Journal of Soil Biology* **38**, 255–259.

Convey, P. and McInnes, S.J. (2005) Exceptional, tardigrade dominated, ecosystems from Ellsworth Land, Antarctica. *Ecology* **86**, 519–527.

Convey, P. and Smith, R.I.L. (2006) Thermal relationships of bryophytes from geothermal habitats in the South Sandwich Islands, maritime Antarctic. *Journal of Vegetation Science* **17**, 529–538.

Convey, P. and Stevens, M.I. (2007) Ecology: Antarctic biodiversity. *Science* **317**, 1877–1878.

Convey, P., Greenslade, P., and Pugh, P.J.A. (2000a) Terrestrial fauna of the South Sandwich Islands. *Journal of Natural History* **34**, 597–609.

Convey, P., Smith, R.I.L., Hodgson, D.A., and Peat, H.J. (2000b) The flora of the South Sandwich Islands, with particular reference to the influence of geothermal heating. *Journal of Biogeography* **27**, 1279–1295.

Convey, P., Pugh, P.J.A., Jackson, C., Murray, A.W., Ruhland, C.T., Xiong, F.S., and Day, T.A. (2002) Response of Antarctic terrestrial arthropods to multifactorial climate manipulation over a four year period. *Ecology* **83**, 3130–3140.

Convey, P., Scott, D., and Fraser, W.R. (2003) Biophysical and habitat changes in response to climate alteration in the Arctic and Antarctic. *Advances in Applied Biodioversity Science* **4**, 79–84.

Cook, A.J., Fox, A.J., Vaughan, D.G., and Ferrigno, J.G. (2005) Retreating glacier fronts on the Antarctic Peninsula over the past half-century. *Science* **308**, 541–544.

Corsetti, F.A., Olcott, A.N., and Bakermans, C. (2006) The biotic response to Neoproterozoic Snowball Earth. *Palaeogeography, Palaeoclimatology, Palaeoecology* **232**, 114–130.

Cotté, C. and Guinet, C. (2007) Historical whaling records reveal major regional retreat of Antarctic sea ice. *Deep Sea Research Part I* **54**, 243–252.

Coulson, S.J. (2007) The terrestrial and freshwater invertebrate fauna of the High Arctic archipelago of Svalbard. *Zootaxa* **1448**, 41–58.

Coulson, S.J. and Resfeth, D. (2004) The terrestrial and freshwater fauna of Svalbard (and Jan Mayen). In Prestrud, P., Strøm, H., and Goldman, H.V. (eds), *A Catalogue of the Terrestrial and Marine Animals of Svalbard*, pp. 57–122. Norwegian Polar Institute, Trömso.

Coulson, S.J., Hodkinson, I.D., Webb, N.R., and Harrison, J.A. (2002) Survival of terrestrial soil-dwelling arthropods on and in seawater: implications for transoceanic dispersal. *Functional Ecology* **16**, 353–356.

Couzin, J. (2007) Opening doors to native knowledge. *Science* **315**, 1518–1519.

Cowan, D.A., Russell, N.J., Mamais, A., and Sheppard, D.M. (2002) Antarctic Dry Valley mineral soils contain unexpectedly high levels of microbial biomass. *Extremophiles* **6**, 431–436.

Craig, P.C. and McCart, P.J. (1975) Classification of stream types in Beaufort Sea drainages between Prudhoe Bay and the Mackenzie Delta., NWT, Canada. *Arctic, Alpine Research* **7**, 183–198.

Crawford, R.M.M. (1995) Plant survival in the High Arctic. *Biologist* **42**, 101–105.

Curren, M.A.J., van Ommen, T.D., Morgan, V.I., Phillips, K.L., and Palmer, A.S. (2003) Ice core evidence for Antarctic sea ice decline since the 1950s. *Science* **302**, 1203–1206.

Czygan, F.-C. (1970) Blood-rain and blood-snow: nitrogen-deficient cells of *Haematococcus pluvialis* and *Chlamydomonas nivalis*. *Archives für Mikrobiologie* **74**, 69–76.

Dalén, L., Fuglei, E., Hersteinsson, P., Kapel, C.M.O., Roth, J.D., Samelius G. *et al.* (2005) Population history and genetic structure of a circumpolar species: the arctic fox. *Biological Journal of the Linnean Society* **84**, 79–89.

Dalén, L., Kvaloy, K., Linnell, J.D.C., Elmhagen, B., Strand, O., Tannerfeldt, M. *et al.* (2006) Population structure in a critically endangered arctic fox population: does genetics matter? *Molecular Ecology* **15**, 2809–2819.

Daly, M.J. (2006) Modulating radiation resistance: insights based on defenses against reactive oxygen species in the radioresistant bacterium *Deinococcus radiodurans*. *Clinics in Laboratory Medicine* **26**, 491–504.

Danks, H.V. (1999) Life cycles in polar arthropods—flexible or programmed? *European Journal of Entomology* **96**, 83–102.

Davey, M.C. and Rothery, P. (1992) Factors causing the limitation of growth of terrestrial algae in maritime Antarctica during later summer. *Polar Biology* **12**, 595–602.

Davidson, A.T. and Marchant, H.J. (1994) The impact of ultraviolet radiation on *Phaeocystis* and selected species of Antarctic marine diatoms. *Antarctic Research Series* **62**, 187–205.

Davidson, A.T. and van der Heijden, A. (2000) Exposure of natural Antarctic marine microbial assemblages to ambient UV radiation: effects on bacterioplankton. *Aquatic Microbial Ecology* **21**, 257–264.

Davis, CH, Li, Y., McConnell, J.R., Frey, M.M., and Hanna, E. (2005) Snowfall-driven growth in East Antarctic Ice Sheet mitigates recent sea-level rise. *Science* **308**, 1898–1901.

Davis, R.C. (1980) Peat respiration and decomposition in Antarctic terrestrial moss communities. *Biological Journal of the Linnean Society* **14**, 39–49.

Davis, R.C. (1981) Structure and function of two Antarctic terrestrial moss communities. *Ecological Monographs* **51**, 125–143.

Day, T.A., Ruhland, C.T., Grobe, C.W., and Xiong, F. (1999) Growth and reproduction of Antarctic vascular plants in response to warming and UV radiation reductions in the field. *Oecologia* **119**, 24–35.

Dayton, P.K., Robilliard, G.A., and Paine, R.T. (1970) Benthic faunal zonation as a result of anchor ice at McMurdo Sound, Antarctica. In Holdgate, M.W. (ed.), *Antarctic Ecology*, vol. **1**, pp. 244–258. Academic Press, London.

de Baar, H.J.W., de Jong, J.T.M., Bakker, D.C.E., Löscher, B.M., Veth, C., Bathmann, U., and Smetacek, V. (1995) Importance of iron for plankton blooms and carbon dioxide drawdown in the Southern Ocean. *Nature* **373**, 412–415.

de Broyer, C., Scailteur, Y., Chapelle, G., and Rauschert, M. (2001) Diversity of epibenthic habitats of gammaridean amphipods in the eastern Weddell Sea. *Polar Biology* **25**, 744–753.

de Broyer, C., Nyssen, F., and Dauby, P. (2004) The crustacean scavenger guild in Antarctic shelf, bathyal and abyssal communities. *Deep-Sea Reserach Part II* **51**, 1733–1752.

de Freitas, C.R. and Symon, L.V. (1987) A bioclimatic index of human survival times in the Antarctic. *Polar Record* **23**, 651–659.

de la Mare, W.K. (1997) Abrupt mid-twentieth-century decline in Antarctic sea-ice extent from whaling records. *Nature* **389**, 57–59.

Dell, R.K. (1972) Antarctic benthos. *Advances in Marine Biology* **10**, 1–216.

DeLong, E.F. (1998) Archaeal means and extremes. *Science* **280**, 542–543.

DeLong, E.F. and Karl, D.M. (2005) Genomic perspectives in microbial oceanography. *Nature* **437**, 336–342

DeLong, E.F., Wu, K.Y., Prezelin, B.B., and Jovine, R.V.M. (1994) High abundance of archaea in Antarctic marine picoplankton. *Nature* **371**, 695–697.

Deming, J.W. (2002) Psychrophiles and polar regions. *Current Opinion in Microbiology* **5**, 301–309.

de Mora, S.J., Demers, S., and Vernet, M. (eds) (2000) *The Effects of UV Radiation in the Marine Environment.* Cambridge University Press, Cambridge.

Derocher, A.E., Lunn, N.J., and Stirling, I. (2004) Polar bears in a warming climate. *Integrative and Comparative Biology* **44**, 163–176.

DeVries, A.L. (1997) The role of antifreeze proteins in survival of Antarctic fishes in freezing environemnts. In Battaglia, B., Valencia, J., and Walton, D.W.H. (eds), *Antarctic Communities. Species, Structure and Survival*, pp. 202–208. Cambridge University Press, Cambridge.

Dickman, M. and Ouellet, M. (1987) Limnology of Garrow Lake. NWT, Canada. *Polar Record* **23**, 531–549.

Dieckmann, G.S., Rohardt, G., Hellmer, H., and Kipfstuhl, J. (1986) The occurrence of ice platelets at 250 m depth near the Filchner Ice Shelf and its significance for sea ice biology. *Deep Sea Research* **33**, 141–148.

Dieckmann, G.S., Spindler, M., Lange, M., Ackley, S.F., and Eicken, H. (1991) Antarctic sea ice: as habitat for the foraminferan *Neogloboquadrina pachyderma. Journal of Foraminiferal Research* **21**, 182–189.

Diez, B., Pedros-Alio, C., and Massana, R. (2001) Study of genetic diversity of eukaryotic picoplankton in different oceanic regions by small-subunit rRNA gene cloning and sequencing. *Applied Environmental Microbiology* **67**, 2932–2941.

Dittmar, T. and Kattner, G. (2003) The biogeochemistry of the river and shelf ecosystem of the Arctic Ocean: a review. *Marine Chemistry* **83**, 103–120.

DiTullio, G., Garrison, D.L., and Mathot, S. (1998) Dimethylsulphoniopropionate in sea ice algae from the Ross Sea polynya. *Antarctic Research Series* **73**, 139–146.

Dixon, D., Mayewski, P.A., Kaspari, S., Kreutz, K., Hamilton, G., Maasch, K. *et al.* (2005) A 200 year sulfate record from 16 Antarctic ice cores and associations with Southern Ocean sea-ice extent. *Annals of Glaciology* **41**, 155–166.

Domack, E., Duran, D., Leventer, A., Ishman, S., Doane, S., McCallum, S. *et al.* (2005) Stability of the Larsen B ice shelf on the Antarctic Peninsula during the Holocene epoch. *Nature* **436**, 681–685.

Doran, P.T., Priscu, J.C., Lyons, W.B., Walsh, J.E., Fountain, A.G., McKnight, D.M. *et al.* (2002) Antarctic climate cooling and terrestrial ecosystem response. *Nature* **415**, 517–520.

Doran, P.T., Fritsen, C.H., McKay, C.P., Priscu, J.C., and Adams, E.E. (2003) Formation and character of an ancient 19-m ice cover and underlying trapped brine in an "ice-sealed" east Antarctic lake. *Proceedings of the National Academy Sciences USA* **100**, 26–31.

Drewry, D.J., Laws, R.M., and Pyle, J.A. (ed.) (1992) Antarctica and environmental change. *Philosophical Transactions of the Royal Society of London Series B* **338**, 199–334.

Ducklow, H.W., Baker, K., Martinson, D.G., Quetin, L.B., Ross, R.M., Smith, R.C. *et al.* (2007) Marine pelagic ecosystems: the West Antarctic. *Philosophical Transactions of the Royal Society of London Series B* **362**, 67–94.

Dunbar, R.B., Leventer, A.R., and Stockon, W.L. (1989) Biogenic sedimentation in McMurdo Sound, Antarctica. Symposium on Glaciomarine Environments, INQUA XII International Congress. *Marine Geology* **85**, 155–179.

Dunton, K.H. (1992) Arctic biogeography: the paradox of the marine benthic fauna and flora. *Trends in Ecological Evolution* **7**, 183–189.

Dunton, K. and Schell, D.M. (1987) Dependence of consumers on macroalgal (*Laminaria solidungula*) carbon in an Arctic kelp community: ^{13}C evidence. *Marine Biology* **93**, 615–625.

EPICA Community Members (2004) Eight glacial cycles from Antarctic ice core. *Nature* **429**, 623–628.

EPICA Community Members (2006) One-to-one coupling of glacial climate variability in Greenland and Antarctica. *Nature* **444**, 195–198.

Edmonds, H.N., Michael, P.J., Baker, E.T., Connelly, D.P., Snow, J.E., Langmuir, C.H. *et al.* (2003) Discovery of abundant hydrothermal venting on the ultraslow-spreading Gakkel ridge in the Arctic Ocean. *Nature* **421**, 252–256.

Ehrenberg, C.G. (1841) Einen Nachtrag zu dem Vortrage über Verbreitung und Einfluß des mikroskopischen Lebens in Süd- und Nordamerika. *Berichte über die zur Bekanntmachung geeigneten Verhandlungen der Königlich-Preussischen Akademie der Wissenschaften zu Berlin, Monatsberichte 1841*, 202–207.

Ehrenberg, C.G. (1853) Über neue Anscauungen des kleinstein nördlichen Polarlebens. *Berichte über die zur Bekanntmachung geeigneten Verhandlungen der Königlich-Preussischen Akademie der Wissenschaften zu Berlin, Monatsberichte 1853*, 522–529.

Eicken, H. (1992) The role of sea ice in structuring Antarctic ecosystems. *Polar Biology* **12**, 3–13.

Eicken, H., Bock, C., Wittig, R., Miller, H., and Poertner H-O (2000) Magnetic resonance imaging of sea-ice pore fluids: methods and thermal evolution of pore microstructure. *Cold Regions Science and Technology* **31**, 207–225.

El-Sayed, S.Z.E. (1994) *Southern Ocean Ecology: The BIOMASS perspective.* Cambridge University Press, Cambridge.

Elvebakk, A. and Hertel, H. (1996) Lichens. In Elvebakk, A. and Prestrud, P. (eds), *A Catalogue of Svalbard Plants, Fungi and Cyanobacteria.* pp. 271–359. Norwegian Polar Institute Skrifter 198. Norwegian Polar Institute, Tromsö.

Ernsting, G., Block, W., MacAlister, H., and Todd, C. (1995) The invasion of the carnivorous carabid beetle *Trechisibus antarcticus* on South Georgia (subantarctic) and its effect on the endemic herbivorous beetle *Hydromedion spasutum. Oecologia* **103**, 34–42.

Ewing, M. and Donn, W.L. (1956) A theory of ice ages. *Science* **123**, 1061–1066.

Fahrbach, E., Rohardt, G., and Krause, G. (1992) The Antarctic Coastal Current in the southeastern Weddell Sea. *Polar Biology* **12**, 171–182.

Feder, H.M. and Jewett, S.C. (1981) Feeding interactions in the eastern bering Sea with emphasis on the benthos. In Hood, D.W. and Calder, J.A. (eds), *The Eastern Bering Sea Self: Oceanography and Resources*, vol. **2**, pp. 1229–1261. University of Washington Press, Seattle, WA.

Feder, H.M., Naidu, A.S., Jewett, S.C., Hameedi, J.M., Johnson, W.R., and Whitledge, T.E. (1994) The northeastern Chukchi Sea: benthosenvironmental interactions. *Marine Ecology Progress Series* **111**, 171–190.

Feder, H.M., Jewett, S.C., and Blanchard, A. (2005) Southeastern Chukchi Sea (Alaska) epibenthos. *Polar Biology* **28**, 402–421.

Fenton, J.H.C. and Smith, R.I.L. (1982) Distribution, composition and general characteristics of the moss banks of the maritime Antarctic. *British Antarctic Survey Bulletin* **51**, 215–236.

Finlay, B.J. (2002) Global dispersal of free-living microbial eukaryote species. *Science* **296**, 1061–1063

Fischbach, A.S., Amstrup, S.C., and Douglas, D.C. (2007) Landward and eastward shift of Alaskan polar bear denning associated with recent sea ice changes. *Polar Biology* **30**, 1395–1405.

Fogg, G.E. and Thake, B. (1987) *Algal cultures and phytoplankton ecology.* University of Wisconsin Press, Madison, WI.

Fountain, A.G., Tranter, M., Nylen, T.H., Lewis, K.J., and Mueller, D.R. (2004) Evolution of cryoconite holes and their contribution to meltwater runoff from glaciers in the McMurdo Dry Valleys, Antarctica. *Journal of Glaciology* **50**, 35–45.

Fountain, A.G., Nylen, T.H., MacClune, K.L., and Dana, G.L. (2006) Glacier mass balances (1993–2001), Taylor Valley, McMurdo Dry Valleys, Antarctica. *Journal of Glaciology* **52**, 451–462.

Freckman, D.W. and Virginia, R.A. (1998) Soil biodiversity and community structure in the McMurdo Dry Valleys, Antarctica. *Antarctic Research Series* **72**, 323–336.

Frenot, Y., Chown, S.L., Whinam, J., Selkirk, P., Convey, P., Skotnicki, M., and Bergstrom, D. (2005) Biological invasions in the Antarctic: extent, impacts and implications. *Biological Reviews* **80**, 45–72.

Fricker, H.A., Scambos, T., Bindschadler, R., and Padman, L. (2007) An active subglacial water system in West Antarctica mapped from space. *Science* **315**, 1544–1548.

Friedman, E.I. (1982) Endolithic microorganisms in the Antarctic cold desert. *Science* **215**, 1045–1053.

Friedman, E.I., Kappen, L., Meyer, M.A., and Nienow, J.A. (1993) Long-term productivity in the cryptoendolithic microbial community of the Ross Desert, Antarctica. *Microbial Ecology* **25**, 51–69.

Frisvoll, A.A. and Elvebakk, A (1996) Bryophytes. In Elvebakk, A. and Prestrud, P. (eds), *A Catalogue of Svalbard Plants, Fungi and Cyanobacteria.* pp. 57–172. Norwegian Polar Institute Skrifter 198. Norwegian Polar Institute, Tromsö.

Gabric, A.J., Qu, B.O., Matrai, P., and Hirst, A.C. (2005) The simulated response of dimethylsulfide production in the Arctic Ocean to global warming. *Tellus B* **57**, 391–403.

Gage, J.D. (2004) Diversity in deep-sea benthic macrofauna: the importance of local ecology, the larger scale, history and the Antarctic. *Deep-Sea Reserach Part II* **51**, 1689–1708.

Garrison, D.L. (1991) Antarctic sea ice biota. *American Zoologist* **31**, 17–33.

Garrison, D.L. and Buck, K.R. (1986) Organism losses during ice melting: a serious bias in sea ice community studies. *Polar Biology* **6**, 237–239.

Garrison, D.L. and Buck, K.R. (1991) Surface-layer sea ice assemblages in Antarctic pack ice during the austral spring: environmental conditions, primary production and community structure. *Marine Ecology Progress Series* **75**, 161–172.

Garrison, D.L. and Close, A.R. (1993) Winter ecology of the sea ice biota in Weddell Sea pack ice. *Marine Ecology Progress Series* **96**, 17–31.

Geider, R.J., Delucia, E.H., Falkowski, P.G., Finzi, A.C., Grime, J.P., Grace, J. *et al.* (2001) Primary productivity of planet earth: biological determinants and physical constraints in terrestrial and aquatic habitats. *Global Change Biology* 7, 849–882.

Gerdel, R.W. and Drouet, F. (1960) The cryoconite of the Thule area, Greenland. *Transactions of the American Microscopical Society* LXXIX, 256–272.

Gerdes, D., Hilbig, B., and Montiel, A. (2003) Impact of iceberg scouring on macrobenthic communities in the high-Antarctic Weddell Sea. *Polar Biology* **26**, 295–301.

Gibson, J.A.E., Trull, T., Nichols, P.D., Summons, R.E., and McMinn, A. (1999) Sedimentation of C-13 rich organic matter from Antarctic sea ice algae: a potential indicator of past sea ice extent. *Geology* 27, 331–334.

Gili, J.M., Coma, R., Orejas, C., López-González, P.J., and Zavala, M. (2001) Are Antarctic suspension feeding communities different from those elsewhere in the world? *Polar Biology* **24**, 473–485.

Gili, J.M., Arntz, W.E., Palanques, A., Orejas, C., Clarke, A., Dayton, P.K. *et al.* (2006) A unique assemblage of epibenthic sessile suspensión feeders with archaic features in the high-Antarctic. *Deep-Sea Research Part II* 53, 1029–1052.

Gilichinsky, D. and Wagener, S. (1995) Microbial life in permafrost: a historical review. *Permafrost and Periglacial Processes* **6**, 243–250.

Gill, P.C. and Thiele, D. (1997) A winter sighting of killer whales (*Orcinus orca*) in Antarctic sea ice. *Polar Biology* 17, 401–404.

Gleitz, M., Rutgers vd Loeff, M., Thomas, D.N., Dieckmann, G.S., and Millero, F.J. (1995) Comparison of summer and winter inorganic carbon, oxygen and nutrient concentrations in Antarctic sea ice brine. *Marine Chemistry* **51**, 81–91.

Glud, R.N., Rysgaard, S., and Kuhl, M. (2002) A laboratory study on O_2 dynamics and ophotosynthesis in ice algal communities: quantification by microsensors, O_2 exchange rates ^{14}C incubations and, P.A.M. fluorometer. *Aquatic Microbial Ecology* 27, 301–311.

Goldman, C.R., Mason, D.T., and Wood, B.J.B. (1963) Light injury and inhibition in in Antarctic freshwater phytoplankton. *Limnology and Oceanography* 8, 313–322.

Goodchild, A, Saunders, N.F.W., Ertan, H., Raftery, M., Guilhaus, M., Curmi, P.M.G., and Cavicchioli, R (2004) A proteomic determination of cold adaptation in the Antarctic archaeon, *Methanoccoides burtonii*. *Molecular Microbiology* 53, 309–321.

Gosselin, M., Levasseur, M., Wheeler, P.A., Horner, R.A., and Booth, BC (1997) New measurements of phytoplankton and ice algal production in the Arctic Ocean. *Deep-Sea Research* 44, 1623–1644.

Gowing, M.M. (2003) Large viruses and infected microeukaryotes in Ross Sea summer pack ice habitats. *Marine Biology* **142**, 1029–1040.

Gowing, M.M., Riggs, B.E., Garrison, D.L., Gibson, A.H., and Jeffries, M.O. (2002) Large viruses in Ross Sea late autumn pack ice habitats. *Marine Ecology Progress Series* **241**, 1–11.

Gradinger, R.R. and Bluhm, B.A. (2004) In situ observations on the distribution and behavior of amphipods and Arctic cod (*Boreogadus saida*) under the sea ice of the high Arctic Canadian Basin. *Polar Biology* 27, 595–603.

Gradinger, R. and Ikävalko J. (1998) Organism incorporation into newly forming Arctic sea ice in the Greenland Sea. *Journal of Plankton Research* 20: 871–886.

Gradinger, R. and Lenz, J. (1995) Seasonal occurrence of picocyanobacteria in the Greenland Sea and central Arctic Ocean. *Polar Biology* **15**, 447–452.

Gradinger, R., Meiners, K., Plumley, G., Zhang, Q., and Bluhm, B.A. (2005) Abundance and composition of the sea ice meiofauna in off-shore pack ice of the Beaufort Gyre in summer 2002 and 2003. *Polar Biology* **28**, 171–181.

Grebmeier, J.M., Overland, J.E., Moore, S.E., Farley, E.V., Carmack, E.C., and Cooper, L.W. (2006) A major ecosystem shift in the northern Bering Sea. *Science* **311**, 1461–1464.

Greenslade, P., Farrow, R.A., and Smith, J.M.B. (1999) Long distance migration of insects to a subantarctic island. *Journal of Biogeography* **26**, 1161–1167.

Gressitt, J.L. (ed.) (1970) Subantarctic entomology, particularly of South Georgia and Heard Island. *Pacific Insects Monograph* **23**, 1–374.

Griffiths, D.J. (2006) Chlorophyll b-containing oxygenic photosynthetic prokaryotes: Oxychlorobacteria (Prochlorophytes). *Botanical Review* **72**, 330–366.

Grossmann, S. (1994) Bacterial activity in sea ice and open waters of the Weddell Sea, Antarctica: a microautoradiographic study. *Microbial Ecology* **28**, 1–18.

Grossmann, S. and Dieckmann, G.S. (1994) Bacterial standing stock, activity, and carbon production during formation and growth of sea ice in the Weddell Sea, Antarctica. *Applied Environmental Microbiology* **60**, 2746–2753.

Grzymski, J.J., Carter, B.J., Delong, E.F., Feldman, R.A., Ghadiri, A., and Murray, A.E. (2006) Comparative genomics of DNA fragments from six Antarctic marine planktonic bacteria. *Applied Environmental Microbiology* **72**, 1532–1541.

Guillard, R.R.L. and Kilham, P. (1977) The ecology of marine planktonic diatoms. In Werner, D. (ed.), *The Biology of Diatoms*, pp. 372–469. Blackwell Scientific Publications, Oxford.

Gulliksen, B. and Lønne, O.J. (1989) Distribution, abundance, and ecological importance of marine sympagic fauna in the Arctic. *Rapports et Process-Verbaux des Réunions. Conseil Permanent International pour l'Exploration de la Mer* **188**, 133–138.

Günther, S. and Dieckmann, G.S. (1999) Seasonal development of algal biomass in snow-covered fast ice and the underlying platelet layer in the Weddell Sea, Antarctica. *Antarctic Science* **11**, 305–315.

Gutt, J. (1995) The occurrence of sub-ice algal aggregations off northeast Greenland. *Polar Biology* **15**, 247–252.

Gutt, J. (2000) Some 'driving forces' structuring communities of the sublittoral Antarctic macrobenthos. *Antarctic Science* **12**, 297–313.

Gutt, J. (2001) On the direct impact of ice on marine benthic communities, a review. *Polar Biology* **24**, 553–564.

Gutt, J. (2007) Antarctic macro-zoobenthic communities: a review and an ecological classification. *Antarctic Science* **19**, 165–182.

Gutt, J. and Starmans, A. (1998) Structure and biodiversity of megabenthos in the Weddell Antarctica and Lazarev Seas (Antarctica): ecological role of physical parameters and biological interactions. *Polar Biology* **20**, 229–247.

Haas, C. (2004) Late-summer sea ice thickness variability in the Arctic Transpolar Drift 1991–2001 derived from ground-based electromagnetic sounding, *Geophysical Research Letters* **31**, L09402, doi:10.1029/2003GL019394.

Haas, C., Thomas, D.N., and Bareiss, J. (2001) Surface properties and processes of perennial Antarctic sea ice in summer. *Journal of Glaciology* **47**, 613–625.

Haave, M., Ropstad, E., Derocher, A.E., Lie, E., Dahl, E., Wiig, O., Skaare, J.U., and Jenssen, B.M. (2003) Polychlorinated biphenyls and reproductive hormones in female polar bears at Svalbard. *Environmental Health Perspectives* **111**, 431–436.

Hall, C.M. and Johnston, M.E. (eds) (1995) *Polar Tourism: Tourism in the Arctic and Antarctic Regions.* John Wiley, Chichester.

Halliday, G. (2002) The British flora in the Arctic. *Watsonia* **24**, 133–144.

Hanalt, D., Wiencke, C., and Bischoff, K. (2007) Effects of UV-radiation on seaweeds. In Ørboek, J.B. (ed) *Arctic Alpine Ecosystems and People in a Changing Environment.* Springer, Berlin, 250–277.

Hart, I. (2006) *Whaling in the Falkland Islands Dependencies 1904–1931.* Pequena, Newton St Margaret.

Hawes, I. (1985) Light climate and phytoplankton photosynthesis in maritime Antarctic lakes. *Hydrobiologia* **123**, 69–79.

Hawes, I. (1993) Photosynthesis in thick cyanobacterial films: a comparison of annual and perennial antarctic mat communities. *Hydrobiologia* **252**, 203–209.

Hawes, I. and Howard-Williams, C. (1998) Primary production processes in streams of the McMurdo Dry Valleys, Antarctica. In Priscu, J.C. (ed.), *Ecosystem Dynamics in a Polar Desert, the McMurdo Dry Valleys, Antarctica*, pp. 129–140. American Geophysical Union, Washington DC.

Hawes, I., Howard-Williams, C., and Vincent, W.F. (1992) Desiccation and recovery of Antarctic cyanobacterial mats. *Polar Biology* **12**, 587–594.

Hawes, I.S., Howard-Williams, C., and Schwarz, A.-M. (1999) Environmental conditions during freezing, and response of microbial mats in ponds of the McMurdo Ice Shelf, Antarctica. *Antarctic Science* **11**, 198–208.

Hayes, P.K., Whitaker, T.M., and Fogg, G.E. (1984) The distribution and nutrient status of phytoplankton in the Southern Ocean between 20° and 70° W. *Polar Biology* **3**, 153–165.

Headland, R.K. (1984) *The Island of South Georgia.* Cambridge University Press, Cambridge.

Hegseth, E.N. (1992) Sub-ice algal assemblages of the Barents Sea: species composition, chemical composition, and growth rates. *Polar Biology* **12**, 485–496.

Heide-Jørgensen, H.S. and Kristensen, R.M. (1999) Puilassoq, the warmest homothermal spring of Disko Island. *Berichte zur Polarforschung* **330**, 32–43.

Heide-Jørgensen, M.P., Dietz, R., Laidre, K.L., and Richard, P. (2002) Autumn movements, home ranges, and winter density of narwhals (*Monodon monoceros*) tagged in Tremblay Sound, Baffin Island. *Polar Biology* **25**, 331–341.

Helbing, E.W., Marguet, E.R., Villafañe, V.E., and Holm-Hansen, O. (1995) Bacterioplnkton viability in Antarctic waters as affected by solar ultraviolet radiation. *Marine Ecology Progress Series* 126 293–298.

Helmke, E. and Weyland, H. (1995) Bacteria in sea ice and underlying water of the eastern Weddell Sea in midwinter. *Marine Ecology Progress Series* **117**, 269–287.

Hempel, G. (2007) Antarctic marine biology—two centuries of research. *Antarctic Science* **19**, 157–164.

Hense, I., Timmermann, R., Beckmann, A., and Bathmann, U.V. (2003) Regional and interannual variability of ecosystem dynamics in the Southern Ocean. *Ocean Dynamics* **53**, 1–10.

Henshaw, T. and Laybourn-Parry, J. (2002) The annual patterns of photosynthesis in two large, freshwater, ultra-oligotrophic Antarctic lakes. *Polar Biology* **25**, 744–752.

Hernandez, E.A., Ferreyra, G.A., and MacCormack, W.P. (2006) Response of two Antarctic marine bacteria to different natural UV radiation doses and wavelengths. *Antarctic Science*,18, 205–212.

Hernando, M., Carreto, J.I., Carignan, M.O., Ferreyra, G.A., and Gross, C. (2002) Effects of solar radiation on growth and mycosporine-like amino acids content in *Thalassiosira* sp, an Antarctic diatom. *Polar Biology* **25**, 12–20.

Hiruki, L.M., Schwartz, M.K., and Boveng, P.L. (1999) Hunting and social behaviour of leopard seals (*Hydrurga leptonyx*) at Seal Island, South Shetland Islands, Antarctica. *Journal of Zoology* **249**, 97–109.

Hobbie, J.E., Bahr, M., and Rublee, P.A. (1999) Controls on microbial food webs in oligotrophic arctic lakes. *Arch Hydrobiologia, Special Issue Advances in Limnology* **54**, 61–76.

Hodgson, D.A., Doran, P.T., Roberts, D., and McMinn, A. (2004) Palaeolimnological studies from the Antarctic and sub-Antarctic islands. In Pienitz, R., Douglas, M.S.V., and Smol, J.P. (eds), *Long-term Environmental Change in Arctic and Antarctic Lakes*, pp. 419–474. Springer, Berlin.

Hodkinson, I.D., Webb, N.R., and Coulson, S.J. (2002) Primary community assembly on land—the missing stages: why are the heterotrophic organisms always there first? *Journal of Ecology* **90**, 569–577.

Hoel, A.H. and Viljamsson, H. (2005) Commercial fisheries. In Nuttall, M. (ed.), *Encyclopedia of the Arctic*, pp. 635–641. Routledge, New York.

Hoffecker, J. (2004) *A Prehistory of the North: Human Settlement of the Higher Latitudes*. Rutgers University Press, New Brunswick, NJ.

Hofmann, E.E. and Lascara, C.M. (2000) Modeling the growth dynamics of Antarctic krill *Euphausia superba*. *Marine Ecology Progress Series* **194**, 219–231.

Hofmann, E.E. and Murphy, E.J. (2004) Advection, krill and Antarctic marine ecosystems. *Antarctic Science* **16**, 487–500.

Hogg, I.D., Cary, S.C., Convey, P., Newsham, K.K., O'Donnell, A.G., Adams, B.J. *et al.* (2006) Biotic interactions in Antarctic terrestrial ecosystems: are they a factor? *Soil Biology and Biochemistry* **38**, 3035–3040.

Hoham, R.W. (1980) Unicellular chlorophytes—snow algae. In Cox, E. (ed.), *Phytoflagellates*, pp. 61–84. Elsevier North Holland, New York.

Hoham, R.W. and Duval, B. (2001) Microbial ecology of snow and freshwater ice with emphasis on snow algae. In Jones, H.G., Pomeroy, J.W., Walker, D.A., and Hoham, R.W. (eds), *Snow Ecology: an Interdisciplinary Examination of Snow-covered Ecosystems*, pp. 168–228. Cambridge University Press, Cambridge.

Hoham, R.W., Schlag, E.M., Kang, J.Y., Hasselwander, A.J., Behrstock, A.F., Blackburn, I.R. *et al.* (1998) The effect of irradiance levels and spectral composition on mating strategies in the snow alga, *Chloromonas* sp.—D, from the Tughill Plateau, New York State. *Hydrological Processes* **12**, 1627–1639.

Holland, D.M. (2001) Explaining the Weddell polynya—a large ocean eddy shed at Maud Rise. *Science* **292**, 1697–1700.

Hooker, J.D. (1847) *The Botany of the Antarctic voyage of H.M. Discovery Ships Erebus and Terror in the Years 1838–1843. Part 1: Flora Antarctica*. Reeve Brothers, London.

Horner, R.A. (1985) *Sea Ice Biota*. CRC Press, Boca Raton, FL.

Howard-Williams, C., Pridmore, R., Downes, M.T., and Vincent, W.F. (1989) Microbial biomass, photosynthesis and chlorophyll-a related pigments in the ponds of the McMurdo Ice Shelf, Antarctica. *Antarctic Science* **1**, 125–131.

Howard-Williams, C., Schwarz A-M and Hawes, I. (1998) Optical properties of the McMurdo Dry Valley lakes, Antarctica. In Priscu, J.C. (ed.), *Ecosystem Dynamics in a Polar Desert, the McMurdo Dry Valleys, Antarctica*, pp. 154–189. American Geophysical Union, Washington DC.

Hughes, K.A. and Lawley, B. (2003) A novel Antarctic microbial endolithic community within gypsum crusts. *Environmental Microbiology* **5**, 555–565.

Huntley, B. (1991) How plants respond to climate change: migration rates, individualism and the consequences for plant communities. *Annals of Botany* **67**, 15–22.

Huntley, M.E., Nordhausen, W., and Lopez, M.D.G. (1994) Elemental composition, metabolic-activity and growth of Antarctic krill *Euphausia superba* during winter. *Marine Ecology Progress Series* **107**, 23–40.

Hutchinson, G.E. (1957) *A Treatise on Limnology*, vol. 1. Wiley, New York.

Ikävalko, J. and Gradinger, R. (1997) Flagellates and heliozoans in the Greenland Sea ice studied alive using light microscopy. *Polar Biology* **17**, 473–481.

Ikävalko, J. and Thomsen, H.A. (1997) The Baltic Sea ice biota (March 1994): A study of the protistan community. *European Journal of Protistology* **33**, 229–243.

Ingold, T. (1980) *Hunters, Pastoralists and Ranchers: Reindeer Economies and their Transformations*. Cambridge University Press, Cambridge.

Inman, M. (2007) The dark and mushy side of a frozen continent. *Science* **317**, 35–36.

Isla, E., Rossi, S., Palanques, A., Gili, J.M., Gerdes, D., and Arntz, W.E. (2006) Biochemical composition of marine sediments from the eastern Weddell Sea (Antarctica): High nutritive value in a high benthic-biomass environment. *Journal of Marine Systems* **60**, 255–267.

Jacobs, S.S. (2004) Bottom water production and its links with the thermohaline circulation. *Antarctic Science* **16**, 427–437.

James, M.R., Pridmore, R.D., and Cummings, V.J. (1995) Planktonic communities of melt ponds on the McMurdo Ice Shelf, Antarctica. *Polar Biology* **15**, 555–567.

Janech, M.G., Krell, A., Mock, T., Kang, J.-S., and Raymond, J.A. (2006) Ice-binding proteins from sea ice diatoms (Bacillariophyceae). *Journal of Phycology* **42**, 410–416.

Janssen, H.H. and Gradinger, R. (1999) Turbellaria (Archoophora: Acoela) from Antarctic sea ice endofauna—examination of their micromorphology. *Polar Biology* **21**, 410–416.

Jarman, S., Elliott, N., Nicol, S., McMinn, A., and Newman S. (1999) The base composition of the krill genome and its potential susceptibility to damage by UV-B. *Antarctic Science* **11**, 23–26.

Jeffrey, W.H. and Mitchell, D.L. (2001), Measurment of UVB induced DNA damage in marine planktonic communities. In Paul, J. (ed.), *Methods in Marine Microbiology*, pp. 469–488. Academic Press, New York,

Jensen, D.B. and Christensen, K.D. (eds) (2003) *The Biodiversity of Greenland—a Country Study*. Technical report 55. Grønlands Naturinstitut, Nuuk.

Jin, M., Deal, C., Wang, J., Alexander, V., Gradinger, R., Saitoh, S. *et al.* (2007) Ice-associated phytoplankton blooms in the southeastern Bering Sea. *Geophysical Research Letters* **34**, L06612.

Jokat, W., Ritzmann, O., Schmidt-Aursch, M.C., Drachev, S., Gauger, S., and Snow, J. (2003) Geophysical evidence for reduced melt production on the Arctic ultraslow Gakkel mid-ocean ridge. *Nature* **423**, 962–965.

Johannessen, O.M., Bengtsson, L., Miles, M.W., Kuzmina, S.I., Semenov, V.A., Alekseev, G.V. *et al.* (2004) Arctic climate change: observed and modelled temperature and sea-ice variability. *Tellus (A)* **56**, 328–341.

Johansen, S. (1998) The origin and age of driftwood on Jan Mayen. *Polar Research* **17**, 125–146.

Jones, A.E. and Shanklin, J.D. (1995) Continued decline of total ozone over Halley, Antarctica, since 1985. *Nature* **376**, 409–411.

Junge, K., Imhoff, F., Staley, T., and Deming, J.W. (2002) Phylogenetic diversity of numerically important Arctic sea-ice bacteria at subzero temperature. *Microbial Ecology* **43**, 315–328.

Junge, K., Eicken, H., and Deming, J.W. (2004) Bacterial activity at –2 to –20°C in Arctic wintertime sea ice. *Applied and Environmental Microbiology* **70**, 550–557.

Kain, J.M. (1989) The seasons in the subtidal. *British Phycological Journal* **24**, 203–215.

Kallio, P. and Valanne, N. (1975) On the effects of continuous light on photosynthesis in mosses. In Wielgolaski, F.E. (ed.), *Fennoscandian Tundra Ecosystems. Part 1: Plants and Microorganisms*, pp. 149–162. Springer, New York.

Kappen, L. and Straka, H. (1988) Pollen and spores transport into the Antarctic. *Polar Biology* **8**, 173–180.

Karentz, D. (1991) Ecological considerations of Antarctic ozone depletion. *Antarctic Science* **3**, 3–11.

Karl, D.M., Bird, D.F., Bjorkman, K., Houlihan, T., Shackelford, R., and Tupas, L. (1999) Microorganisms in the accreted ice if Lake Vostok, Antarctica. *Science* **286**, 2144–2147.

Kattner, G., Thomas, D.N., Haas, C., Kennedy, H.A., and Dieckmann, G.S. (2004) Surface ice and gap layers in Antarctic sea ice: highly productive habitats. *Marine Ecology Progress Series* **277**, 1–12.

Kennedy, A.D. (1993) Water as a limiting factor in the Antarctic terrestrial environment: a biogeographical synthesis. *Arctic and Alpine Research* **25**, 308–315.

Kennedy, A.D. (1995) Simulated climate change: are passive greenhouses a valid microcosm for testing the biological effects of environmental perturbations? *Global Change Biology* **1**, 29–42.

Kennedy, H., Thomas, D.N., Kattner, G., Haas, C., and Dieckmann, G.S. (2002) Particulate organic carbon in Antarctic summer sea ice: Concentration and stable carbon isotopic composition. *Marine Ecology Progress Series* **238**, 1–13.

Kepner, R.L., Wharton, R.A., and Coats, D.W. (1999) Ciliated protozoa of two Antarctic lakes: analysis by qualitative protargol staining and examination of artificial substrates. *Polar Biology* **21**, 285–294.

Kerry, K.R. and Hempel, G. (eds) (1990) *Antarctic Ecosystems. Ecological Change and Conservation*. Springer, Berlin.

King, J.E. (1983) *Seals of the World*, 2nd edn. British Museum (Natural History) and Oxford University Press, Oxford.

Kirchman, D.L., Dittel, A.I., Malmstrom, R.R., and Cottrell, M.T. (2005) Biogeography of major bacterial groups in the Delaware estuary. *Limnology and Oceanography* **50**, 1697–1706.

Kirk, J.T.O. (1994) *Light and Photosynthesis in Aquatic Ecosystems*, 2nd edn. Cambridge University Press, Cambridge.

Kirst, G.O. and Wiencke, C. (1995) Ecophysiology of polar algae. *Journal of Phycology* **31**, 181–199.

Kol, E. (1972) Snow algae from Signy Island (South Orkney Island, Antarctica). *Annales Historico-Naturales Musei Nationalis Hungarici* **64**, 63–70.

Kol, E. and Eurola, S. (1974) Red snow algae from Spitsbergen. *Astarte* 7, 61–66.

Krajick, K. (2007) Race to plumb the frigid depths. *Science* **315**, 1525–1528.

Krembs, C., Gradinger, R., and Spindler, M. (2000) Implications of brine channel geometry and surface area for the interaction of sympagic organisms in Arctic sea ice. *Journal of Experimental Marine Biology and Ecology* **243**, 55–80.

Krembs, C., Eicken, H., Junge, K., and Deming, J.W. (2002) High concentrations of exopolymeric substances in Arctic winter sea ice: implications for the polar ocean carbon cycle and cryoprotection of diatoms. *Deep Sea Research Part I* **49**, 2163–2181.

Laurion, I., Demers, S., and Vézina, A.F. (1995) The microbial food web associated with the ice algal assemblage: biomass and bacterivory of nanoflagellate protozoans in Resolute Passage (High Canadian Arctic). *Marine Ecology Progress Series* **120**, 77–87.

Law, K.S. and Stohl, A. (2007) Arctic air pollution: origins and impacts. *Science* **315**, 1537–1540.

Laws, R.M. (1977) The significance of vertebrates in the Antarctic marine ecosystem. In Llano, G.A. (ed.), *Adaptations Within Antarctic Ecosystems. Proceedings of the Third SCAR Symposium in Antarctic Biology*, pp. 411–438. Smithsonian Institution, Washington DC.

Lawley, B., Ripley, S., Bridge, P., and Convey, P. (2004) Molecular analysis of geographic patterns of eukaryotic diversity in Antarctic soils. *Applied and Environmental Microbiology* **70**, 5963–5972.

Lawson, J., Doran, P.T., Kenig, F., Des Marais, D.J., and Prsicu, J.C. (2004) Stable carbon and nitrogen isotopic composition of benthic and pelagic organic matter in lakes of the McMurdo Dry Valleys, Antarctica. *Aquatic Geochemistry* **10**, 269–301.

Laybourn-Parry, J. (2002) Survival strategies in Antarctic lakes. *Philosophical Transactions of the Royal Society London Series B* **357**, 863–869.

Laybourn-Parry, J. and Marshall, W.A. (2003) Photosynthesis, mixotrophy and microbial plankton dynamics in two high Arctic lakes during summer. *Polar Biology* **26**, 517–524.

Laybourn-Parry, J. and Pearce, D.A. (2007) The biodiversity and ecology of Antarctic lakes—models for evolution. *Philosophical Transaction of the Royal Society London Series B* **362**, doi:10.1098/rstb.2006.1945.

Laybourn-Parry, J., Marshall, W.A., and Marchant, H.J. (2005) Nutritional versatility as a key to survival in Antarctic phytoflagellates in two contrasting saline lakes. *Freshwater Biology* **50**, 830–838.

Laybourn-Parry, J., Madan, N.J., Marshall, W.A., Marchant, H.J., and Wright, S.W. (2006) Carbon dynamics in an ultr-oligotrophic epishelf lake (Beaver Lake), Antarctica) in summer. *Freshwater Biology* **51**, 1116–1130.

Le Romancer, M., Gaillard, M., Geslin, C., and Prieur, D. (2007) Viruses in extreme environments. *Reviews in Environmental Science and Biotechnology* **6**, 17–31.

Lee, P.A., de Mora, S.J., Gosselin, M., Levasseur, M., Bouillon, R.-C., Nozais, C., and Michel, C. (2001) Particulate dimethysulfoxide in Arctic sea-ice algal communities: the cryoprotectant hypothesis revisited. *Journal of Phycology* **37**, 488–499.

Lee, R.F., Hagen, W., and Kattner, G. (2006) Lipid storage in marine zooplankton. *Marine Ecology Progress Series* **307**, 273–306.

Lee, S.H. and Whitledge, T.H. (2005) Primary and new production in the deep Canada Basin during summer 2002. *Polar Biology* **28**, 190–197.

Legendre, L., Ackley, S.F., Dieckmann, G.S., Gulliksen, B., Horner, R., Hoshai, T. et al. (1992) Ecology of sea ice biota—2. Global significance. *Polar Biology* **12**, 429–444.

Ling, H.U. and Seppelt, R.D. (1990) Snow algae of the Windmill Islands, continental Antarctica. *Mesotaenium berggrenii* (Zygnematales, Chlorophyta) the alga of grey snow. *Antarctic Science* **2**, 143–148.

Lister, A., Block, W., and Usher, M.B. (1988) Arthropod predation in an Antarctic terrestrial community. *Journal of Animal Ecology* **57**, 957–971.

Livingstone, D.A. (1963) Alaska, Yukon, Northwest Territories and Greenland. In Frey, D.G. (ed.), *Limnology in North America*, pp. 559–574. University of Wisconsin Press, Madison, WI.

Lizotte, M.P. and Priscu, J.C. (1992) Photosynthesis-irradiance relationships in phytoplankton from thre physically stable water column of a perennially ice-covered lake (Lake Bonney, Antarctica). *Journal of Phycology* **28**, 179–185.

Lizotte, M.P., Sharp, T.R., and Priscu, J.C. (1996) Phytoplankton dynamics in the stratified water column of Lake Bonney, Antarctica. 1. Biomass and productivity during the winter-spring transition. *Polar Biology* **16**, 155–162.

Loeb, V., Siegel, V., Holm-Hansen, O., Hewitt, R., Fraser, W., Tivelpiece, W., and Trivelpiece, S. (1997) Effects of sea-ice extent and krill or salp dominance on the Antarctic food web. *Nature* **387**, 897–900.

Longton, R.E. (1988) *The Biology of Polar Bryophytes and Lichens*. Cambridge University Press, Cambridge.

Lønne, O.J. and Gulliksen, B. (1991) On the distribution of sympagic macro-fauna in the seasonally ice covered Barents Sea. *Polar Biology* **11**, 457–469.

Lorius, C. (1991) *Glaces de l'Antarctique: une mémoire, des passions*. Éditions Odile Jacob, Paris.

Lovejoy, C., Massana, R., and Pedros-Alio, C. (2006) Diversity and distribution of marine microbial eukaryotes in the Arctic Ocean and adjacent Seas. *Applied and Environmental Microbiology* **72**, 3085–3095.

Lunine, J.I. (2005) *Astrobiology—A Multidisciplinary Approach*. Pearson, Addison Wesley, San Francisco, CA.

Lunn, N.J., Stirling, I., Andriashek, D., and Richardson, E. (2004) Selection. of maternity dens by female polar bears in western Hudson. Bay. *Polar Biology* **27**, 350–356.

Lydersen, C. and Kovacs, K. (1999) Behaviour and energetics of ice-breeding, North Atlantic phocid seals during the lactation period. *Marine Ecology Progress Series* **187**, 265–281.

Lyons, W.B., Tyler, S.W., Wharton, J.R., McKnight, D.M., and Vaughn, B.H. (1998) A late Holocene desiccation of Lake Hoare and Lake Fryxell, McMurdo Dry Valleys, Antarctica. *Antarctic Science* **10**, 247–256.

Lyons, W.B., Frape, S.K., and Welch, K.A. (1999) History of the McMurdo Dry Valley lakes, Antarctica, from stable chlorine isotope data. *Geology* **27**, 527–530.

Mader, H.M. (1992) Observations of the water-vein system in polycrystalline ice. *Journal of Glaciology* **38**, 333–347.

Malin, G. and Kirst, G.O. (1997) Algal production of dimethyl sulfide and its atmospheric role. *Journal of Phycology* **33**, 889–896.

Mancuso-Nichols, C.A., Garon, S., Bowman, J., Raguénès, G., and Guezennec, J. (2004) Production of exopolysaccharides by Antarctic marine bacterial isolates. *Journal of Applied Microbiology* **96**, 1057–1066.

Mancuso-Nichols, C.A., Guezennec, J., and Bowman, J.P. (2005) Bacterial exopolysaccharides from extreme marine environments, with special with special consideration of the Southern Ocean, sea ice and hydrothermal vents a review. *Marine Biotechnology* **7**, 253–271.

Maranger, R., Bird D.F., and Juniper, S.K. (1994) Viral and bacterial dynamics in Arctic sea ice during the spring algal bloom near Resolute, NWT, Canada. *Marine Ecology Progress Series* **111**, 121–127.

Marion, G.M., Fritsen, C.H., Eicken, H., and Payne, M.C. (2003) The search for life on Europa: Limiting environmental factors, potential habitats, and earth analogues. *Astrobiology* **3**, 785–811.

Markager, S., Vincent, W.F., and Tang, E.P.Y. (1999) Carbon fixation by phytoplankton in high arctic lakes: implications of low temperature for photosynthesis. *Limnology and Oceanography* **44**, 597–607.

Marshall, D.J. and Pugh, P.J.A. (1996) Origin of the inland Acari of continental Antarctica, with particular reference to Dronning Maud Land. *Zoological Journal of the Linnean Society* **118**, 101–118.

Marshall, D.J. and Convey, P. (2004) Latitudinal variation in habitat specificity of ameronothroid mites. *Experimental and Applied Acarology* **34**, 21–35.

Marshall, J., Kushnir, Y., Battisti, D., Chang, P., Czaja, A., Dickson, R. *et al.* (2001) North Atlantic climate variability; Phenomena, impacts and mechanisms. *International Journal of Climatology* **21**, 1863–1898.

Marshall, W.A. (1996) Biological particles over Antarctica. *Nature* **383**, 680.

Marshall, W. and Laybourn-Parry, J. (2002) The balance between photosynthesis and grazing in Antarctic mixotrophic cryptophytes. *Freshwater Biology* **47**, 2060–2070.

Martin, A.R., Hall, P., and Richard, P.R. (2001) Dive behaviour of belugas (*Delphinapterus leucas*) in the shallow waters of western Hudson Bay. *Arctic* **54**, 276–283.

Martin, J.H., Coale, K.H., Johnson, K.S., Fitzwater, S.E., Gordon, R.M., Tanner, S.J. *et al.* (2002) Testing the iron hypothesis in ecosystems of the equatorial Pacific Ocean. *Nature* **371**, 123–129.

Maslen, N.R. and Convey, P. (2006) Nematode diversity and distribution in the southern maritime Antarctic—clues to history? *Soil Biology and Biochemistry* **38**, 3141–3151.

Mathot, S., Dandois, J.-M., and Lancelot, C. (1992) Gross and net primary production in the Scotia–Weddell Sea sector of the Southern Ocean during spring 1988. *Polar Biology* **12**, 321–332.

Matthews, J.A. (1992) *The Ecology of Recently-Deglaciated Terrain: a Geoecological Approach to Glacier Forelands and Primary Succession.* Cambridge University Press, Cambridge.

McClintock, J.B. and Karentz, D. (1997) Mycosporine-like amino acids in 38 species of subtidal marine organisms from McMurdo Sound, Antarctica. *Antarctic Science* **9**, 392–398.

McGhee, R. (2006) *The Last Imaginary Place:a Human History of the Arctic World.* Oxford University Press, Oxford.

McKenna, K.C., Moorhead, D.L., Roberts, E.C., and Laybourn-Parry, J. (2006) Simulated patterns of carbon flow in the pelagic food web of Lake Fryxell, Antarctica: little evidence of top-down control. *Ecological Modelling* **192**, 457–472.

McKnight, D.M., Alger, A., Tate, C.M., Shupe, G., and Spaulding, S.A. (1998) Longitudinal patterns in algal abundance and species distribution in meltwater streams in Taylor Valley, Southern Victoria Land, Antarctica. In Priscu, J.C. (ed.), *Ecosystem Dynamics in a Polar Desert, the McMurdo Dry Valleys, Antarctica*, pp. 109–128. American Geophysical Union, Washington DC.

McMinn, A., Skerratt, J., Trull, T., Ashworth, C., and Lizotte, M.P. (1999) Nutrient stress gradient in the bottom 5 cm of fast ice, McMurdo Sound, Antarctica. *Polar Biology* **21**, 220–227.

McMinn, A., Ashworth, C., and Ryan, K.G. (2000) In situ net primary productivity of an Antarctic fast ice bottom algal community. *Aquatic Microbial Ecology* **21**, 177–185.

Medlin, L.K., Lange, M., and Baumann, M.E.M. (1994) Genetic differentiation among three-colony-forming species of *Phaeocystis* further evidence for the phylogeny of the Prymnesiophyta. *Phycologia* **33**, 199–212.

Meiners, K., Brinkmeyer, R., Granskog, M.A., and Lindfors, A. (2004) Abundance, size distribution and bacterial colonization of exopolymer particles in Antarctic sea ice (Bellingshausen Sea). *Aquatic Microbial Ecology* **35**, 283–296.

Meurk, C.D., Foggo, M.N., and Wilson, J.B. (1994) The vegetation of subantarctic Campbell Island. *New Zealand Journal of Ecology* **18**, 123–168.

Michel, C., Nielsen, T.G., Nozais, C., and Gosselin, M. (2002) Significance of sedimentation and grazing by ice micro- and meiofauna for carbon cycling in annual sea ice (northern Baffin Bay). *Aquatic Microbial Ecology* **30**, 57–68.

Mikucki, J.A., Foreman, C.M., Sattler, B., Lyons, W.B., and Priscu, J.C. (2004) Geomicrobiology of Blood Falls: an iron-rich saline discharge at the terminus of the Taylor Glacier, Antarctica. *Aquatic Geochemistry* **10**, 199–220.

Mincks, S.L., Smith, C.R., and, De Master, D.J. (2005) Persistence of labile organic matter and microbioal biomass in Antarctic shelf sediments: evidence of a sediment 'food bank'. *Marine Ecology Progress Series* **300**, 3–19.

Mitchell, A.D., Meurk, C.D., and Wagstaff, S.J. (1999) Evolution of Stilbocarpa, a megaherb from New Zealand's sub–antarctic islands. *New Zealand Journal of Botany* **37**, 205–211.

Miteva, V.I., Sheridan, P.P., and Brenchley, J.E. (2004) Phylogenetic and physiological diversity of microorganisms isolated from a deep Greenland glacier ice core. *Applied Environmental Microbiology* **70**, 202–213.

Moberg, A., Sonechkin, D.M., Holmgren, K., Datsenko, N.M., and Karlén, W. (2005) Highly variable Northern Hemisphere temperatures reconstructed from low- and high-resolution proxy data. *Nature* **433**, 613–617.

Mock, T. (2002) In situ primary production in young Antarctic sea ice. *Hydrobiologia* **470**, 127–132.

Mock, T. and Thomas, D.N. (2005) Sea ice—recent advances in microbial studies. *Environmental Microbiology* **7**, 605–619.

Molau, U. and Molgaard, P. (1996) *ITEX Manual*, 2nd edn. Danish Polar Center, Copenhagen.

Moline, M.A., Claustre, H., Frazer, T.K., Schofield, O., and Vernet, M. (2004) Alteration of the food web along the Antarctic Peninsula in response to a regional warming trend. *Global Change Biology* **10**, 1973–1980.

Monaghan, A.J., Bromwich, D.H., Fogt, R.L., Wang, S.-H., Mayewski, P.A., Dixon, D.A. *et al.* (2006) Insignificant change in Antarctic snowfall since the International Geophysical Year. *Science* **313**, 827–831.

Moore, J.K. and Abbott, M.R. (2000) Phytoplankton chlorophyll distributions and primary production in the Southern Ocean. *Journal of Geophysical Research* **105**, 28709–28722.

Moran, D.M., Anderson, O.R., Dennett, M.R., Caron, D.A., and Gast, R.J. (2007) A description of seven Antarctic marine Gymnamoebae including a new sub-species, two new species and a new genus: *Neoparamoeba aestuarina antarctica* n. subsp., *Platyamoeba oblongata* n. sp., *Platyamoeba contorta* n. sp. and *Vermistella antarctica* n. gen. n. sp. *The Journal of Eukaryotic Microbiology* **54**, 169–183.

Mortimer, E. and Jansen van Vuuren, B. (2006) Phylogeography of *Eupodes minutus* (Acari: Prostigmata) on sub-Antarctic Marion Island reflects the impact of historical events. *Polar Biology* doi:10.1007/s00300–006-0205–7.

Mostajir, B., Gosselin, M., Gratton, Y., Booth, B., Vasseur, C., Garneau, M.V. *et al.* (2001) Surface water distribution of pico- and nanophytoplankton in relation to two distinctive water masses in the North Water, northern Baffin Bay, during fall. *Aquatic Microbial Ecology* **23**, 205–212.

Mountfort, D.O., Kaspar, H.F., Downes, M., and Asher, R.A. (1999) Partitioning effects during terminal carbon and electron flow in sediments of a low-salinity meltwater pond near Bratina Island, McMurdo ice shelf, antarctica. *Applied and Evnironmental Microbiology* **65**, 5493–5499.

Mueller, D. (2001) *A Bipolar Comparison of Glacial Cryoconite Ecosystems*. McGill University, Montreal.

Mueller, D.R. and Pollard, W.H. (2004) Gradient analysis of cryoconite ecosystems from two polar glaciers. *Polar Biology* **27**, 66–74.

Müller, C.H. (1952) Plant succession in Arctic heath and tundra in northern Scandinavia. *Bulletin of the Torrey Botanical Club* **79**, 296–309.

Müller, T., Bleiß, W., Martin, C.-D., Rogaschewski, S., and Fuhr, G. (1998) Snow algae from northwest Svalbard: their identification, distribution, pigment and nutrient content. *Polar Biology* **20**, 14–32.

Muñoz, J., Felicísimo, A.M., Cabezas, F., Burgaz, A.R., and Martínez, I. (2004) Wind as a long-distance dispersal vehicle in the Southern Hemisphere. *Science* **304**, 1144–1147.

Murphy, E.J., Clarke, A., Symon, C., and Priddle, J. (1995) Temporal variation in Antarctic sea-ice: analysis of a long-term fast-ice record from the South Orkney Islands. *Deep-Sea Research* **42**, 1045–1062.

Murphy, E.J., Boyd, P.W., Leakey, R.J.G., Atkinson, A., Edwards, E.S., Robinson, C. *et al.* (1998) Carbon flux in ice-ocean-plankton systems of the Bellingshausen Sea during a period of ice retreat. *Journal of Marine Systems* **17**, 207–227.

Nansen, F. (1897) *Farthest North: Being a Record of a Voyage of Exploration of the Ship 'Fram' 1893–96 and of a Fifteen Months's Sleigh Journey*. Westminster Archibald Constable and Co. London.

Newman, S.J., Dunlap, W.C., Nicol, S., and Ritz, D. (2000) Antarctic krill (*Euphausia superba*) acquire a UV-absorbing mycosporine-like amino acid from dietary algae. *Journal of Experimental Marine Biology and Ecology* **255**, 93–110.

Nghiem, S.V., Chao, Y., Neumann, G., Li, P., Perovich, D.K., Street, T., and Clemente-Colon, P. (2006) Depletion of perennial sea ice in the East Arctic Ocean. *Geophysical Research Letters* **33**, L17501.

Nicol, S. (2006) Krill, currents, and sea ice: *Euphausia superba* and its changing environment. *BioScience* **56**, 111–120.

Nichols, D.S. (2003) Prokaryotes and the input of polyunsaturated fatty acids to the marine food web. *FEMS Microbiology Letters* **219**, 1–7.

Nichols, D.S., Nichols, P.D., and McKeen, T.A. (1995) Ecology and physiology of psychrophilic bacteria from Antarctic saline lakes and sea ice. *Science Progress* **78**, 311–347.

O'Brien, W.J., Hershey, A.E., Hobbie, J.E., Hullar, M.A., Kipphut, G.W., Miller, M.C. *et al.* (1992) Control mechanisms of arctic lake ecosystems: a limnocorral experiment. *Hydrobiologia* **240**, 143–188.

O'Brien, W.J., Bahr, M., Hershey, A.E., Hobbie, J.E., Kipphut, G.W., Kling, H. *et al.* (1997) The limnology of Toolik Lake. In Milner, A.M. and Oswood, M.W. (eds), *Freshwaters of Alaska Ecological Synthesis*. Springer, New York.

Oppenheimer, M. and Alley, R.B. (2005) Ice sheets, global warming, and article 2 of the UNFCCC. *Climatic Change* **68**, 257–267.

Orejas, C., Gili, J.M., Arntz, W.E., Ros, J.D., López-González, P., Teixidó, N., and Pinto, P. (2000) Benthic suspension feeders, key players in Antarctic marine ecosystems. *Contributions to Science* **1**, 299–311.

Orejas, C., Gili, J.M., and Arntz, W.E. (2003) The role of small-plankton communities in the diet of two Antarctic octocorals (*Primnoisis antarctica* and *Primnoella* sp.). *Marine Ecology Progress Series* **250**, 105–116.

Palmisano, A.C. and Sullivan, C.W. (1985) Pathways of photosynthetic carbon assimilation in sea-ice microalgae from McMurdo Sound, Antarctica. *Limnology and Oceanography* **30**, 674–678.

Papadimitriou, S., Thomas, D.N., Kennedy, H., Haas, C., Kuosa, H., Krell, A., and Dieckmann, G.S. (2007) Biogeochemical composition of natural sea ice brines from the Weddell Sea during early austral summer. *Limnology and Oceanography* **52**, 1809–1823.

Parkinson, C.L. (2004) Southern Ocean sea ice and its wider linkages: insights revealed from models and observations. *Antarctic Science* **16**, 387–400.

Pearse, J.S. (1994) Cold-water echinoderms break Thorson's rule. In Young, C.M. and Eckelbarger, K.J. (eds), *Reproduction, Larval Biology, and Recruitment of the Deep-Sea Benthos*, pp. 26–39. Columbia University Press, New York.

Peat, H.J., Clarke, A., and Convey, P. (2007) Diversity and biogeography of the Antarctic flora. *Journal of Biogeography* **34**, 132–146.

Peck, L.S., Convey, P., and Barnes, D.K.A. (2006) Environmental constraints on life histories in Antarctic ecosystems: tempos, timings and predictability. *Biological Reviews* **81**, 75–109.

Perovich, D.K., Grenfell, T.C., Light, B., and Hobbs, P.V. (2002) The seasonal evolution of Arctic sea ice albedo. *Journal of Geophysical Research* doi:10.1029/2000JC000438.

Petit, J.R., Alekhina, I., and Bulat, S.A. (2005) Lake Vostok, Antarctica: exploring a subglacial environment and searching life in an extreme environment. In Gargaud, M., Barbier, B., Martin, H., and Reisse, J. (eds), *Lessons for Exobiology.* pp. 227–288. Springer, Berlin.

Petz, W., Song, W., and Wilbert, N. (1995) Taxonomy and ecology of the ciliate fauna (Protozoa, Ciliophora) in the endopagial and pelagial of the Weddell Sea, Antarctica. *Stapfia* **40**, 1–223.

Piepenburg, D. (2000) Arctic brittle stars (Echinodermata: Ophiuroidea). *Oceanography and Marine Biology: an Annual Review* **38**, 189–256.

Piepenburg, D. (2005) Recent research on Arctic benthos: common notions need to be revised. *Polar Biology* **28**, 733–755.

Plettner, I. (2002) *Streßphysiologie bei antarktischen Diatomeen: Ökophysiologische Untersuchungen zur Bedeutung von Prolin bei der Anpassung an hohe Salinitäten und tiefe Temperaturen.* http://elib.suub.uni-bremen.de/publications/dissertations/E-diss445-plettner.pdf. Dissertation, University of Bremen.

Plötz, J., Bornemann, H., Knust, R., Schroder, A., and Bester, M. (2001) Foraging behaviour of Weddell seals, and its ecological implications. *Polar Biology* **24**, 901–909.

Pomeroy, L.R. and Wiebe, W.J. (2001) Temperature and substrates as interactive limiting factors for marine heterotrophic bacteria. *Aquatic Microbial Ecology* **23**, 187–204.

Pisek, A. (1960) Pflanzen der Arktis und des Hochgebirges. In Pirson, A. (ed.), *Encyclopedia of Plant Physiology*, pp. 376–414. Springer, Berlin.

Podgorny, I.A. and Grenfell, T.C. (1996) Absorption of solar energy in a cryoconite hole. *Geophysical Research Letters* **23**, 2465–2468.

Porazinska, D.L., Fountain, A.G., Nylen, T.H., Tranter, M., Virginia, R.A., and Wall, D.H. (2004) The Biodiversity and biogeochemistry of cryoconite holes from McMurdo Dry Valley glaciers, Antarctica. *Arctic Antarctic and Alpine Research* **36**, 84–91.

Prézelin, B., Moline, M.A., and Matlick, H.A. (1998) ICECOLORS '93: Spectral UV radiation effects on Antarctic frazil ice algae. *Antarctic Research Series* **73**, 45–83.

Price, P.B. (2000) A habitat for psychrophiles in deep Antarctic ice. *Proceedings of the National Academy of Sciences USA* **97**, 1247–1251.

Price, P.B. and Sowers, T. (2004) Temperature dependence of metabolic rates for microbial growth, maintenance, and survival. *Proceedings of the National Academy of Sciences USA* **101**, 4631–4636.

Priscu, J.C. (2003) An international plan for Antarctic subglacial lake exploration. *Polar Geography* **27**, 69–83.

Priscu, J.C. and Christner, B.C. (2004) Earth's icy biosphere. In Bull, A.T. (ed.), *Microbial Diversity and Bioprospecting*, pp. 130–145. American Society for Microbiology, Washington DC.

Priscu, J.C., Fritsen, C.H., Adams, E.E., Giovannoni, S.J., Paerl, H.W., McKay, C.P. *et al.* (1998) Perennial Antarctic lake ice: an oasis for life in a polar desert. *Science* **280**, 2095–2098.

Priscu, J.C., Adams, E.E., Lyons, W.B., Voytek, M.A., Mogk, D.W., Brown, R.L. *et al.* (1999) Geomicrobiology of subglacial ice above Lake Vostok, Antarctica. *Science* **286**, 2141–2144.

Pugh, P.J.A. (1997) Spiracle structure in ticks (Ixodida: Anactinotrichida: Arachnida): resume, taxonomic and functional significance. *Biological Reviews* **72**, 549–564.

Pugh, P.J.A. (2003) Have mites (Acarina: Arachnida) colonized Antarctica and the islands of the Southern Ocean via air currents? *Polar Record* **39**, 239–244.

Quayle, W.C., Peck, L.S., Peat, H., Ellis-Evans, J.C., and Harrigan, P.R. (2002) Extreme responses to climate change in Antarctic lakes. *Science* **295**, 645.

Quetin, L.B. and Ross, R.M. (1991) Behavioural and physiological characteristics of the Antarctic krill, *Euphausia superba. American Zoologist* **31**, 49–63.

Quetin, L.B., Ross, R.M., Frazer, T.K., and Haberman, K.L. (1996) Factors affecting distribution and abundance of zooplankton, with an emphasis on Antarctic krill, *Euphausia superba. Antarctic Research Series* **70**, 357–371.

Raven, J.A. (1984) *Energetics and Transport in Aquatic Plants.* AR Liss, New York.

Raymond, J.A. (2000) Distribution and partial characterisation of ice-active molecules associated with sea ice diatoms. *Polar Biology* **23**, 721–729.

Raymond, J.A. and Knight, C.A. (2003) Ice binding, recrystallization inhibition, and cryoprotective properties of ice-active substances associated with Antarctic sea ice diatoms. *Cryobiology* **46**, 174–181.

Reay, D.S., Nedwell, D.B., Priddle, J., and Ellis-Evans, J.C. (1999) Temperature dependence of inorganic nitrogen uptake: reduced affinity for nitrate at suboptimal temperatures in both algae and bacteria. *Applied and Environmental Microbiology* **65**, 2577–2584.

Rees, W.G. (1993) A new wind-chill nomogram. *Polar Record* **29**, 229–234.

Remias, D., Lutz-Meindl, U., and Lutz, C. (2005) Photosynthesis, pigments and ultrastructure of the alpine snow alga *Chlamydomonas nivalis. European Journal of Phycology* **40**, 259–268.

Reynolds, C.S. (1992) The role of fluid motion in the dynamics of phytoplankton in lakes and rivers. In Giller, P.S., Hildrew, A.G., and Raffaelli, D.G. (eds), *Aquatic Ecology Scale, Patterns and Process*, pp. 141–187. Blackwell Scientific Publications, London.

Reynolds, J.F. and Tenhunen, J.D. (eds) (1996) *Landscape function and disturbance in Arctic tundra.* Springer, Berlin.

Richard, P.R., Martin, A.R., and Orr, J.R. (2001) Summer and autumn movements of Belugas of the Eastern Beaufort Sea Stock. *Arctic* **54**, 223–236.

Riebesell, U., Schloss, I., and Smetacek, V. (1991) Aggregation of algae released from melting sea ice: implications for seeding and sedimentation. *Polar Biology* **11**, 239–248.

Riegger, L. and Robinson, D. (1997) Photoinduction of UV-absorbing compounds in Antarctic diatoms and *Phaeocystis antarctica. Marine Ecology Progress Series* **160**, 13–25.

Riemann, F. and Sime-Ngando, T. (1997) Note on sea ice nematodes (Monhysteroidea) from Resolute Passage, Canadian High Arctic. *Polar Biology* **18**, 70–75.

Rivkin, R.B. and Putt, M. (1987) Heterotrophy and photoheterotrophy ny Antarctic microalgae: Light-dependent incorporation of amino acids and glucose. *Journal of Phycology* **23**, 442–452.

Roberts, E.C. and Laybourn-Parry, J. (1999) Mixotrophic cryptophytes and their predators in the Dry Valley lakes of Antarctica. *Freshwater Biology* **41**, 737–746.

Roberts, E.C., Priscu, J.C., Wolf, C., Lyons, W.B., and Laybourn-Parry, J. (2004a) The distribution of microplankton in the McMurdo Dry valley lakes, Antarctica: response to ecosystem legacy or present-day climatic controls. *Polar Biology* **27**, 238–250.

Roberts, E.C., Priscu, J.C., and Laybourn-Parry, J. (2004b) Microplankton dynamics in a perennially ice-covered Antarctic Lake—Lake Hoare. *Freshwater Biology* **49**, 853–869.

Røen, U. (1994) A theory for the origin of the Arctic freshwater fauna. *Verhandlungen der Internationale Vereinigung für Limnologie* **25**, 2409–2412.

Rønning, O.I. (1996) *The Flora of Svalbard*. Norsk Polarinstitut, Oslo.

Rosswall, T. and Heal, O.W. (eds) (1975) *Structure and Function of Tundra Ecosystems*. Swedish Natural Science Research Council, Stockholm.

Rothrock, D.A. and Zhang, J. (2005) Arctic Ocean sea ice volume: what explains its recent depletion? *Journal of Geophysical Research—Oceans* **110**, C01002.

Ryan, P.G. and Watkins, B.P. (1989) The influence of physical factors and ornithogenic products on plant and arthropod abundance at an inland nunatak group in Antarctica. *Polar Biology* **10**, 151–160.

Sage, B. (1986) *The Arctic and its Wildlife*. Croom Helm, London.

Sakshaug, E. (2004) Primary and secondary production in the Arctic seas. In, R. Stein and, R.W. Macdonald, (eds), *The Organic Carbon Cycle in the Arctic Ocean*, pp. 57–81. Springer, New York.

Sambrotto, R.N., Goering, J.J., and McRoy, C.P. (1984) Large yearly production of phytoplankton in western Bering Strait. *Science* **225**, 1147–1150.

Sarmiento, J.L., Gruber, N., Brzezinski, M., and Dunne, J.P. (2004) High-latitude controls of thermocline nutrients and low latitude biological productivity. *Nature* **427**, 56–60.

Säwström, C., Laybourn-Parry, J., Granéli, W., and Anesio, A.M. (2007) Heterotrophic bacterial and viral dynamics in Arctic Freshwaters: Results from a field study of nutrient temperature—manipulation experiments. *Polar Biology*, **30** 1407–1416.

Scharek, R. and Nöthig, E.M. (1995) Das einzellige Plankton im Ozean der Arktis und Antarktis. In Hempel, G. (ed.), *Biologie der Polarmeere*, pp. 116–127. Gustav Fischer, Jena.

Schiermeier, Q. (2007) The new face of the Arctic. *Nature* **446**, 133–135.

Schlensog, M., Pannewitz, S., Green, T.G.A., and Schroeter, B. (2004) Metabolic recovery of continental Antarctic cryptogams after winter. *Polar Biology* **27**, 399–408.

Schlichting, H.E., Speziale, B.J., and Zink, R.M. (1978) Dispersal of algae and protozoa by Antarctic flying birds. *Antarctic Journal of the USA* **13**, 147–149.

Schnack-Schiel, S.B. and Isla, E. (2005) The role of zooplankton in the pelagic-benthic coupling of the Southern Ocean. *Scientia Marina* **69** (suppl. 2), 39–55.

Schnack-Schiel, S.B., Thomas, D.N., Dieckmann, G.S., Eiken, H., Gradinger, R., Spindler, M. *et al.* (1995) Life cycle strategy of the Antarctic calanoid copepod *Stephos longipes*. *Progress in Oceanography* **36**, 45–75.

Schnack-Schiel, S.B., Dieckmann, G.S., Kattner, G., and Thomas, D.N. (2004) Copepods in summer platelet ice in the eastern Weddell Sea. *Polar Biology* **27**, 502–506.

Schreiber, A., Eisinger, M., and Storch, V. (1996) Allozymes characterize sibling species of bipolar Priapulida (*Priapulis, Priapulopsis*). *Polar Biology* **16**, 521–526.

Schultes, S., Verity, P.G., and Bathmann, U. (2006) Copepod grazing during an iron-induced diatom bloom in the Antarctic Circumpolar Current (EisenEx): I. Feeding patterns and grazing impact on prey populations. *Journal of Experimental Marine Biology and Ecology* **338**, 16–34.

Schulze-Makuch, D. and Irwin, L.N. (2004) *Life in the Universe: Expectations and Constraints*. Springer, Berlin.

Schumacher, J.D., Bond, N.A., Brodeur, R.D., Livingston, P.A., Napp, J.M., and Stabeno, P.J. (2003) Climate change in the Southeastern Bering Sea and some consequences for biota. In Hempel, G. and Sherman, K. (eds), *Large Marine Ecosystems of the World—Trends in Exploitation, Protection and Research*, pp. 17–40. Elsevier Science, Amsterdam.

Scott, F.J., Davidson, A.T., and Marchant, H.J. (2001) Grazing by the Antarctic sea-ice ciliate *Pseudocohnilembus*. *Polar Biology* 24 127–131.

Serreze, M.C., Holland, M.M., and Stroeve, J. (2007) Perspectives on the Arctics shrinking sea ice cover. *Science* **315**, 1533–1536.

Shain, D.H., Masson, T.A., Farrell, A.H., Michalewicz, L.A. (2001) Distribution and behavior of ice worms (Mesenchytraeus solifugus) in south-central Alaska. *Canadian Journal of Zoology* **79**, 1813–1821.

Shepard, A. and Wingham, D.J. (2007) Recent sea level contributions of the Antarctic and Greenland ice sheets. *Science* **315**, 1529–1532.

Sheridan, P.P., Miteva, V.I., Brenchley, J.E. (2003) Phylogenetic analysis of anaerobic psychrophilic enrichment cultures obtained from a Greenland ice core. *Applied and Environmental Microbiology* **69**, 2153–2160.

Shick, M.J. and Dunlap, W.C. (2002) Mycosporine like amino acids and related gadusols: biosynthesis, accumulation, and UV-protective functions in aquatic organisms. *Annual Review of Physiology* **64**, 223–262.

Shreeve, R.S. and Peck, L.S. (1995) Distribution of pelagic larvae of benthic marine invertebrates in the Bellingshausen Sea. *Polar Biology* **15**, 369–374.

Siegel, V. (2000) Krill (Euphausiacea) life history and aspects of population dynamics. *Canadian Journal of Fisheries and Aquatic Sciences* 57, 130–150.

Siegel, V. and Loeb, V. (1995) Recruitment of Antarctic krill *Euphausia superba* and possible causes for its variability. *Marine Ecology Progress Series* **123**, 45–56.

Siegert, M.J., Dowdeswell, J.A., Gorman, M.R., and McIntyre, N.F. (1996) An inventory of antarctic sub-glacial lakes. *Antarctic Science* **8**, 281–286.

Siegert, M.J., Carter, S.P., Tabacco, I.E., Popov, S., and Blankenship, D.D. (2005) A revised inventory of Antarctic subglacial lakes. *Antarctic Science* **17**, 453–460.

Simó, R. and Vila-Costa, M. (2006) Ubiquity of algal dimethylsulfoxide in the surface ocean: geographic and temporal distribution patterns. *Marine Chemistry* **100**, 136–146.

Sjoling, S. and Cowan, D.A. (2003) High 16S rDNA bacterial diversity in glacial meltwater lake sediment, Bratina Island, Antarctica. *Extremophiles* 7, 275–282.

Skidmore, M.L., Foght, J.M., and Sharp, M.J. (2000) Microbial life beneath a high arctic glacier. *Applied and Environmental Microbiology* **66**, 3214–3220.

Slabber, S. and Chown, S.L. (2002) The first record of a terrestrial crustacean, *Porcellio scaber* (Isopoda, Porcellionidae), from sub-Antarctic Marion Island. *Polar Biology* **25**, 855–858.

Smetacek, V. and Passow, U. (1990) Spring bloom initiation and Sverdrup's critical-depth model. *Limnology and Oceanography* **35**, 228–234.

Smetacek, V. and Nicol, S. (2005) Polar ocean ecosystems in a changing world. *Nature* **437**, 362–368.

Smetacek, V., Scharek, R., and Nöthig, E.M. (1990) Seasonal and regional variation in the pelagial and its relationship to the life history cycle of krill. In Kerry, K.R. and Hempel, G. (eds), *Antarctic Ecosystems: Ecological Change and Conservation*, pp. 103–114. Springer, Berlin.

Smith, C.R., Minks, S., and De Master, D.J. (2006) A synthesis of bentho-pelagic coupling on the Antarctic shelf: food banks, ecosystem inertia and global climate change. *Deep-Sea Research Part II* **53**, 875–894.

Smith, Jr, KL, Robison, B.H., Helly, J.J., Kaufmann, R.S., Ruhl, H.A., Shaw, T.J. *et al.* (2007) Free-drifting icebergs: hot spots of chemical and biological enrichment in the Weddell Sea. *Science* **317**, 478–482.

Smith, R., Prézelin, B., Baker, K., Bidigare, R., Boucher, N., Coley, T. *et al.* (1992) Ozone depletion: ultraviolet radiation and phytoplankton biology in antarctic waters. *Science* **255**, 952–959.

Smith, R.C., Prezelin, B.B., Baker, K.S., Bidigare, R.R., Boucher, N.P., Coley, T. *et al.* (1992) Ozone depletion: ultraviolet radiation and phytoplankton biology in antarctic waters. *Science* **255**, 952–959.

Smith, R.I.L. (1972) Vegetation of the South Orkney Islands with particular reference to Signy Island. *British Antarctic Survey Scientific Reports* no. 68.

Smith, R.I.L. (1988) Destruction of Antarctic terrestrial ecosystems by a rapidly increasing fur seal population. *Biological Conservation* **45**, 55–72.

Smith, R.I.L. (1990) Signy Island as a paradigm of biological and environmental change in Antarctic terrestrial ecosystems. In Kerry, K.R. and Hempel, G. (eds), *Antarctic Ecosystems, Ecological Change and Conservation*, pp. 32–50. Springer, Berlin.

Smith, R.I.L. (2005) The thermophilic bryoflora of Deception Island: unique plant communities as a criterion for designating an Antarctic Specially Protected Area. *Antarctic Science* **17**, 17–27.

Sømme, L. (1995) *Invertebrates in Hot and Cold Arid Environments*. Springer, Berlin.

Sommer, U. (1988) The species composition of Antarctic phytoplankton interpreted in terms of Tilman's competition theory. *Oecologia* **77**, 464–467.

Song, W. and Wilbert, N. (2000) Ciliates from Antarctic sea ice. *Polar Biology* **23**, 212–222.

Southwell, C., De la Mer, W.K., Borchers, D., and Burt, L. (2004) Shipboard line transect of crabeater seal abundance in the pack ice off East Antarctica: evaluation of assumptions. *Marine Mammal Science* **20**, 602–620.

Spaulding, S.A., McKnight, D.M., Smith, R.I., and Dufford, R. (1994) Phytoplankton population dynamics in perennial ice-covered Lake Fryxell, Antarctica. *Journal of Plankton Research* **16**, 527–541.

Spindler, M. and Dieckmann, G.S. (1986) Distribution and abundance of the planktic foraminfer *Neogloboquadrina pachyderma* in sea ice of the Weddell Sea (Antarctica). *Polar Biology* **5**, 185–191.

Staley, J.T. and Gosink, J.J. (1999) Poles apart: biodiversity and biogeography of sea ice bacteria. *Annual Review of Microbiology* **53**, 189–215.

Steig, E.J. (2006) Climate change—the south-north connection. *Nature* **444**, 152–153.

Stevens, M.I., Greenslade, P., Hogg, I.D., and Sunnucks, P. (2006) Southern hemisphere springtails: could any have survived glaciation of Antarctica? *Molecular Biology and Evolution* **23**, 874–882.

Stirling, I. and Parkinson, C.L. (2006) Possible effects of climate warming on selected populations of polar bears (*Ursus maritimus*) in the Canadian Arctic. *Arctic* **59**, 261–275.

Stoecker, D.K., Buck, K.R., and Putt, M. (1993) Changes in the sea ice brine community during the spring-summer transition, McMurdo Sound, Antarctica. I. Photosynthetic protists. *Marine Ecology Progress Series* **95**, 103–113.

Stoecker, D.K., Gustafson, D.E., Merrell, J.R., Black, M.M.D., and Baier, C.T. (1997) Excystment and growth of chryophytes and dinoflagellates at low temperatures and high salinities in Antarctic sea-ice. *Journal of Phycology* **33**, 585–595.

Stoecker, D.K., Gustafson, D.E., Black, M.M.D., and Baier, C.T. (1998) Population dynamics of microalgae in the upper land-fast sea ice at a snow free location. *Journal of Phycology* **34**, 60–69.

Stoecker, D.K., Gustafson, D.E., Baier, C.T., and Black, M.M.D. (2000) Primary production in the upper sea ice. *Aquatic Microbial Ecology* **21**, 275–287.

Stroeve, J., Holland, M.M., Meier, W., Scambos, T., and Serreze, M. (2007) Arctic sea ice decline: Faster than forecast. *Geophysical Research Letters* **34**, L09501.

Stokstad, E. (2007) Boom and bust in a Polar hot zone. *Science* **315**, 1522–1523.

Sunda, W., Kieber, D.J., Kiene, R.P., and Huntsman, S. (2002) An antioxidant function for DMSP and, D.M.S. in marine algae. *Nature* **418**, 317–320.

Suren, A. (1990) Microfauna associated with algal mats in melt ponds of the Ross Ice Shelf. *Polar Biology* **10**, 329–335.

Suttle, C.A. (2005) Viruses in the sea. *Nature* **437**, 356–361.

Suydam, R.S., Lowry, L.F., Frost, K.J., O'Corry-Crowe, G.M., and Pikok, D. (2001) Satellite tracking of eastern Chukchi Sea beluga whales into the Arctic Ocean. *Arctic* **54**, 237–243.

Tedrow, J.C.F. (1977) *Soils of the Polar Landscapes.* Rutgers University Press, New Brunswick, NJ.

Teixidó, N., Garrabou, J., Gutt, J., and Arntz, W.E. (2004) Recovery in Antarctic benthos after iceberg disturbance: trends in benthic composition, abundance and growth forms. *Marine Ecology Progress Series* **278**, 1–16.

Teixidó, N., Gili, J.M., Uriz, M.J., Gutt, J., and Arntz, W.E. (2006) Observations of sexual reproductive strategies in hexactinellid sponges from ROV video records. *Deep Sea Research Part II* **53**, 972–984.

Thayer, C.W. (1983) Sediment-mediated biological disturbance and the evolution of marine benthos. In Taevesz, M.J.S. and McCall, P.L. (eds), *Biotic Interactions in Recent and Fossil Benthic Communities,* pp. 479–625. Plenum Press, New York.

Thomas, D.N. and Dieckmann, G.S. (2002) Antarctic sea ice—a habitat for extremophiles. *Science* **295**, 641–644.

Thomas, D.N., Lara, R.J., Haas, C., Schnack-Schiel, S.B., Nöthig, E.-M., Dieckmann, G.S. *et al.* (1998) Biological soup within decaying summer sea ice in the Amundsen Sea, Antarctica. *Antarctic Research Series* **73**, 161–171.

Thomas, D.N., Engbrodt, R., Giannelli, V., Kattner, G., Kennedy, H., Haas, C., and Dieckmann, G.S. (2001) Dissolved organic matter in Antarctic sea ice. *Annals of Glaciology* **33**, 297–303.

Thomas, W.H. and Duval, B. (1995) Sierra Nevada, California, USA, snow algae: snow albedo changes, algal-bacterial interrelationships, and ultraviolet radiation effects. *Arctic and Alpine Research* **27**, 389–399.

Thrush, S., Dayton, P., Cattaneo-Vietti, R., Chiantore, M., Cummings, V., Andrew, N. *et al.* (2006) Broad-scale factors influencing the biodiversity of coastal benthic communities of the Ross Sea. *Deep-Sea Research Part II* **53**, 959–971.

Tilzer, M., Elbrächter, M., Gieskes, W.W., and Beese, B. (1986) Light temperature interactions in the control of photosynthesis in Antarctic phytoplankton. *Polar Biology* **5**, 105–111.

Tong, S., Vørs, N., and Patterson, D.J. (1997) Heterotrophic flagellates, centrohelid heliozoa and filose amoebae from marine and freshwater sites in the Antarctic. *Polar Biology* **18**, 91–106.

Tønesson, J.N. and Johnsen, A.O. (1982) *The History of Modern Whaling*. C. Hurst, London.

Tranter, M., Fountain, A.G., Fritsen, C.H., Lyons, W.B., Priscu, J.C., Statham, P.J., and Welch, K.A. (2004) Extreme hydrochemical conditions in natural microcosms entombed within Antarctic ice. *Hydrological Processes* **18**, 379–387.

Trenerry, L.J., McMinn, A., and Ryan, K.G. (2001) In situ oxygen microelectrode measurements of bottom –ice algal production in mcMurdo Sound, Antarctica. *Polar Biology* **25**, 72–80.

Trevena, A.J. and Jones, G.B. (2006) Dimethylsulphide and dimethylsulphoniopropionate in Antarctic sea ice and their release during sea ice melting. *Marine Chemistry* **98**, 210–222.

Trevena, A.J., Jones, G.B., Wright, S.W., and van den Enden, R.L. (2003) Profiles of dimethylsulphonioproprionate (DMSP), algal pigments, nutrients, and salinity in the fast ice of Prydz Bay, Antarctica. *Journal of Geophysical Research* 108 (C5), **3145**, doi:10.1029/2002JC001369.

Turner, J., Colwell, S.R., Marshall, G.J., Lachlan-Cope, T.A., Carleton, A.M., Jones, P.D. *et al.* (2005) Antarctic climate change during the last 50 years. *International Journal of Climatology* **25**, 279–294.

van de Poll, W.H., Alderkamp, A.C., Janknegt, P.J., Roggeveld, J., and Buma, A.G.J. (2006) Photoacclimation modulates excessive photosynthetically active and ultraviolet radiation effects in a temperate and an Antarctic marine diatom. *Limnology and Oceanography* **51**, 1239–1248.

Vaughan, S. (2000) Can Antarctic sea-ice extent be determined from whaling records? *Polar Record* **36**, 345–346.

Vaughan, D.G., Marshall, G.J., Connolley, W.M., King, J.C., and Mulvaney, R. (2001) Climate change—Devil in the detail. *Science* **293**, 1777–1779.

Vaughan, D.G., Marshall, G.J., Connolley, W.M., Parkinson, C.L., Mulvaney, R., Hodgson, D.A. *et al.* (2003) Recent rapid regional climate warming on the Antarctic Peninsula. *Climatic Change* **60**, 243–274.

Villafañe, V., Sundbäck, K., Figueroa, F., and Helbling, E. (2003) Photosynthesis in the aquatic environment as affected by UVR. In Ew, H. and He, Z. (eds), *UV Effects in Aquatic Organisms and Ecosystems*, pp. 357–397. Royal Society of Chemistry, Cambridge.

Vincent, W.F. (1987) Antarctic limnology. In Viner, A.B. (ed.), *Inland Waters of New Zealand*, pp. 379–412. SIPC, Wellington.

Vincent, W.F. (1988) *Microbial Ecosystems of Antarctica*. Cambridge University Press, Cambridge.

Vincent, W.F. and Vincent, C.L. (1982) Factors controlling phytoplankton production in Lake Vanda (77°S). *Canadian Journal Fisheries and Aquatic Science* **39**, 1602–1609.

Vincent, W.F. and Roy, S. (1993) Solar ultraviolet-B radiation and aquatic primary production: damage, protection and recovery. *Environment Review* **1**, 1–12.

Vincent, W.F. and Hobbie, J.E. (1999) Ecology of lakes and rivers. In Nuttall, M. and Callaghan, T.V. (eds), *The Arctic: a Guide to Research in the Natural and Social Sciences*. Harwood Academic Publishers, Amsterdam.

Vincent, W.F. and Howard-Williams, C. (2000) Life on snowball Earth. *Science* **287**, 2421.

Vincent, W.F., Castenholz, R.W., Downes, M.T., and Howard-Williams, C. (1993) Antarctic cyanobacteria: light, nutrients and photosynthesis in the microbial mat environment. *Journal of Phycology* **29**, 745–755.

Vincent, W.F., Rae, R., Laurion, I., Howard-Williams, C., and Priscu, J.C. (1998) Transparency of Antarctic ice-covered lakes to solar UV radiation. *Limnology and Oceanography* **43**, 618–624.

Vincent, W.F., Gibson, J.A.E., Pienitz, R., and Villeneuve, V. (2000) Ice shelf microbial ecosystems in the high arctic and implications for life on snowball earth. *Naturwissenschaften* **87**, 137–141.

Vincent, W.F., Mueller, D.R., and Bonilla, S. (2004) Ecosystems on ice: the microbial ecology of Markham Ice Shelf in the high Arctic. *Cryobiology* **48**, 103–112.

Vitousek, P.M., D'Antonio, C.M., Loope, L.L., Rejmanek, M., and Westbrooks, R. (1997) Introduced species: a significant component of human-caused global change. *New Zealand Journal of Botany* **21**, 1–16.

Vopel, K. and Hawes, I. (2006) Photosynthetic performance of benthic microbial mats in Lake Hoare, Antarctica. *Limnology and Oceanography* **51**, 1801–1812.

Wadhams, P., Dowdswell, J.A., and Schofield, A.N. (eds) (1995) The Arctic and environmental change. *Philosophical Transactions of the Royal Society of London Series A* **352**, 197–385.

Waleron, M., Waleron, K., Vincent, W.F., and Wilmotte, A. (2007) Allochthonous inputs of riverine picocyanobacteria to coastal waters in the Arctic Ocean. *FEMS Microbiology Ecology* **59**, 356–365.

Walker, G. (2003) *Snowball Earth: The Story of the Global Catastrophe that Spawned Life as we Know It*. Bloomsbury Publishing, London.

Walther, G.-R., Post, E., Convey, P., Menel, A., Parmesan, C., Beebee, T.J.C. *et al.* (2002) Ecological responses to recent climate change. *Nature* **416**, 389–395.

Walton, D.W.H. (ed.) (1987) *Antarctic Science*. Cambridge University Press, Cambridge.

Weissenberger, J. and Grossmann, S. (1998) Experimental formation of sea ice: importance of water circulation and wave action for incorporation of phytoplankton and bacteria. *Polar Biology* **20**, 178–188.

Weissenberger, J., Dieckmann, G.S., Gradinger, R., and Spindler, M. (1992) Sea ice: a cast technique to examine and analyse brine pockets and channel structure. *Limnology and Oceanography* **37**, 179–183.

Werner, I. (2006) Seasonal dynamics, cryo-pelagic interactions and metabolic rates of Arctic pack ice and under ice fauna—a review. *Polarforschung* **75**, 1–19.

Weykam, G., Thomas, D.N., and Wienke, C. (1997) Growth and photosynthesis of the Antarctic red algae, *Palmaria decipiens* (Palmariales) and *Iridaea cordata* (Gigartinales) during and following extended periods of darkness. *Phycologia* **36**, 395–405.

Wharton, J.R.A., Vinyard, W.C., Parker, B.C., Simmons, J.G.M., and Seaburg, K.G. (1981) Algae in cryoconite holes on Canada Glacier in Southern Victorialand, Antarctica. *Phycologia* **20**, 208–211.

Whitaker, T.M. (1977) Sea ice habitats of Signy Island (South Orkneys) and their primary productivity. In Llano, G.A. (ed.), *Adaptations within Antarctic Ecosystems*, pp. 75–82. Smithsonian Institution, Washington DC.

Whitehead, K., Karentz, D., and Hedges, J.I. (2001) Mycosporine-like amino acids (MAAs) in phytoplankton, a herbivorous pteropod (*Limacina helicina*), and its pteropod predator (*Clione antarctica*) in McMurdo Bay, Antarctica. *Marine Biology* **139**, 1013–1019.

Whitton, B.A. and Potts, M. (eds) (2000) *The Ecology of Cyanobacteria; their Diversity in Time and Space*. Kluwer Academic Publishers, Dordrecht.

Wiebe, W.J., Sheldon, Jr, W.M., and Pomeroy, L.R. (1992) Bacterial growth in the cold: evidence for an enhanced substrate requirement. *Applied and Environmental Microbiology* **58**, 359–364.

Wielgolaski, F.E. (1975) Primary production of tundra. In Cooper, J.P. (ed.), *Photosynthesis and Productivity in Different Environments*, pp. 75–106. Cambridge University Press, Cambridge.

Wiencke, C. (1996) Recent advances in the investigation of Antarctic macroalgae. *Polar Biology* **16**, 231–240.

Wiencke, C., Clayton, M.N., Gōmez, I., Iken, K., Lüder, V.-H., Amoler, C.D., Karsten, V., Hanelt, D., Bischoff, K., and Dunton, K. (2007) Life strategy, eco-physiology and ecology of seaweeds in polar waters. *Review In Environmental Science and Biotechnology* **6**, 95–126.

Willerslev, E., Hansen, A.J., Christensen, B., Steffensen, J.P., and Arctander, P. (1999) Diversity of holocene life forms in fossil glacier ice. *Proceedings of the National Academy of Sciences USA* **96**, 8017–8021.

Willerslev, E., Hansen, A.J., Brand, T., Binladen, J., Gilbert, T.M.P., Shapiro, B.A. *et al.* (2003) Diverse plant and animal DNA from Holocene and Pleistocene sedimentary records *Science* **300**, 791–795.

Willerslev, E., Hansen, A.J., Ronn, R., Brand, T.B., Barnes, I., Wiuf, C. *et al.* (2004) Long-term persistence of bacterial DNA. *Current Biology* **14**, R9–R10.

Willerslev, E., Cappellini, E., Boomsma, W., Nielson, R., Hebsgaard, M.B., Brand, T.B. *et al.* (2007) Ancient biomolecules from deep ice cores reveal a forested Southern Greenland. *Science* **317**, 111–114.

Williams, T.D. (1995) *The Penguins*. Oxford University Press, Oxford.

Wingham, D.J., Siegert, M.J., Shepard, A., and Muir, A.S. (2006) Rapid discharge connects Antarctic subglacial lakes. *Nature* **440**, 1033–1036.

Winter, C., Moesenerer, M.M., and Herndl, G.J. (2001) Impact of UV radiation on bacterioplancton community composition. *Applied and Environmental Microbiology* **67**, 665–672.

Wöhrmann, A.H.A. (1997) Freezing resistance in Antarctic and Arctic fishes: its relation to mode of life, ecology and evolution, *Cybium* **21**, 423–442.

Worland, M.R. and Lukesová, A. (2000) The effect of feeding on specific soil algae on the cold-hardiness of two Antarctic micro-arthropods (*Alaskozetes antarcticus* and *Cryptopygus antarcticus*). *Polar Biology* **23**, 766–774.

World Meteorological Organization (2007) *Scientific Assessment of Ozone Depletion: 2006, Global Ozone Research and Monitoring Project*. Report no. 50. World Meteorological Organization, Geneva.

Wortmann, H. (1995) Medizinische Untersuchungen zur Circadian-rhythmik und zum Verhalten bei Überwinterern auf einer antarktischen Forschungsstation. *Berichte zur Polarforschung* **169**, 1–261.

Wynn-Williams, D.D. (1996) Antarctic microbial diversity: the basis of polar ecosystem processes. *Biodiversity and Conservation* **5**, 1271–1293.

Yager, P.L. and Deming, J.W. (1999) Pelagic microbial activity in an arctic polynya: testing for temperature and substrate interactions using a kinetic approach. *Limnology and Oceanography* **44**, 1882–1893.

Zwally, H.J., Comiso, J.C., Parkinson, C.L., Cavalieri, D.J., and Gloersen, P. (2002) Variability of the Antarctic Sea Ice Cover. *Journal of Geophysical Research* **107**, 1029–1047.

Index